D1558442

SURVEYING

Board of Advisors, Engineering

SURVEYING

Jack B. Evett
University of North Carolina at Charlotte

John Wiley & Sons
New York Chichester Brisbane Toronto

Library of Congress Cataloging in Publication Data
Evett, Jack B., 1942-
Surveying.

Bibliography: p.
Includes index.
1. Surveying. I. Title.
TA545.E93 526.9 78-8332
ISBN 0-471-03132-1

Printed in the United States of America

10 9 8 7 6 5 4 3 2

To my wife, Linda,
my children, Susan, Scott, and Sarah,
and my parents, "Dutch" and Lennie.

PREFACE

My principal purpose in writing this book was to prepare a more basic surveying text that would be useful to anyone seeking his or her first formal study of surveying. I believe this book would be well suited for students of forestry, architecture, geography, agriculture, building and construction technology, architectural technology, civil engineering technology, and related areas. I also believe this book will be well suited for a first surveying course in a civil engineering curriculum when a general, introductory course in surveying is desired. In nonacademic contexts, I believe this book would be useful for surveying assistants who want to learn more about their work and for persons such as builders and lawyers who deal with surveying and surveyors and would like to learn more about the subject.

In preparing this book, I have tried to be as practical as possible, rather than highly theoretical. That is, I have tried to cover the "how to" of surveying. The first three chapters are preparatory in nature. Chapter 2 gives a review of the fundamental mathematics needed in surveying. I suspect this will be helpful to many who use this book, and for some this chapter may be more than a "review." Chapter 3 introduces the various pieces of equipment used in surveying. In the next four chapters, I have attempted to cover in some detail, but in the simplest language possible, the use of basic surveying equipment (primarily the tape, transit, and level) to make the basic measurements of surveying (primarily horizontal distances, vertical distances, and angles, bearings, and azimuths). In the remaining chapters, I have tried to show how data so obtained are displayed, utilized, and interpreted.

Several changes that have taken place recently or are about to take place have had some influence on the preparation of this book. One is the advent of the hand-held calculator. Virtually anyone who needs one can afford a calculator that will at least add, subtract, multiply, and divide. For this reason I have omitted any reference to logarithms–both in the text and with regard to listing a table of logarithms in the appendix. Although some will have calculators with values of trigonometric functions available at the touch of a button, I have included trigonometric tables in Appendix A.

Also available are calculators, including hand-held ones as well as desk types for office use, that are programmable or preprogrammed to perform extended surveying calculations by simply entering the necessary data. Using such preprogrammed calculators, one can, for example, compute the closure and the area of a closed traverse without knowing what a latitude, departure, or DMD is. Since such calculators are not so readily available, and since I believe anyone claiming to be doing surveying should have some understanding of what he or she is doing, rather than just entering data into a calculator and getting answers therefrom, I have included explanations of all procedures involved (i.e., in the step-by-step manual-surveying calculations). For illustration, however, I have included one traverse problem worked using a preprogrammed calculator.

Also widely available–particularly on college and university campuses–is the high-speed digital computer. To facilitate reduction of field data for a closed traverse, I have written a computer program and included it in Chapter 8. I have included an explanation as to how to utilize this program along with an example problem. Thus after obtaining field data for a closed traverse, the student has three options for computing closure, adjusting the traverse, and computing the area. These are: (1) doing the computations by hand, (2) using a preprogrammed calculator (including hand-held calculators), and (3) punching the data on data cards and processing on a digital computer, utilizing the given program.

Another change that is about to take place in the United States is the conversion to the "International System of Units" (SI). This conversion will have a significant effect on the art of surveying. In order to help effect a smooth transition to SI, I have used SI units in approximately half the problems herein.

When I started writing this book, I wanted to avoid giving the impression that surveying is man's work. I tried to avoid the use of all nouns and pronouns that indicate male or female. This, however, proved troublesome, with sentences like this one: "One does not want to place a specified foot marker at one's point; one wants to pull the tape tightly and see which marker lines up with one's point". Also, I personally dislike words such as "tape-person" and "instrumentperson." Then I decided to go back and use "he/she," "him/her," and "his/her." Using this scheme the first part of the sample sentence above would be: "He/she does not want to place a specified foot marker at his/her point". I tried to do this for a while, but each time I read back through what I had written, I felt that the terms "he/she" etc. intefered with my train of thought and seemed unnatural. Therefore, I decided to use "he or she" or to pluralize in some cases and to use the masculine nouns and pronouns in other cases. But let me assure you that when I use the masculine nouns and pronouns alone I mean for these words to refer to both men and women and certainly not to the male sex in particular.

I have included a number of problems at the end of most chapters. Answers are given in the back of the book for most of the even-numbered problems.

Finally, I wish to express publicly my sincere appreciation to my colleagues and friends, Carlos G. Bell, Richard R. Phelps, and Richard P. Pinckney, who read my manuscript and offered many helpful suggestions. Also, I thank my father, W. W. "Dutch" Evett, who, in addition to reading the entire manuscript and offering many helpful suggestions, taught me in the first place much of what I know about surveying. I would also like to express my appreciation to Keuffel & Esser Company, Hewlett-Packard, and the U. S. Geological Survey for furnishing illustrative material for use in this book.

Jack B. Evett

CONTENTS

INTRODUCTION

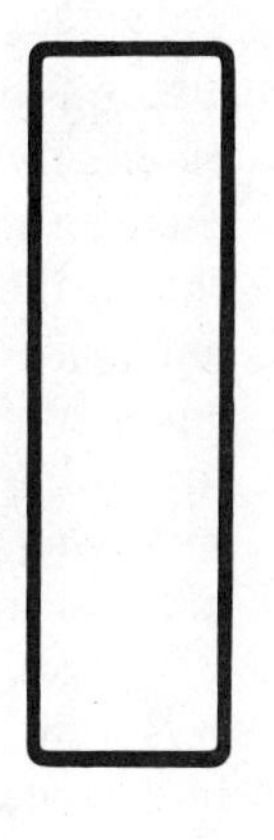

Most surveying books begin by defining the word "surveying," and this one is no exception. Surveying might be defined as "the act of determining the relative positions of points above, on, or below the earth's surface, or of establishing such points, by the measurements of distances (both horizontal and vertical) and directions." That is certainly a very general definition and does not attempt to relate all the techniques, computations, and procedures involved in the art of surveying. These should, however, become more lucid as the reader studies these aspects of surveying throughout the remainder of this book.

It should be pointed out that the surveying presented in this book is "plane" surveying, as opposed to "geodetic" surveying. In simple terms, in geodetic surveying, the spheroidal shape of the earth is taken into consideration; in plane surveying, it is not. Thus in plane surveying it is assumed that horizontal distances and angles are projected onto a horizontal plane; while in geodetic surveying it is assumed that they are projected onto the surface of a spheroid. The assumption of plane surveying is reasonable as far as the scope of this book is concerned, because the type of surveying involved will likely cover such a small portion of the earth's surface that the assumption of a "flat earth" is acceptable.

1-1 Units of Measure

The unit for measuring length most used in surveying in recent years in the United States is the foot (ft), with distances being given in feet and decimal fractions of a foot (as opposed to feet and inches). Sometimes, particularly on older surveys, distances are expressed in a unit of length called the Gunter's chain, which is equal to 66 ft. Such distances may be expressed in Gunter's chains and decimal fractions thereof or in Gunter's chains and links, where a link is $\frac{1}{100}$ of a chain. Thus a distance of 5 chains 42 links is 5.42 chains and is equivalent to 358 ft (obtained by multiplying 5.42 chains by 66 ft per chain). In the International System of Units (SI), the unit of linear measure used in surveying is the meter (m), or for longer distances, the kilometer (km), which is equal to 1000 m. The relationship between meters and feet is as follows: 1 ft = 0.3048 m, or 1 m = 3.281 ft.[1]

Another important measurement made in surveying is that of surface area. If distances have been measured in feet, then a bounded area would likely be computed in square feet (sq ft). Land is frequently bought and sold by a unit of area called the acre, with the relationship: 1 acre = 43,560 sq ft. (Also, 1 acre is equal to 10 square chains.) In SI if distances have been measured in meters, then a bounded area could be computed in square meters (sq m). A unit of area in SI analogous to the acre is the hectare (ha), with the relationship: 1 ha = 10,000 sq m. The relationship between hectares and acres is as follows: 1 ha = 2.471 acres, or 1 acre = 0.4047 ha.

Sometimes, particularly when dealing with large areas and with some government maps, area is expressed in square miles (sq mi). A square mile is, of course, the area equal to that of a square parcel of land with sides equal to 1 mile (5280 ft) in length. (Also, 1 square mile is equal to 640 acres.) An analogous unit of area in SI is the square kilometer (sq km). A square kilometer is the area equal to that of a square parcel of land with sides equal to one kilometer in length. The relationship between square miles and square kilometers is as follows: 1 sq km = 0.3861 sq mi, or 1 sq mi = 2.590 sq km.

Another measurement made in surveying is that of volume. If distances are measured in feet and area in square feet, then it follows that volume would be measured in cubic feet (cu ft). Certainly volume can be measured in cubic feet, but in most phases of surveying where volume is concerned—such as earthwork computations—volumes are usually expressed in cubic yards (cu yd). A cubic yard is the volume equal to that of a cube, each side of which is 1 yard (yd) in length. One yard is equal to exactly 3 ft, and 1 cubic yard is equal to exactly 27 cu ft. In SI volume is expressed in cubic meters (cu m).

The usual units of angular measurement used in surveying are degrees, minutes, and seconds. One degree is $\frac{1}{360}$ of the total angle measured about a point. One minute is $\frac{1}{60}$ of a degree, and one second is $\frac{1}{60}$ of a minute and also $\frac{1}{3600}$ of a degree. Degrees, minutes, and seconds are usually indicated by the symbols °, ′, and ″ (respectively). Thus 42 degrees, 29 minutes, 15 seconds is written as 42°29′15″.

[1] These, and many other conversion factors here, are approximate values, rounded off to four significant figures.

1-2 Recording Field Data

In general terms, surveying consists of two phases—(1) gathering data in the field, and (2) analyzing and utilizing these data in specific ways at various later times. One of the most important aspects of surveying, which is necessary if the two phases above are to be carried out effectively, is the recording of field data for later use. The importance of recording field data accurately, legibly, and without ambiguity cannot be overemphasized.

Field data should be recorded in a permanent manner and filed for future reference. For the professional surveyor, a set of field notes constitutes a legal document, which could conceivably end up as evidence in a court of law. Consequently, each set of field notes should be documented by listing the type of survey and location in titular form, followed by the date, the weather, members of the survey party, list of equipment used, and time involved (for pricing jobs).

Field data should be recorded legibly and without ambiguity. It is entirely possible—and frequently happens—that the person using field data (as, for example, to draw a map) may not be the same person who gathered and recorded the field data. It is difficult to think of something more frustrating than trying to follow field notes made by someone else (or even made by oneself, for that matter) and not being able to decipher the recorded numbers or not being able to determine what angle in the survey corresponds to a value of an angle written in the margin of a page, or otherwise trying to interpret something recorded in an inexplicable manner. A good principle to follow in recording field data is always to assume that the notes will be utilized by someone else who has never even visited the site of the survey.

Field notes are usually recorded on standard ruled sheets in a field book. This field book may be either loose-leaf or bound. A loose-leaf field book is preferable if it is desired to remove each individual set of notes for filing. On the other hand, a bound field book is preferable if it is desired to leave all field notes intact and then file the book when all its pages have been used.

Examples of note keeping for specific types of surveying are deferred until later in the book when the specific types of surveying are discussed. However, some general remarks are presented at this point.

1. A hard lead pencil (3H or harder) should be used to record field notes. The reason for this is that field notes are often subject to moisture (rain, perspiration, creek, etc.), which can cause the lead (graphite) to smear. When a hard lead pencil is used, the user is forced to press down hard, making an indentation in the paper. Thus if the notes become smeared, the data can still be ascertained by examining the indentations. Also, harder lead is less likely to smear.

2. Erasures should never be made in a field book. If a measured value is recorded incorrectly, a single horizontal line should be drawn through the incorrect value, and the correct value recorded above it. There are several reasons for this. An indication of an erasure in a field book (and replacement with another value) compromises the integrity of the notes. At least the suspicion arises that someone may have tampered with the data. If, instead, a line is drawn through the incorrect number, and the correct value written above it, the implication is more that of an honest mistake that was admittedly detected and corrected at the time. Such considerations might be of great importance if, for example, the notes became evidence in a court case. Another reason is that if a mistake was made and a corrected value recorded, the implication is that at the time of the measurement there was apparently some confusion as to what the correct value was. If, later on, the survey does not check out, the place where the questionable measurement was made may be a good place to start checking to find a mistake. In some cases, it may be useful to look back and see what that questionable value really was. Occasionally, it may turn out that the original value that was thought to be incorrect may have been correct after all. In summary, it is just good practice not to erase any data after they have been written down.

3. In general, the left page of a field book is used for recording data, while the right page is used for sketches.
4. Field data should be recorded in the field book at the time the measurement is made. To try to memorize data or write them on a slip of paper or matchbook for later permanent recording in a field book is an abhorrent practice.

1-3 Some Practical Considerations

In a way, surveying is like golf. A person could read all the books ever written about golf, but the probability that he or she could go out the first time and shoot par golf is virtually zero. With sufficient field practice—probably several years—and under the supervision of a professional, a novice could possibly become so proficient that he or she could shoot par golf. Similarly, a person could read all the books ever written about surveying, but the probability that he or she could go out the first time and do a perfect job of surveying is very low. With sufficient field practice—probably several years—and under the supervision of a professional surveyor, the novice could become so proficient that he could do quality land surveying on his own.

In golf, much of the practice is necessary to develop and perfect one's ability and skill. The same is true in surveying, but another important purpose of practice (supervised experience) is to learn some of the applications of surveying—things not normally covered in the classroom.

For example, in the classroom and laboratory field period a person can learn how to measure linear distances and angles and/or bearings. These are measures that would be utilized in establishing a property line between adjacent landowners. It is one thing, however, to know how to measure linear distance and angles, but something else to establish a property line between adjacent landowners. In learning to measure linear distance in the laboratory field period, the student is generally directed to measure distance between two given points. However, in establishing a property line, the points defining the property line may not be "given," that is, the property corners may have to be established—a procedure often requiring practical experience and common sense.

The purpose of relating the preceding discussion is to point out the need for actual field experience after academic study in order to really become a surveyor. Most (if not all) states require evidence of both academic study and field experience, followed by successfully passing a comprehensive examination, in order to be licensed as a "registered land surveyor." Any serious student of surveying should set early the goal of becoming licensed as a registered land surveyor.

1-4 Problems

1-1 What is the difference between plane surveying and geodetic surveying?

1-2 On an old plat, a distance was recorded as 14 chains 12 links. How many feet is this? How many meters?

1-3 A rectangular piece of property has sides of 101.77 m and 50.08 m. What is the area of the property in square meters? In hectares? In square feet? In acres?

1-4 A rectangular piece of property has sides of 9400.0 ft and 12,208.2 ft. What is the area of the property in acres? In square miles? In square kilometers? In hectares?

1-5 A box-shaped volume of earth has sides of 13.36 m, 25.00 m, and 100.00 m. What is the volume in cubic meters? In cubic yards?

1-6 What is the area in acres of a square tract of land, each side of which is 100 chains in length?

1-7 What is the length of the side of a square tract of land, the area of which is 1.000 acre?

1-8 What is the length of the radius of a circle, the area of which is 1.000 acre?

1-9 What is the length of a rectangle, the area and width of which are 1.000 acre and 99.15 ft, respectively?

1-10 A flat area of 2.20 acres is to be excavated to a depth of 2.5 ft. How many cubic yards of earth will be removed?

1-11 A straight highway segment is 7.25 km long. What is the length in miles? In feet?

1-12 If the right-of-way for the highway segment of Problem 1-11 is 30.0 m wide, what is the area of the right-of-way? Give answer in square meters and in hectares.

1-13 If the right-of-way described in Problems 1-11 and 1-12 must be filled in an average of 2.0 m, what volume of earth will be required? Give answer in cubic meters and in cubic yards.

1-14 A 100-story skyscraper is 125.0 ft wide and 175.0 ft long. If each story is 9.8 ft tall, calculate the volume of the building.

1-15 The U.S. Interstate Highway System consists of about 42,500 miles of highway. If the average right-of-way is 200 ft wide, what is the total area of right-of-way? How does this compare with the area of Rhode Island?

MATHEMATICS USED IN SURVEYING

Frankly, there is no tremendous amount of highly sophisticated mathematics used in this book. However, there is a certain amount of mathematics needed by the user, and it is important that he or she have a thorough understanding of that which is employed. Undoubtedly, some users of this book are already prepared, and these are invited to skip directly to Chapter 3. It is believed, however, that many who use this book will either lack this knowledge or will at least need to review the following material.

2-1 Significant Figures

Prior to attempting to carry out mathematical manipulations, an understanding of significant figures is of great importance. Significant figures refer to those digits in a given number whose values are known. For example, if a distance of about 50 ft is measured accurately to the nearest hundredth of a foot (say 50.22 ft), then the result has four significant figures. If, however, a distance of about 150 ft is measured accurately to the nearest hundredth of a foot (say 150.22 ft), then the result has five significant figures. Generally, all digits in a number except for leading zeroes are significant figures (leading zeroes are used only to place the decimal). Consider the following examples:

7.405 contains four significant figures
0.0074 contains two significant figures

10.0074 contains six significant figures
10.000 contains five significant figures
10.0 contains three significant figures
0.00001 contains one significant figure

In the case of a number ending in one or more zeroes and no decimal, there may be some question as to the number of significant figures intended. For example, a distance of 3000 ft has four significant figures if the measurement was in fact determined to the nearest foot. If, however, the measurement was made only to the nearest 100 ft, then 3000 ft has only two significant figures. Confusion can be avoided by using scientific notation. Thus 3.000×10^3 has four significant figures, while 3.0×10^3 has two.

It is important in surveying (as well as in some other enterprises) that the proper number of significant figures be recorded for every measured value. If this is done, the precision with which the measurement was made is clearly indicated. If a distance is read to the nearest hundredth of a foot as 58.00 ft, then it should be recorded as 58.00 ft—not 58.0 ft or 58 ft. If 58.0 ft is recorded, the implication is that the distance was measured only to the nearest tenth of a foot; if only 58 ft is recorded, the implication is that the distance was measured only to the nearest foot.

Care must be taken with regard to significant figures in computations. In addition and subtraction, the number of significant figures is controlled by the last column *full* of significant figures. For example, if the three measured quantities, 62.417 cm, 12.0 cm, and 55.08 cm, are to be added, the result upon entering these numbers into a calculator is 129.497 cm. This number implies that the sum is accurate to the nearest thousandth of a centimeter. However, two of the three addends were not measured to the nearest thousandth; hence it is improper to imply that the sum is correct to the nearest thousandth. In this case, since the value 12.0 cm was read only to the nearest tenth of a centimeter, then the sum is no better than this and should be reported only to the nearest tenth—in this case 129.5 cm. Arranging this summation in column form shows that the last column full of significant figures is the tenths column.

$$
\begin{array}{r}
\downarrow \\
62.417 \\
12.0 \\
\underline{55.08} \\
129.5
\end{array}
$$

In multiplication and division, the number of significant figures is the same as the least number of significant figures in any of the factors. For example, consider the following fraction made up of measured quantities:

$$\frac{(25.62)(127.18)}{(144)(0.011222)}$$

If these numbers are entered into a calculator, the result is 2016.346611, which has ten significant figures. The two numbers in the numerator of the fraction have four and five significant figures, respectively; the two in the denominator have three and five. In accord with the first sentence of this paragraph, the number 144 with three significant figures controls the result, and the answer should be reported to three significant figures—that is, 2020 or (better) 2.02×10^3. Suppose, however, that the 144 in the denominator is a conversion factor for converting square feet to square inches. In this case, 144 is an exact number (not a measurement); in effect, it has as many significant figures as one wants. In this case, the previous answer (2016.346611) is controlled by the number 25.62 (in the numerator), which has four significant figures, and thus the answer should be reported to four significant figures—that is, 2016.

In practice it is common to carry one or two more significant figures than is necessary for intermediate results and then to report the final answer to the proper number of significant figures. If this is done, there is little chance that precision is lost merely through calculation.

2-2 Algebra

The surveyor must be able to carry out simple algebraic manipulations and to solve simple equations. Considered here are (1) solution of a linear equation, (2) solution of a quadratic equation, and (3) solution of simultaneous equations. Let us begin with the solution of a linear equation.

The general procedure for solving a linear equation is to move terms so that the unknown remains alone on one side of the equation and everything else appears on the other side. The "everything else" can then be solved mathematically to obtain the value of the unknown. The movement of terms in the equations can be effected by either adding (or subtracting) the same value to (or from) both sides of the equation or by multiplying (or dividing) both sides of the equation by the same value. Consider the following examples.

Example 2-1

$$x = a + b^2 + \frac{c}{d}$$

Solve for x when $a = -5.6$, $b = -2.0$, $c = 0.8$, and $d = 0.4$.

Solution

Substituting these values into the equation gives

$$x = (-5.6) + (-2.0)^2 + \frac{0.8}{0.4}$$

In this example, the unknown is already alone on one side of the equation; thus the solution involves the algebraic combination of the terms on the right. Solving in this manner gives $x = 0.4$.

Example 2-2

$$\frac{2}{y} + 20.2 = \frac{6}{y} + 10.2$$

Solve for y.

Solution

As a first step, the y's can be eliminated in the denominator by multiplying both sides of the equation by y. The result of this operation is

$$2 + 20.2y = 6 + 10.2y$$

In order to get the unknown (y) to one side of the equation, the value "$10.2y$" can be subtracted from both sides, giving

$$2 + 10.0y = 6$$

In order to get everything other than the unknown to the other side of the equation, the value "2" can be subtracted from both sides, giving

$$10.0y = 4$$

In order to isolate y completely on one side of the equation, both sides of the equation can now be divided by 10.0, giving $y = 0.4$.

Consider now the quadratic equation, of the form

$$ax^2 + bx + c = 0 \tag{2-1}$$

where a, b, and c are constants and x is the unknown. This equation differs from the linear equation in that there is an "x squared" term. In the special case where b equals zero, this equation can be solved by isolating x^2 on one side of the equation and then taking the square root of both sides. In the general case, however, a solution can always be obtained by substituting into the quadratic formula

$$x = \frac{-b \pm \sqrt{b^2 - 4ac}}{2a} \tag{2-2}$$

Actually there are two values of x (roots) obtainable from this equation—one using the plus (+) sign and one using the minus (−) sign just in front of the radical. If the value under the radical is zero, there is only one numerical solution. Technically, however, there are two roots, each having the same numerical value. (Mathematicians call these "repeated roots.") If the value under the radical is negative, the roots are "imaginary." Such a solution (imaginary roots) generally should not occur in surveying applications.

Example 2-3

$$3x^2 - 40 = x - 20 + x^2$$

Solve for x.

Solution

In solving this type of equation, it is helpful to get the equation in the form of Eq. (2-1). This can be accomplished in this example by subtracting x^2 and x from both sides of the equation and adding 20 to both sides. If this is done, the result is

$$2x^2 - x - 20 = 0$$

This is now in the form of Eq. (2-1), with $a = 2$, $b = -1$, and $c = -20$. The equation can be solved by substituting these values of a, b, and c into Eq. (2-2), giving

$$x = \frac{-(-1) \pm \sqrt{(-1)^2 - 4(2)(-20)}}{2(2)}$$

Upon simplifying,

$$x = \frac{1 \pm 12.69}{4}$$

Using the plus sign gives $x_1 = 3.42$. Using the minus sign gives $x_2 = -2.92$. Both these values of x are solutions of the equation.

Some quadratic equations can be solved by factoring and setting each factor equal to zero. However, since not all equations can be solved readily in this manner, the procedure is not presented here. The use of Eq. (2-2) will always give a solution.

Finally, consider the solution of simultaneous equations. Simultaneous equations are two or more equations containing a number of unknowns equal to the number of equations. While there may be any number of equations with a corresponding number of unknowns, the discussion here is limited to two equations with two unknowns. It is rare in elementary surveying to need to solve more than two simultaneous equations.

Two general methods for solving two simultaneous equations are considered. One is to add (or subtract) the two equations in such a manner that one of the unknowns is eliminated while the other remains in a single equation. The single equation can then be solved for the remaining unknown, and that value substituted into one of the original equations to solve for the other unknown. The other general method is to use one of the equations to solve for one of the unknowns in terms of the other. This value can then be substituted into the other equation, resulting in one equation with one unknown. The following example illustrates both these methods.

Example 2-4

Solve for x and y in the following two simultaneous equations:

(1) $2x-3y=11$

(2) $3x+4y=8$

Solution 1 (by first method related above):

Multiply (1) by 4 and (2) by 3, giving

(3) $8x-12y=44$

(4) $9x+12y=24$

Adding (3) and (4) gives

(5) $17x=68$

(6) $x=4$

Substituting this value in (1) gives

(7) $2(4)-3y=11$

(8) $-3y=3$

(9) $y=-1$

Thus the solution is $x=4$ and $y=-1$.

Solution 2 (by second method related above):

Solve (1) for x, giving

(10) $x=\frac{11}{2}+\frac{3y}{2}$

Substituting this value of x into (2) gives

(11) $3\left(\frac{11}{2}+\frac{3y}{2}\right)+4y=8$

(12) $\frac{33}{2}+\frac{9y}{2}+4y=8$

(13) $33+9y+8y=16$

(14) $17y=-17$

(15) $y=-1$

Substituting this value of y into (10) gives

(16) $x=\frac{11}{2}+\frac{3(-1)}{2}$

(17) $x=4$

Thus the solution is (again) $x=4$ and $y=-1$.

2-3 Geometry

In order to deal with problems of a geometric nature, it is convenient to use a rectangular coordinate system. A rectangular coordinate system is made up simply ot two lines that intersect at right angles. One of the lines may be thought of as oriented horizontally, and the other vertically. The horizontal line is referred to as the abscissa, or the x axis. The vertical line is referred to as the ordinate, or the y axis. A typical x/y rectangular coordinate system is shown in Figure 2-1*a*. If the point of intersection of the two axes (called the origin) is assumed to be zero for both axes, then positive values of x are measured toward the right and negative values toward the left. Positive values of y are measured upward and negative values downward. The x/y rectangular coordinate system shown in Figure 2-1*a* is commonly used in engineering and mathematics. In surveying, however, the coordinate system is frequently designated according to the four compass points, with the abscissa

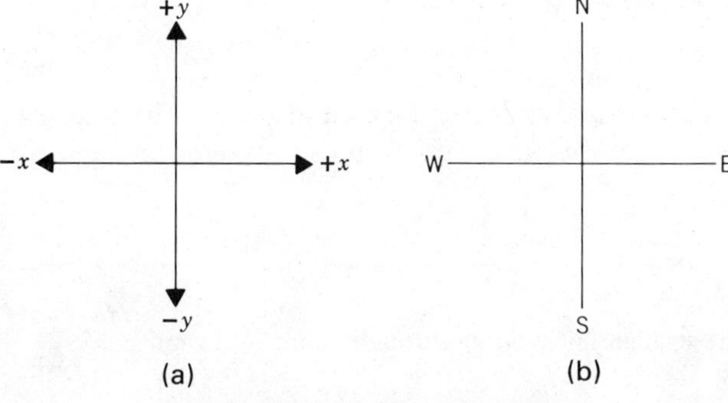

Figure 2-1 Rectangular coordinate systems.

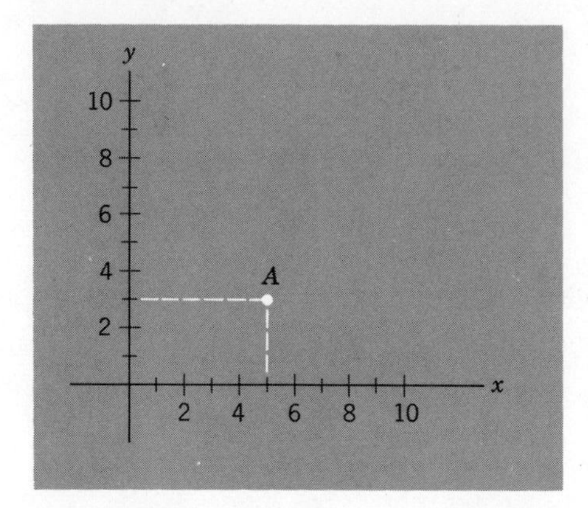

Figure 2-2

designating East (E) toward the right and West (W) toward the left and the ordinate designating North (N) upward and South (S) downward. Such a coordinate system is shown in Figure 2-1*b*.

The location of any point may be specified by indicating a distance from the origin along the x axis and a distance from the origin along the y axis. For example, in Figure 2-2, point A is specified as being 5 units along the x axis and 3 units along the y axis. This point would be designated as (5, 3), where the first number specifies distance along the x axis, and the second specifies distance along the y axis.

In surveying it is sometimes necessary to find the equation of a straight line passing through two given points and to find the equation of a circle with a given center and radius. The general equation of a straight line is

$$y = mx + b \tag{2-3}$$

where m and b are constants for a specific line. (Actually, m is the slope of the line, and b is the y intercept—the value of y when x equals zero.) The general equation of a circle is

$$(x-a)^2 + (y-b)^2 = r^2 \tag{2-4}$$

where the circle has a radius r and a center located at (a, b). The following examples illustrate how to determine the equation of a straight line and that of a circle.

Example 2-5

Find the equation of the straight line passing through points (4, 1) and (−1, −2), as shown in Figure 2-3.

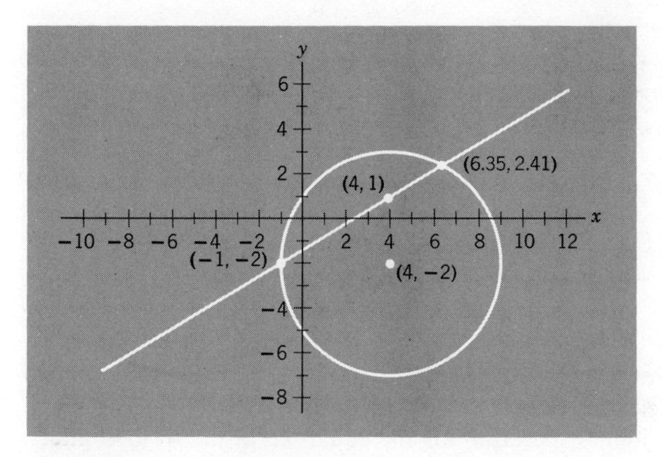

Figure 2-3

Solution

$$y = mx + b \tag{2-3}$$

Substituting the coordinates (4, 1) into Eq. (2-3) gives

(1) $1 = m(4) + b$

Substituting the coordinates (−1, −2) into Eq. (2-3) gives

(2) $-2 = m(-1) + b$

The result is two equations with two unknowns (m and b), and these may be solved simultaneously to determine the values of m and b, which define the straight line passing through the given points. If (2) is subtracted from (1), the result is

(3) $3 = 5m$

Thus $m = \frac{3}{5}$.
This value of m can be substituted into (1) to determine the value of b.

(4) $1 = \frac{3}{5}(4) + b$

Thus $b = -\frac{7}{5}$.

By substituting $m = \frac{3}{5}$ and $b = -\frac{7}{5}$ into Eq. (2-3), the equation of the line is found to be

$y = \frac{3}{5}x - \frac{7}{5}$ or $5y = 3x - 7$

Example 2-6
Find the equation of a circle with center at (4, −2) and radius 5.

Solution

$$(x-a)^2+(y-b)^2=r^2 \tag{2-4}$$

In this case, $a=4$, $b=-2$, and $r=5$. Thus the equation of the circle is

$$(x-4)^2+(y+2)^2=5^2$$

Sometimes we want to find the point or points of intersection of two straight lines, two circles, or a circle and a straight line (all, of course, lying in a plane). This can be done by determining the equations of the straight line(s) and/or circle(s) and solving them simultaneously to determine the values of x and y that will satisfy both equations.

Example 2-7
Find the points of intersection of the straight line determined in Example 2-5 and the circle determined in Example 2-6.

Solution

(1) $y=\frac{3}{5}x-\frac{7}{5}$

(2) $(x-4)^2+(y+2)^2=5^2$

To determine the points of intersection, these two equations must be solved simultaneously. Since (1) is already solved for y in terms of x, this value of y may be substituted into (2), giving

$$(x-4)^2+(\tfrac{3}{5}x-\tfrac{7}{5}+2)^2=5^2$$

$$(x-4)^2+(\tfrac{3}{5}x+\tfrac{3}{5})^2=5^2$$

$$x^2-8x+16+\tfrac{9}{25}x^2+\tfrac{18}{25}x+\tfrac{9}{25}=25$$

$$25x^2-200x+400+9x^2+18x+9=625$$

$$34x^2-182x-216=0$$

Substituting into Eq. (2-2) with $a=34$, $b=-182$, and $c=-216$ gives

$$x=\frac{-(-182)\pm\sqrt{(-182)^2-4(34)(-216)}}{2(34)}$$

$$x=\frac{182\pm 250.0}{68}$$

$$x_1=6.35$$

$$x_2=-1.00$$

These values of x can be substituted into (1) to determine the corresponding values of y.

$$y_1 = \tfrac{3}{5}(6.35) - \tfrac{7}{5}$$

$$y_1 = 2.41$$

$$y_2 = \tfrac{3}{5}(-1.00) - \tfrac{7}{5}$$

$$y_2 = -2.00$$

Thus the two points of intersection are (6.35, 2.41) and (−1.00, −2.00). These are shown in Figure 2-3.

Note: In Example 2-7, if the value under the radical had been negative, it would mean that the straight line and circle do not intersect.

2-4 Trigonometry

Of all phases of mathematics, trigonometry (trig, for short) is certainly one of the most important (if not the most important) in regard to surveying. Generally speaking, trigonometry is the study of triangles where a triangle is a closed figure comprised of three straight sides.

Of special interest is the right triangle, a triangle containing one right angle (i.e., a 90° angle). For every right triangle, there is a special relationship among the lengths of the sides, as given by the following equation:

$$r^2 = x^2 + y^2 \tag{2-5}$$

where r (called the hypotenuse) is the side opposite the right angle and x and y (called the legs) are the other two sides of the triangle (see Figure 2-4). This important relationship is known as the Pythagorean Theorem.

There are also some interesting and important relationships among the sides and angles of a right triangle. These are called natural functions or trigonometric (trig) functions. Consider the right triangle shown in Figure 2-4.

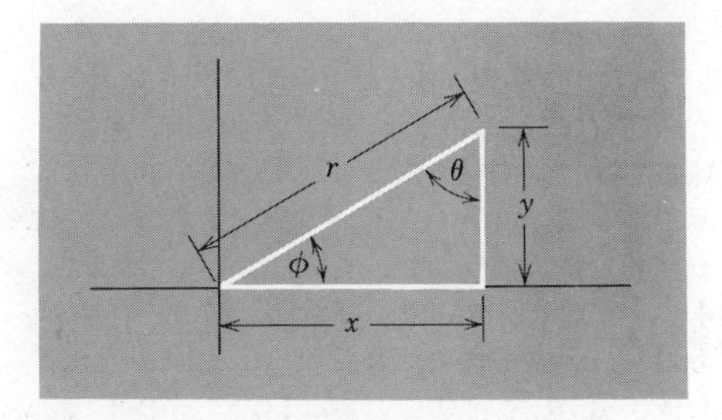

Figure 2-4 The right triangle

The sine (abbreviated sin) of ϕ is equal to the leg opposite ϕ divided by the hypotenuse of the right triangle. Or

$$\sin \phi = \frac{y}{r} \tag{2-6}$$

The cosine (cos) of ϕ is equal to the leg adjacent to ϕ divided by the hypotenuse of the right triangle. Or

$$\cos \phi = \frac{x}{r} \tag{2-7}$$

The tangent (tan) of ϕ is equal to the leg opposite ϕ divided by the leg adjacent to ϕ in the right triangle. Or

$$\tan \phi = \frac{y}{x} \tag{2-8}$$

The tangent of ϕ can also be determined by dividing the sine of ϕ by the cosine of ϕ. Or

$$\tan \phi = \frac{\sin \phi}{\cos \phi} = \frac{y/r}{x/r} = \frac{y}{x}$$

These three functions are the ones most used in surveying; however, there are three additional ones, related to these first three, that should be mentioned here. They are cotangent (cot), secant (sec), and cosecant (csc). The relationships are as follows:

$$\cot \phi = \frac{1}{\tan \phi} = \frac{x}{y} \tag{2-9}$$

$$\sec \phi = \frac{1}{\cos \phi} = \frac{r}{x} \tag{2-10}$$

$$\csc \phi = \frac{1}{\sin \phi} = \frac{r}{y} \tag{2-11}$$

These functions are important because they relate the unique relationships that exist among the sides and angles of a right triangle. For a given angle, there is one (and only one) value of the sine of that angle. This value of the sine of a given angle may be obtained by looking in trig tables, or it can be obtained at the punch of a button using the modern hand calculator. For example, the sine of 56°26′ is 0.83324, to five significant figures. Thus in any right triangle, one angle of which is 56°26′, the ratio of the length of the side opposite this angle to the length of the hypotenuse is 0.83324. This ratio is the same regardless of how large or how small the triangle is. The same considerations apply to the other functions (cosine, tangent, etc.).

Example 2-8
If the length of the side of a right triangle opposite an angle of 56°26′ is 100.00 m, what is the length of the hypotenuse?

Solution
Referring to Figure 2-4 with $\phi = 56°26'$ and $y = 100.00$ m, use Eq. (2-6).

$$\sin \phi = \frac{y}{r} \qquad (2\text{-}6)$$

$$\sin 56°26' = \frac{100.00 \text{ m}}{r}$$

$$r = \frac{100.00 \text{ m}}{\sin 56°26'} = \frac{100.00 \text{ m}}{0.83324}$$

$$r = 120.01 \text{ m}$$

Example 2-9
If the lengths of the two legs of a right triangle are 526 cm and 138 cm, what is the size of the angle opposite the shorter leg?

Solution
Referring to Figure 2-4 with $y = 138$ cm and $x = 526$ cm, use Eq. (2-8).

$$\tan \phi = \frac{y}{x} \qquad (2\text{-}8)$$

$$\tan \phi = \frac{138 \text{ cm}}{526 \text{ cm}} = 0.26236$$

$$\phi = \arctan 0.26236$$

$$\phi = 14°42'$$

The third step in Example 2-9 above involves the arctan, which means "the angle whose tangent is." Thus in this case it was determined that the arctan of 0.26236, or the angle whose tangent is 0.26236, is 14°42′. This can be determined by using trig tables or by using the hand calculator. "Arc" can be used in the same manner with the other trig functions.

Example 2-10

For the conditions given in Example 2-9, determine the length of the hypotenuse and the size of the angle opposite the other (longer) leg.

Solution

Since the lengths of the two legs are known, the length of the hypotenuse can be determined by utilizing Eq. (2-5).

$$r^2 = x^2 + y^2 \qquad (2\text{-}5)$$

$$r^2 = (526 \text{ cm})^2 + (138 \text{ cm})^2$$

$$r^2 = 295{,}720 \text{ cm}^2$$

$$r = 544 \text{ cm}$$

The length of the hypotenuse could have been determined using either the sine or cosine function. For example,

$$\cos \phi = \frac{x}{r} \qquad (2\text{-}7)$$

$$\cos 14°42' = \frac{526 \text{ cm}}{r}$$

$$r = \frac{526 \text{ cm}}{\cos 14°42'} = \frac{526 \text{ cm}}{0.96727}$$

$$r = 544 \text{ cm}$$

The size of the angle opposite the longer leg can be determined using any of the functions, since all sides are now known. For example,

$$\tan \theta = \frac{x}{y}$$

$$\tan \theta = \frac{526 \text{ cm}}{138 \text{cm}} = 3.81159$$

$$\theta = \arctan 3.81159$$

$$\theta = 75°18'$$

Note that in Example 2-10, $\tan \theta$ is equal to x/y. This seems to be in disagreement with Eq. (2-8), which says that $\tan \phi$ is equal to y/x. It should be noted, however, that Eq. (2-8) refers to the angle "ϕ" (phi) in Figure 2-4; whereas the equation involving the tangent function in Example 2-10 refers to the angle "θ" (theta) in Figure 2-4. Recall that the tangent of an angle is equal to the leg opposite the angle divided by the leg adjacent to the angle in a right triangle. Thus the leg opposite angle ϕ (Figure 2-4) is y and the leg adjacent to angle ϕ is x, giving tan ϕ equal to y/x. The leg opposite angle θ is x and the leg adjacent to angle θ is y, giving tan θ equal to x/y.

In Examples 2-9 and 2-10, the values of the three angles are 14°42′, 75°18′, and 90°00′. If these three angles are added, the sum is 180°00′. This will always be true because the sum of the angles of a triangle is equal to 180° (exactly).

In Figure 2-4, the angle ϕ is an acute angle (i.e., less than 90°). Suppose, however, that an obtuse angle (i.e., greater than 90° and less than 180°) is under consideration. How are the sine, cosine, and tangent of such an angle determined? If a hand calculator containing trig functions is available, the solution is easy. For example, consider the angle 162°15′. Using a calculator, sin 162°15′ is 0.30486, cos 162°15′ is −0.95240, and tan 162°15′ is −0.32010. Why is the value of sine positive, while those of cosine and tangent are negative? Also, how are these values determined if trig tables (which normally only give values of the functions for angles between zero and 90°) are used and no hand calculator is available?

To answer these questions, refer to Figure 2-5. The values of the functions of an obtuse angle are determined by looking up the values of the functions for the *supplement* (i.e., 180° minus the obtuse angle). The supplement of 162°15′ is 17°45′. Looking up the functions of this angle, one finds sin 17°45′ is 0.30486, cos 17°45′ is 0.95240, and tan 17°45′ is 0.32010. It will be noted that these values are equal in absolute value to the corresponding ones for an angle of 162°15′. The only difference is in the signs. The sine of 162°15′ and the sine of 17°45′ are both positive; whereas the cosine and tangent of 162°15′ are negative, while the cosine and tangent of 17°45′ are positive. Thus to look up the trig functions for an obtuse angle, one must look up the functions for the supplement and place a minus sign in front of values of cosine and tangent (also cotangent and secant) of the supplementary angle. The sine (and cosecant) will have the same sign (i.e., positive) for the angle and its supplement.

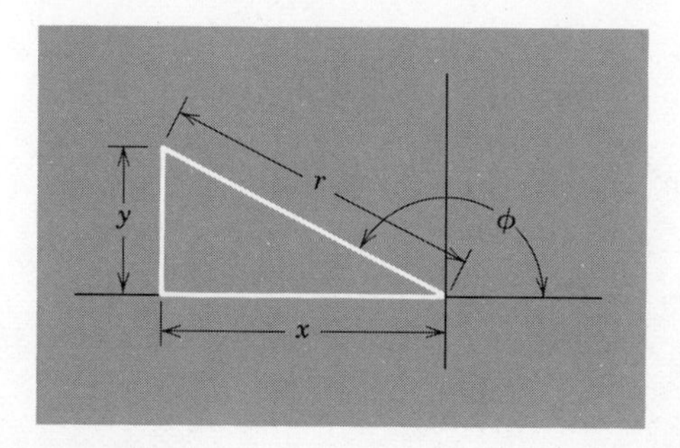

Figure 2-5

Example 2-11
Determine from trig tables the values of the trig functions for an angle of 177°16′.

Solution
Supplement of 177°16′ is 2°44′. From tables

$$\sin 2°44' = 0.04769$$
$$\cos 2°44' = 0.99886$$
$$\tan 2°44' = 0.04774$$
$$\cot 2°44' = 20.9460$$
$$\sec 2°44' = 1.00114$$
$$\csc 2°44' = 20.9698$$

In accord with the previous discussion,

$$\sin 177°16' = 0.04769$$
$$\cos 177°16' = -0.99886$$
$$\tan 177°16' = -0.04774$$
$$\cot 177°16' = -20.9460$$
$$\sec 177°16' = -1.00114$$
$$\csc 177°16' = 20.9698$$

All the discussion thus far in this section has concerned the right triangle. Frequently, however, triangles other than right triangles are encountered, and it is necessary to solve for unknown sides and angles in these. This can be accomplished by utilizing two laws—the law of sines and the law of cosines.

The law of sines states that the ratios of each side of any triangle to the sine of the angle opposite the side are equal. Referring to Figure 2-6, the law of sines may be stated in equation form as

$$\frac{a}{\sin A} = \frac{b}{\sin B} = \frac{c}{\sin C} \tag{2-12}$$

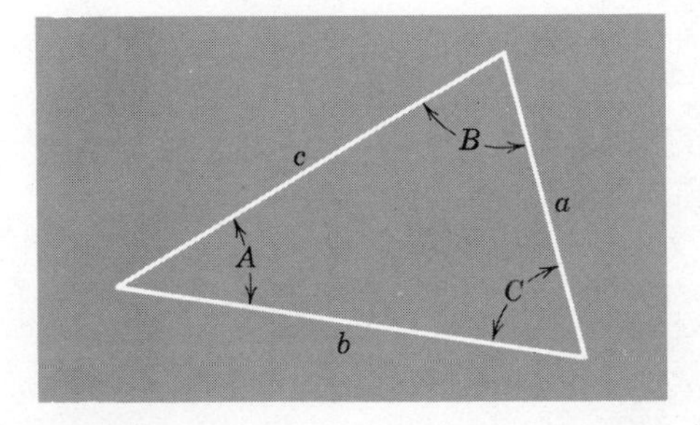

Figure 2-6

The law of cosines may be stated (referring to Figure 2-6) as

$$c^2 = a^2 + b^2 - 2ab \cos C \tag{2-13}$$

It should be emphasized that, in Eq. (2-13), the side c on the left side of the equation must be the side opposite angle C, which appears on the right side of the equation.

The law of sines and the law of cosines apply to all triangles. Considering every triangle to consist of six parts (three sides and three angles), it is possible to solve for all six parts of a triangle if any three are known—provided at least one side is known.

Example 2-12

Referring to Figure 2-6, if $C = 25°03'$, $b = 250$ cm, and $a = 266$ cm, calculate the remaining side and angles.

Solution

Using the law of cosines,

$$c^2 = a^2 + b^2 - 2ab \cos C \tag{2-13}$$

$$c^2 = (266 \text{ cm})^2 + (250 \text{ cm})^2 - 2(266 \text{ cm})(250 \text{ cm}) \cos 25°03'$$

$$c = 113 \text{ cm}$$

Using the law of sines,

$$\frac{a}{\sin A} = \frac{b}{\sin B} = \frac{c}{\sin C} \tag{2-12}$$

$$\frac{250 \text{ cm}}{\sin B} = \frac{113 \text{ cm}}{\sin 25°03'}$$

$$\sin B = 0.93675$$

$$B = 69°31'$$

The angle A can be determined by subtracting the sum of angles B and C from 180°.

$$A = 180° - (69°31' + 25°03')$$

$$A = 85°26'$$

2-5 Probability and Statistics

Surveying practice is replete with physical measurements, including distances, angles, elevations, and so on. Strictly speaking, there is some error involved in every measurement that is made. In order to understand and deal with these errors, the surveyor should have some knowledge of probability and statistics.

The term error, as used here, refers to a difference between a measured value and the "true" value, the difference resulting from circumstances more or less beyond the control of the individual making the measurement. (Since the "true" value is generally not known, the amount of the error is not known.) Suppose, for example, an angle, the true value of which is 26°36′, is measured; but, because the transit is slightly out of adjustment, the angle is measured as 26°37′. The difference of 01′ between the measured value and the true value is an *instrumental error*, arising from imperfection or faulty adjustment of the measuring device. Or, suppose that, in the process of measuring the angle with the transit, the individual inadvertently sights the cross hair slightly off the target, resulting in a difference of 01′ between the measured value and the true value. This would constitute a *personal error*. Finally, suppose a true distance of 98.18 ft is being measured with a 100-ft tape, but the temperature is cold enough that the tape has contracted by 0.01 ft. The result would be a tape reading of 98.19 ft, which would constitute a *natural error* of 0.01 ft.

Errors are also classified as systematic errors and accidental errors. *Systematic errors* have the same sign for a given situation. For example, using the tape that is 0.01 ft too short (previous paragraph) would result in a systematic error of +0.01 ft for each tape length measured. Systematic errors are cumulative, and thus their effect can generally be computed and corrections applied. *Accidental errors* are such that the magnitude and sign of each error are matters of chance. Since these will be both positive and negative, they may tend in part to average out, or compensate. For example, in taping a long distance, each time a full tape length is marked off with a taping pin, the possibility exists that the taping pin may be set slightly to one side or the other of the graduation mark on the tape. Thus, sometimes the pin would be set slightly too short and other times it would be set slightly too long.

It was stated initially in this section that there is some error in every measurement that is made. To elaborate on this point, consider that a measurement of 12.82 ft is made with a tape that is calibrated to the nearest tenth of a foot, with interpolation to the nearest hundredth of a foot estimated visually by the tapeman. A recorded value of 12.82 ft implies that it is believed that the true value lies somewhere between 12.815 ft and 12.825 ft. The third digit to the right of the decimal cannot be read on the indicated tape; thus the exact true value between 12.815 and 12.825 cannot be determined, and some error exists. Suppose now the tape is replaced by another that is calibrated to the nearest hundredth of a foot. The same measurement with this tape might result in a value of 12.823 ft, with interpolation to the nearest thousandth of a foot estimated visually by the tapeman. A recorded value of 12.823 ft implies that it is believed that the true value lies somewhere between 12.8225 ft and 12.8235 ft. The fourth digit to the right of the decimal cannot be read; thus the exact true value between 12.8225 ft and 12.8235 ft cannot be determined, and some error exists. Theoretically, this scheme could be repeated to use a tape calibrated to the nearest thousandth of a foot with interpolation to the nearest ten-thousandth of a foot. Practically speaking, however, it is virtually impossible for the human eye to see divisions of a foot into 10,000 parts. Thus, no matter how many digits are read in making a given measurement, there is always another digit that cannot be read, and there is always some error of unknown magnitude in every measurement.

It should be noted that the term "measurement" refers to a quantity that has fractional parts possible, such as the measurement of distance in the previous example. The discussion does not refer to counts of whole number quantities, for these, of course, have no fractional part. For example, suppose one asks how many students are registered for a given class, and a check of the class roll indicates that there are 29 students registered. Assuming no mistake is made in counting the number of students registered, then 29 is the "true" value; and there is no error. This is because the number 29 is an exact whole number and no fractional part is possible. (It would make no sense to say there are 29.36 students registered.)

In the previous example, even if the number of students registered in the class had been miscounted as 28, there would still not be any error. Instead, the discrepancy would be a "mistake." The distinction between an error and a mistake should be understood. As was previously explained, an *error* results from circumstances more or less beyond the control of the individual making the measurement. A *mistake* results from circumstances more or less within the control of the individual. A mistake generally is the result of carelessness or poor judgment. In other words, a mistake is a "blunder" or a "goof." For example, if a person reads a tape distance as 76.4 but records 74.6 in the field notes, then this is a mistake. Mistakes should not occur in surveying; however, human beings are not infallible, and everyone is going to make a mistake from time to time. The surveyor must make every effort to minimize the occurrence of mistakes, and when they are discovered (many mistakes are quite conspicuous), corrections should be made. This frequently means that the measurement in question must be repeated. It should be emphasized that, in the discussion in the remainder of this chapter concerning adjustment of measured values, the adjustments refer to errors—not mistakes.

If systematic errors occur, their effects should be calculated and corrections applied to the measurements in question. In the case of accidental errors, however, since some are positive and some are negative, and since their respective causes may not be known, it is generally not possible to calculate their effects and apply corrections to the measurements. Instead, measurements containing accidental errors are dealt with on the basis of probability theory to determine the "best" value to use based on available data.

Probability theory is complex, and a complete treatment is far beyond the scope of this book. However, a few simple applications in surveying will be presented here, without going into much theoretical detail.

The term "most probable value" (*MPV*) refers to a quantity which, based on available data, has the best chance of being correct (i.e., the "true" value). Suppose a given quantity has been measured more than one time, resulting in more then one value. Obviously not all the different values are correct, so what should be reported as the correct value? According to probability theory, if all measured values are thought to be equally reliable, the most probable value is the mean of the measured values. In equation form

$$MPV_m = \frac{\sum_{i=1}^{n} y_i}{n} \tag{2-14}$$

where MPV_m = most probable value (mean)
y_i = individual measured value
n = number of measured values

(The symbol "$\sum$" (sigma) means "summation." Thus the designation $\sum_{i=1}^{n} y_i$ in the numerator of Eq. (2-14) means to find the sum: $y_1 + y_2 + y_3 + \cdots + y_n$.)

Example 2-13

Suppose each of seven students in a class is given a meter stick and a piece of chalk and is asked to measure the width of a classroom. What is the most probable value of the width of the classroom if the seven students report the following (respective) results: 914.4 cm, 914.3 cm, 914.0 cm, 914.9 cm, 915.1 cm, 914.8 cm, 914.5 cm?

Solution

With no other information given, it will be assumed that all measurements are equally reliable. Hence, the most probable value is the mean of the seven values.

$$MPV_m = \frac{\sum_{i=1}^{n} y_i}{n} \tag{2-14}$$

$$MPV_m = \frac{914.4 + 914.3 + 914.0 + 914.9 + 915.1 + 914.8 + 914.5}{7}$$

$$MPV_m = 914.6 \text{ cm}$$

One other term (in addition to most probable value) that is important here is "probable error" (PE). Probable error is a quantity which, when added to and subtracted from a most probable value, defines a range within which there is a 50 percent chance that the true value lies. There is, of course, a 50 percent chance that the true value lies outside this range. In the previous case where a number of measurements of the same quantity are made, each of equal reliability, and the most probable value is computed as the mean, the probable error of the most probable value (mean) is computed from the following formula:

$$PE_m = \pm 0.6745 \sqrt{\frac{\sum_{i=1}^{n} (\bar{y} - y_i)^2}{n(n-1)}} \tag{2-15}$$

where PE_m = probable error of the most probable value (mean)
$\bar{y}$ = the mean value [same as MPV_m in Eq. (2-14)]
y_i = individual measured value
n = number of measured values

Example 2-14
Calculate the probable error of the most probable value for the data given in Example 2-13.

Solution
$\bar{y} = 914.6$ cm (MPV_m computed in Example 2-13)
$n = 7$

Measured Value (y_i)	$\bar{y} - y_i$	$(\bar{y} - y_i)^2$
914.4	0.2	0.04
914.3	0.3	0.09
914.0	0.6	0.36
914.9	−0.3	0.09
915.1	−0.5	0.25
914.8	−0.2	0.04
914.5	0.1	0.01
		0.88

$$PE_m = \pm 0.6745 \sqrt{\frac{\sum_{i=1}^{n} (\bar{y} - y_i)^2}{n(n-1)}} \tag{2-15}$$

$$PE_m = \pm 0.6745 \sqrt{\frac{0.88}{7(7-1)}}$$

$$PE_m = \pm 0.1 \text{ cm}$$

The results of Examples 2-13 and 2-14 are usually expressed in the form

$MPV = 914.6 \pm 0.1$ cm

There is a 50 percent chance that the true distance lies within the range 914.5 to 914.7 cm. There is also a 50 percent chance that the true distance lies outside that range.

In the previous examples, the seven measurements of the same quantity were assumed to have an equal reliability. Sometimes there may be reason to think that some measurements may be more reliable than others. It may be, for example, that some of the measurements were made under poor weather conditions or some were made by less experienced crew members. In such cases, one might choose to assign weights to each measured value to reflect the relative reliabilities. In this case, the most probable value is the weighted mean, which is calculated by multiplying each measured value by its weight, adding these products, and dividing the resulting sum by the sum of the weights. In equation form

$$MPV_{wm} = \frac{\sum_{i=1}^{n} (W_i \times y_i)}{\sum_{i=1}^{n} W_i} \qquad (2\text{-}16)$$

where MPV_{wm} = most probable value (weighted mean)
W_i = individual weight
y_i = individual measured value
n = number of measured values

The probable error of the weighted mean is computed from the following formula:

$$PE_{wm} = \pm 0.6745 \sqrt{\frac{\sum_{i=1}^{n} W_i(\bar{y} - y_i)^2}{\left(\sum_{i=1}^{n} W_i\right)(n-1)}} \qquad (2\text{-}17)$$

where PE_{wm} = probable error of the most probable value (weighted mean)
W_i = individual weight
$\bar{y}$ = the weighted mean value [same as MPV_{wm} in Eq. (2-16)]
y_i = individual measured value
n = number of measured values

Example 2-15

Using the data of Example 2-13, calculate the most probable value and the probable error, assuming weights of 1:1:1:2:2:3:3 are applied respectively to the given measured distances.

Solution

Measured Value (y_i)	Weight (W_i)	$W_i \times y_i$	$\bar{y} - y_i$	$W_i(\bar{y} - y_i)^2$
914.4	1	914.4	0.3	0.09
914.3	1	914.3	0.4	0.16
914.0	1	914.0	0.7	0.49
914.9	2	1,829.8	−0.2	0.08
915.1	2	1,830.2	−0.4	0.32
914.8	3	2,744.4	−0.1	0.03
914.5	3	2,743.5	0.2	0.12
	13	11,890.6		1.29

$$MPV_{wm} = \frac{\sum_{i=1}^{n} (W_i \times y_i)}{\sum_{i=1}^{n} W_i} \tag{2-16}$$

$$MPV_{wm} = \frac{11,890.6}{13}$$

$$MPV_{wm} = 914.7 \text{ cm}$$

$$PE_{wm} = \pm 0.6745 \sqrt{\frac{\sum_{i=1}^{n} W_i(\bar{y} - y_i)^2}{\left(\sum_{i=1}^{n} W_i\right)(n-1)}} \tag{2-17}$$

$$PE_{wm} = \pm 0.6745 \sqrt{\frac{1.29}{13(7-1)}}$$

$$PE_{wm} = \pm 0.1 \text{ cm}$$

Thus, $MPV = 914.7 \pm 0.1$ cm.

Note: In Example 2-15, $\bar{y}$ is the weighted mean (MPV_{wm}), computed to be 914.7 cm. This value must be determined (using the third column) before the fourth column ($\bar{y} - y_i$) can be determined.

Another situation arises when several related quantities are measured, but because of error of measurement, the known relationship is not sustained precisely. For example, it is a fact that the sum of the three angles of a (plane) triangle is exactly 180°. However, if the three angles of a triangle are measured in the field, their sum might not be exactly 180° because of error in measurement. It is desirable to adjust the values so that they satisfy the required relationship.

If several related quantities of equal reliability are measured and the required relationship is not sustained, then each measured quantity should be adjusted by the equal amount that will sustain the required relationship.

Example 2-16

Suppose the three angles of a triangle are measured with the following results:

$A = 116°40'$
$B = 40°04'$
$C = 23°07'$

Determine the most probable value of each angle.

Solution

The sum of these measured angles is 179°51′. It should, of course, be 180° exactly. According to the previous discussion, each angle should be adjusted by an equal amount to make the sum equal to 180°. The difference between the true value (180°) and the sum obtained from measured values (179°51′) is 09′. If this is divided into three equal parts, the result will be 03′, which must be added to each angle. The most probable values of the three angles are

$A = 116°43'$
$B = 40°07'$
$C = 23°10'$

These three angles now add to 180° exactly. If the computed sum had been larger than 180°, the adjustments would have been subtracted from the measured angles.

Just as in the case of several measurements of the same quantity, in the case of related measurements, it is possible that the related measurements may not have equal reliability. In this case, instead of dividing the difference between the true value and the measured value into equal parts to be applied to the individual measurements, the difference is divided inversely according to the respective weights. The division is made inversely because the more reliable values will have a larger weight, but since they are more reliable, less adjustment should be made to them.

Example 2-17

Using the data of Example 2-16, calculate the most probable value of each angle, assuming weights of 1:2:3 are applied respectively to the given angles.

Solution

Angle	Weight	Adjustment Factor	Adjustment
$A = 116°40'$	1	$1 = \frac{6}{6}$	$\frac{(6/6)}{(11/6)}(09') = 04.9'$
$B = 40°04'$	2	$\frac{1}{2} = \frac{3}{6}$	$\frac{(3/6)}{(11/6)}(09') = 02.5'$
$C = 23°07'$	3	$\frac{1}{3} = \frac{2}{6}$ $\frac{11}{6}$	$\frac{(2/6)}{(11/6)}(09') = 01.6'$

The most probable values of the three angles are

$$
\begin{aligned}
A &= 116°40' + 04.9' = 116°44.9' \text{ or } 116°45' \\
B &= 40°04' + 02.5' = 40°06.5' \text{ or } 40°06' \\
C &= 23°07' + 01.6' = 23°08.6' \text{ or } \underline{23°09'} \\
& \qquad\qquad\qquad\qquad\qquad\qquad 180°00'
\end{aligned}
$$

In this case, the difference of 09′ was apportioned among the three angles inversely with the weights. In order to get adjustment factors inversely proportional to the weights, the reciprocal of each weight is taken. The sum of these reciprocals (adjustment factors) is determined and the difference of 09′ is apportioned according to the ratio of the adjustment factor to the sum of the adjustment factors.

Note : In Example 2-1.7, the adjustments for the three angles (fourth column) were computed to one place beyond the decimal. When these adjustments were added to the original angles, the results were rounded off to the nearest minute. Since original values were read to the nearest minute, adjusted values should not imply greater precision.

One final consideration concerning errors is errors in computed quantities. These may be summarized as follows:

1. If several measured values are added, the probable error of the sum is given by

$$PE_s = \pm\sqrt{PE_1^2 + PE_2^2 + \cdots + PE_n^2} \tag{2-18}$$

where PE_s = probable error of the sum
PE_1, PE_2, etc. = probable error of each respective measured value
n = the number of values added

If all probable errors of the respective measured values are equal, then

$$PE_s = \pm PE_1\sqrt{n} \tag{2-19}$$

2. If one measured value is subtracted from another, the probable error of the difference is given by

$$PE_d = \pm\sqrt{PE_1^2 + PE_2^2} \tag{2-20}$$

3. If one line of length L_1 and probable error PE_1 is multiplied by another of length L_2 and probable error PE_2, the probable error of the product is given by

$$PE_p = \pm\sqrt{(L_1PE_2)^2 + (L_2PE_1)^2} \tag{2-21}$$

Example 2-18

Given the three measurements and corresponding probable errors below, what is the most probable value and the probable error of the sum of these three?

$A = 116.25 \pm 0.02$ m
$B = 222.18 \pm 0.04$ m
$C = 175.75 \pm 0.02$ m

Solution

The most probable value is the sum, or 514.18 m. The probable error of the sum may be computed from Eq. (2-18).

$$PE_s = \pm\sqrt{PE_1^2 + PE_2^2 + \cdots + PE_n^2} \tag{2-18}$$

$$PE_s = \pm\sqrt{(0.02)^2 + (0.04)^2 + (0.02)^2}$$

$$PE_s = \pm 0.05 \text{ m}$$

Thus the sum would be expressed as 514.18 ± 0.05 m.

Example 2-19

Assume that A and B in Examplc 2-18 are the sides of a rectangle. Determine the area of the rectangle and the probable error of that area.

Solution

Area $= A \times B$
Area $= (116.25 \text{ m})(222.18 \text{ m})$
Area $= 25{,}828$ sq m

The probable error may be computed using Eq. (2-21).

$$PE_p = \pm\sqrt{(L_1PE_2)^2 + (L_2PE_1)^2} \tag{2-21}$$

$$PE_p = \pm\sqrt{[(116.25 \text{ m})(0.04 \text{ m})]^2 + [(222.18 \text{ m})(0.02 \text{ m})]^2}$$

$$PE_p = \pm 6 \text{ sq m}$$

Thus the area would be expressed as $25{,}828 \pm 6$ sq m.

2-6 Problems

Solve for y in Problems 2-1, 2-2, and 2-3.

2-1 $25y-17=6.4y+11$

2-2 $y^2-16y=2.5+3y^2+18.1$

2-3 $23.4+y=16y^2-13y$

Solve for a and b in Problems 2-4, 2-5, and 2-6.

2-4 $a+b=10$

$a-b=9$

2-5 $a^2+b=21$

$a-b=-1$

2-6 $b=a+1$

$2a=b-1$

2-7 Find the equation of the straight line passing through points (4,2) and (−6,4).

2-8 Find the equation of the circle with center at (3, −1) and radius 12.

2-9 Find the equation of a circle passing through points (4,4) and (8,5) with radius 12.

2-10 Referring to Problem 2-9, find the coordinates of the point(s) on the circle where x is 6.5.

2-11 Find the point(s) of intersection of the circle $(x-2)^2+(y-3)^2=49$ and the straight line $y=x+1$.

In Problems 2-12 through 2-15, find the unknown value. (All refer to the triangle in Figure 2-4.)

2-12 $y=3.47$ ft; $r=5.47$ ft; $x=$ __?__

2-13 $x=222$ cm; $r=400$ cm; $\phi=$ __?__

2-14 $y=25.5$ m; $x=5.5$ m; $\phi=$ __?__

2-15 $x=155.7$ ft; $\phi=33°33'$; $y=$ __?__

In Problems 2-16 and 2-17, find the three unknown parts of the triangle. (All refer to the triangle in Figure 2-6.)

2-16 $a=75.5$ m; $b=65.5$ m; $c=50.0$ m; $A=$ __?__; $B=$ __?__; $C=$ __?__

2-17 $A=46°09'$; $b=100.0$ ft; $c=82.2$ ft; $a=$ __?__; $B=$ __?__; $C=$ __?__

2-18 For the ten measurements of the same quantity given below, determine the most probable value and the probable error, assuming equal reliability.

121.75; 121.85; 121.68; 121.66; 121.82; 121.65; 121.88; 121.73; 121.71; 121.77

2-19 Determine the most probable value and the probable error for the data of Problem 2-18, assuming respective weights of 1:2:3:3:3:1:4:1:1:4.

2-20 For the four values (given below) of the interior angles of a four-sided figure, determine the most probable values. Assume equal reliability. (The sum of the four angles should be 360°, exactly.)

69°13′; 82°02′; 131°22′; 77°36′

2-21 Determine the most probable values of the angles of Problem 2-20, assuming respective weights of 1:2:4:4.

Problems 2-22 through 2-25 refer to the following measurements.

$R = 83.12 \pm 0.08$ ft
$S = 70.05 \pm 0.04$ ft
$T = 99.14 \pm 0.10$ ft
$U = 66.04 \pm 0.03$ ft

2-22 Determine the most probable value and the probable error of the sum of R, S, T, and U.

2-23 Determine the most probable value and the probable error of the difference between T and U.

2-24 Determine the most probable value and the probable error of $(T + U - S)$.

2-25 Determine the most probable value and the probable error of the area of a rectangle with sides S and T.

2-26 It will be shown in Chapter 7 that the horizontal distance between two points can be determined by stadia using the following formula:

$$H = Ks \cos^2 \alpha + C \cos \alpha$$

where H = horizontal distance, in feet
s = stadia interval, in feet
K = constant for transit
α = vertical angle
C = constant for transit

Compute the horizontal distance between two points if $s = 6.64$ ft, $K = 100$, $\alpha = 13°30'$, and $C = 1.0$.

2-27 Referring to the information given in Problem 2-26, what vertical angle will give a horizontal distance of 587 ft? Assume that values of s, K, and C remain the same as in Problem 2-26.

2-28 The four sided figure shown in Figure 2-7 represents a parcel of property owned by Scott. The line EF represents the center line of a right-of-way for a pipeline that is to cross Scott's parcel. Determine the coordinates of the points of intersection of line EF and side AB and of line EF and side DC. Coordinates are indicated in feet.

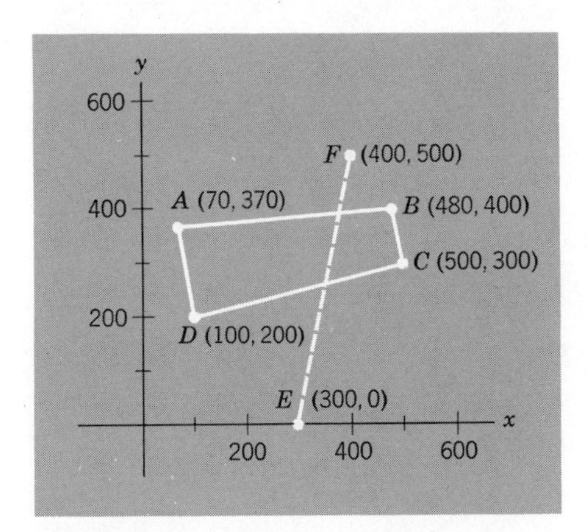

Figure 2-7

2-29 The city limits for a small town are defined as a circle of radius 1.00 mile. A right-of-way for a railroad is to pass through points A and B. If the right-of-way is a straight line, determine the points of intersection of the city limits and the railroad right-of-way. The coordinates of A and B are (0, −700) ft and (900, 0) ft [these assuming the center of the circle at (0, 0)]. See Figure 2-8.

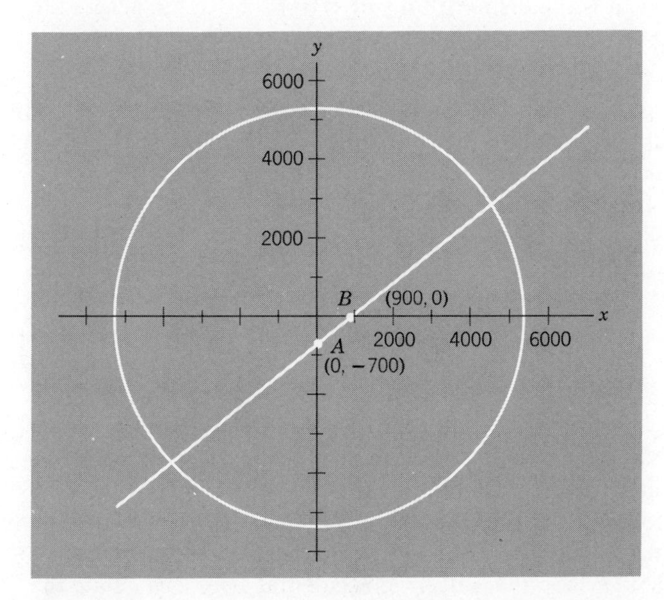

Figure 2-8

2-30 Dutch is a golfer. On a certain lake hole, he keeps hitting the ball into the lake. Feeling that the yardage for the hole is incorrectly stated on the

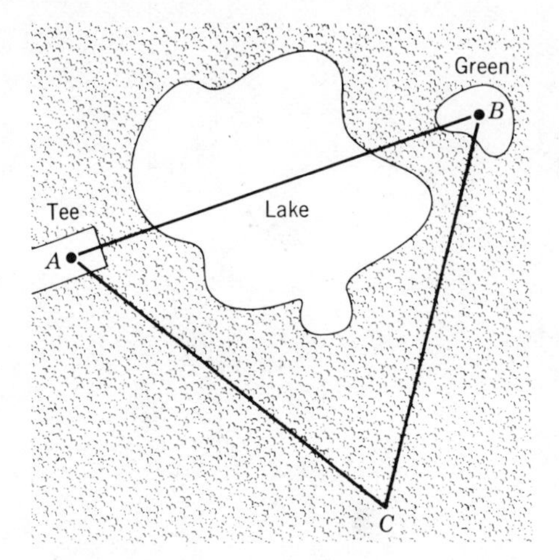

Figure 2-9

score card, he borrowed some equipment from his surveying class and set out to determine the correct yardage. In the sketch (Figure 2-9), A is the tee and B is the pin. He established point C arbitrarily and, using the borrowed equipment, measured AC as 472 ft, BC as 508 ft, and angle ACB as 62°25′. Based on these data, what is the distance (in yards) from tee to pin across the lake (i.e., AB)?

2-31 Linda wanted to know how far it is from her house to a television tower visible in the distance. She laid out a distance AB of 80.00 m in her front yard. She then used a transit to measure angles CAB and CBA (Figure 2-10), obtaining 88°38′ and 89°50′, respectively. If C represents the tower, calculate the distance from B to the tower.

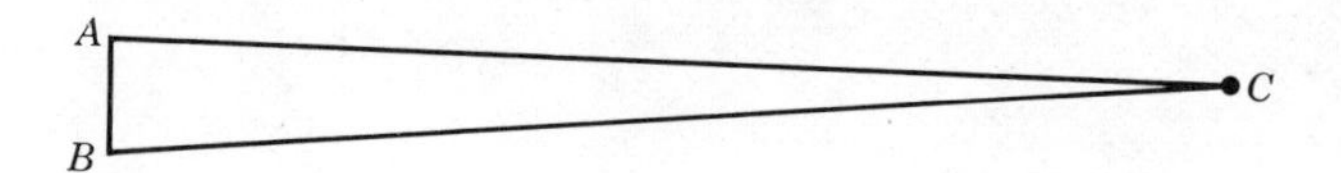

Figure 2-10

2-32 Susan, a surveying instructor, sent the 20 students in her class out to measure a distance between two nails driven into a highway. The highway was over rugged terrain, and the 20 students working in pairs came up with 10 different answers as follows: 420.2 m, 420.8 m, 421.4 m, 420.0 m, 419.9 m, 421.6 m, 421.7 m, 420.2 m, 420.9 m, and 421.1 m. Assuming these values are equally reliable and that variations result from accidental errors, calculate the most probable value and the probable error of the most probable value.

2-33 Referring to Problem 2-32, suppose Susan feels that the work of "A" and "B" students is twice as reliable as that of "C," "D," and "F" students.

Assuming the first five measurements in Problem 2-32 were made by "A" and "B" students and the last five, by "C," "D," and "F" students, calculate the most probable value and the probable error of the most probable value.

2-34 The sides of a rectangle were measured as 416.92 ft and 518.32 ft. If the area must be correct within ± 100 sq ft, what is the maximum permissible probable error in the measurement of each side? Assume the probable error is the same for each side.

SURVEYING EQUIPMENT

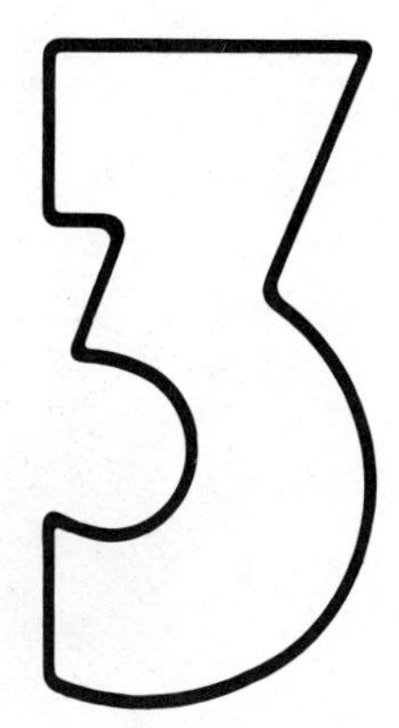

Surveying requires the use of fairly specialized equipment. Some surveying equipment, such as the 100-ft tape for measuring linear distance, is fairly commonplace; whereas some, such as the transit, is known well only by the surveyor. It is the purpose of this chapter to introduce equipment used in surveying and to discuss the use and care of the individual items.

3-1 Taping Equipment

Included in this section are the surveyor's tape, plumb bobs, range poles, taping pins, and the hand level. These items, shown in Figure 3-1, are ordinarily used by taping parties to measure linear distances.

The standard surveyor's tape is made of steel about $\frac{1}{4}$ to $\frac{3}{8}$ inch wide. The usual length of a tape is 100 ft (or 30 m for SI measurement), although shorter and longer ones are available. When not in use, a tape is prepared for storage either by rolling it on a reel or by taking the tape up in successive loops of 5 ft and then "throwing" it into a circle.

Surveyor's tapes are commonly divided into 1 ft intervals, with each interval properly marked. One foot at one or both ends of the tape is graduated in tenths and (sometimes) hundredths of a foot to permit measurement to the nearest tenth or hundredth of a foot. [Tapes graduated in meters are divided and marked at meter intervals, with each meter graduated into centimeters (hundredths of a meter) and a meter at the end graduated into millimeters

Figure 3-1 Equipment used in taping.

(thousandths of a meter).] This graduated foot may be between the zero and 1 ft mark (called a "subtracting" tape) or beyond the zero mark—that is, between the zero mark and the end of the tape—(called an "adding" tape). While tapes of a differing nature are available, the ones described here are the most common. The individual using a tape for the first time should examine it carefully to see how it is graduated in order to avoid mistakes.

Tapes are surprisingly fragile in that they can easily be snapped into two pieces. This can occur if the tape is allowed to kink or if it is run over by an automobile or other vehicle; thus care should be taken to see that these things do not happen. If it is necessary to stretch a tape across a highway and traffic is approaching, the tape should be snatched off the highway immediately. If time does not permit this, the tape can be partially protected by stretching it out flat against the highway surface. A broken tape can be mended; but it introduces a "weak link" in the tape, and a spliced tape is not as accurate as the original. A spliced tape can also be dangerous to the hands if the mending is poorly done. If a tape gets wet during use, it should be wiped dry and lightly oiled prior to storing.

Taping pins are used primarily in marking tape lengths when a distance longer than one tape length is being measured. A taping pin is a thin metal rod a foot or so in length, with a loop formed at one end and a point at the other. The looped end serves more or less as a handle, while the pointed end is inserted into the ground to mark a tape length. Pins are brightly painted—usually red and white—so they can be spotted easily. Eleven pins make up a set, and they are carried on a ring while not actually being used. The ring containing the pins can be inserted under the belt, and carrying the ring and pins in this manner leaves both hands free to carry other equipment.

A plumb bob is simply a weight—generally shaped somewhat like a solid cone—attached to a string. When a plumb bob is lowered by its string, it hangs in a vertical direction, as a result of gravity.

The hand level can be used to ensure that the tape is level when a measurement is being made with the aid of a plumb bob. In simple terms, it consists of a tube or barrel containing a leveling bubble. The device is manufactured such that the individual looking through the barrel can hold it level by observing the leveling bubble.

Range poles are round (or hexagonal) poles ranging from 6 to 8 ft in length; each has a pointed end for sticking it into the ground. Made of metal, wood, or fiberglass, they are painted red and white so they can be easily seen. Normally they are painted with alternating lengths of one foot white, one foot red, and so on. This allows the range pole to be used to measure short distances approximately when the tape is not readily available. Range poles are used in running lines and in taping to help maintain proper alignment.

3-2 The Transit

Of all pieces of surveying equipment, certainly the transit is the one most associated by the layman with surveying, and the one most important to the surveyor. Using a transit, one can measure horizontal distance (by stadia), measure vertical distance (by running levels), measure angles in a horizontal plane, measure angles in a vertical plane, measure direction (bearings and azimuths), and align points on a straight line. These include all of the basic surveying operations.

A typical transit is shown in Figure 3-2. When in use, the transit is mounted on (screwed onto) a tripod (Figure 3-3). The first step in utilizing a transit is to set it up and level it. In most instances it is necessary to set up the transit directly above a reference point on the ground. In order to do this, a plumb bob is suspended from a hook on the under side of the transit. The transit is then set up over the point, by adjusting the tripod legs so that the plumb bob is directly over the reference point. This is not particularly difficult if the transit is being set up on fairly level ground; it is more difficult if it is being set up on uneven ground. In the latter case, it may be necessary to have one of the tripod legs at an odd angle. In any case, the end of each tripod leg should be firmly planted (inserted) in the ground to prevent movement of the instrument.

Figure 3-2 Transit. (Courtesy of Keuffel & Esser Co.)

Once the tripod is set up with the plumb bob directly over the reference point, the next step before making any measurement is to level the base of the transit. (For much of the discussion of the transit in the remainder of this section, frequent reference to Figure 3-2 should aid the reader's understanding. It would be even better to have an actual transit available.) Leveling the base is accomplished by rotating the four leveling screws at the bottom of the transit. Two opposite screws must be rotated simultaneously, with one hand rotating one screw and the other rotating the *opposite* screw in the opposite

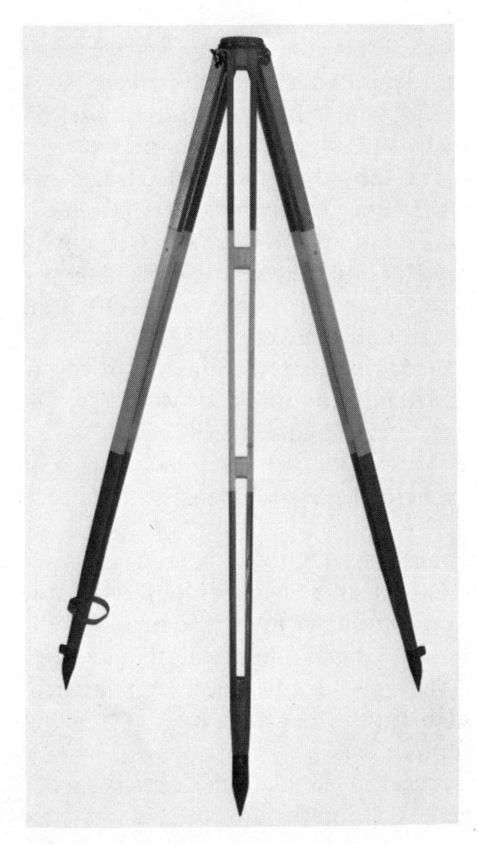

Figure 3-3 Tripod. (Courtesy of Keuffel & Esser Co.)

direction. Stated another way, the two thumbs should both be moving toward the center, or both away from the center. As this action is taken, the base of the transit is adjusted until it is level, as determined by observing the *base leveling bubble*. (The base leveling bubble should be lined up initially with the two leveling screws being rotated.) After the base is level in the direction indicated by the first two leveling screws, the instrumentman moves to the remaining two leveling screws (opposite each other) and repeats the procedure to level the base (as determined by observing the *other* base leveling bubble) in a direction perpendicular to that of the previous step.

When both leveling bubbles are centered, the instrument should be level; however, it is good practice to rotate the transit head to see if the bubbles remain level. If the bubbles do not remain level, the instrument may be out of adjustment. Minor rotation of one or more leveling screws should make it level. If the bubbles do remain level, the instrumentman is ready to proceed with the task at hand.

Two additional remarks should be made at this point. The first is that, when the transit has been properly leveled as described above and is ready to be used, the leveling screws should be fairly "tight." If they are too loose, the transit may tip to one side. On the other hand, they should not be too tight, for it is possible to strip the threads. The second remark is that when two adjacent leveling screws are loosened, the entire transit head is free to move approximately one-quarter inch in a horizontal plane (i.e., without moving the tripod legs). This technique can be useful in setting up the transit initially, when one is trying to set up very precisely over a reference point. Often the reference point is the head of a tack driven into a stake.

Once the transit is set up and leveled, the instrumentman is ready to measure or lay off an angle, determine distances by stadia, or whatever. The actual techniques of measuring an angle, determining distances by stadia, and so on, will be discussed later when angles and stadia are discussed. Discussion continues at this point, however, with regard to physical manipulation of the transit once it is set up and leveled.

In almost every application of the transit the telescope is used. Like any telescope, the one on the transit allows clear viewing of objects in the distance. A *focusing screw* is located on top of the telescope barrel. When looking through the telescope, four cross hairs (one vertical and three horizontal) should be clearly visible. These are used in determining alignment, running levels, and reading stadia. If the cross hairs are out of focus, they can be focused by rotating the eyepiece.

The telescope can be rotated in a vertical plane when the *vertical motion clamp* is loose. Tightening it will prevent such rotation. When it is tightened (and rotation prevented) the telescope can be rotated slightly by turning the *vertical motion tangent screw*. This feature is useful when setting a (horizontal) cross hair precisely on a given target. Associated with the telescope are a vertical circle and vernier (usually graduated into degrees and minutes), used in measuring vertical angles (Chapter 6). There is a leveling bubble attached to the bottom of the telescope that can be used when it is necessary to have the telescope level.

The telescope can also be rotated in a horizontal plane. There are actually two means of doing this. The base of the transit consists of two circular plates—a lower plate and an upper plate. Rotation of the lower plate is controlled by a *lower plate clamp*. When the lower plate clamp is tightened, rotation of the lower plate is prevented; when it is loosened, rotation is possible. There is a *lower plate tangent screw* used in final adjustment of the lower plate after the lower plate clamp is tightened. The upper plate consists of the entire top of the transit, including the telescope, leveling bubble tubes, and the like. Rotation of the upper plate is controlled by an *upper plate clamp* and *upper plate tangent screw*.

When both the upper plate clamp and the lower plate clamp are loose, both upper and lower plates are free to rotate independently. Obviously when both are tightened, rotation of both plates (and thus the entire transit) is prevented. When the upper plate clamp is tight and the lower plate clamp is loose, the lower plate is free to rotate. In this situation the upper plate is, in effect, rigidly attached to the lower plate, and thus they both move together as a unit. Finally, when the lower plate clamp is tight and the upper plate clamp is loose, the upper plate is free to rotate, but rotation of the lower plate is prevented. As will be explained in Chapter 6, proper manipulation of the upper and lower plates and upper and lower plate clamps is necessary in order to measure and lay off horizontal angles.

The typical transit is also equipped with a compass needle which, when allowed to float freely, points toward the magnetic North pole. The needle rotates on a sharp point. A *needle lifting screw* can be tightened to clamp the needle tightly against the glass cover above and loosened to allow the needle to rotate freely. The needle should always be clamped except when actually in use. The needle should never be loosened unless the transit is properly set up in a level position, and it should always be clamped prior to picking the transit up and moving it. Compass readings are made by means of a circular scale divided into four quadrants (Figure 2-1*b*) and located on the upper plate near the ends of the needle. By clamping the upper plate and allowing the lower plate (and telescope) to rotate freely, the instrument can be used as a surveyor's compass when only bearings and azimuths are recorded. In a compass survey the needle readings can be read only to the nearest 15 or perhaps 10 minutes.

More precise readings of angles can be made using a scale and vernier arrangement between upper and lower plates. There is a circular scale on the lower plate. The scale is often graduated in degrees and half degrees, going from zero to 360 degrees both clockwise and counterclockwise. The upper plate contains two verniers. These are short scales adjacent to the circular scale on the lower plate, used to measure fractional parts of a degree. The verniers are often graduated in minutes, going from zero to 30 minutes to both sides of a zero mark (see Figure 3-4).

A scale reading is made by first noting the number of degrees and then noting the number of minutes. The number of degrees is determined by

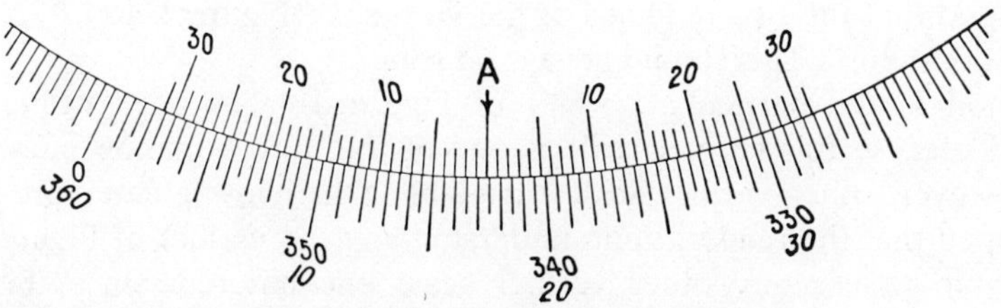

Figure 3-4 Vernier. (Courtesy of Keuffel & Esser Co.)

noting the location of the zero graduation on the vernier with respect to the lower plate scale. For example, in Figure 3-4, the zero graduation on the vernier falls between 17 and 18 degrees; hence, the degrees part of the scale reading is 17. To determine the minutes part of the scale reading, the vernier scale must be examined to find the graduation on the vernier that coincides exactly with a graduation on the lower plate scale. (A magnifying glass is helpful in reading a vernier.) This graduation on the vernier indicates the number of minutes. In Figure 3-4, the only graduation on the vernier that lines up exactly opposite a graduation on the lower plate scale is the 25-minute graduation. Hence, the reading is 17°25′.

Two additional points must be made. First, since the scale on the lower plate is marked in both directions, a reading may be made either in the clockwise direction or the counterclockwise direction. In either case, however, the vernier must always be read in the same direction that the scale on the lower plate is read. Hence, in Figure 3-4, since the value of 17 degrees was read on the scale on the lower plate in a counterclockwise direction, the value of 25 minutes was read on the side of the vernier where the calibration numbers proceed counterclockwise. Second, at first glance there appears to be a problem in that the vernier scale only goes to 30 minutes in each direction, while an angle reading can certainly have more than 30 minutes. It will be noted, however, that the scale on the lower plate is graduated in degrees and half degrees. If the zero graduation on the vernier falls within the lower half of a degree, the vernier reading is the actual number of minutes. But, if the zero graduation on the vernier falls within the upper half of a degree, the vernier reading must be added to 30 to get the actual number of minutes. In Figure 3-4, if a reading is made in the clockwise direction, the number of degrees read is 342 and the number of minutes read is 05. However, since the zero graduation on the vernier falls within the upper half of the degree (i.e., it is nearer 343 than 342), the vernier reading of 05 must be added to 30, giving 35, the actual number of minutes. The complete reading in this case is 342°35′. This necessity of having to add 30 minutes to approximately half of all vernier readings is a potential source of mistake by the surveyor. Needless to say, every time a vernier reading is made, a careful check should be made to determine whether or not 30 minutes must be added to the vernier reading.

It will be noted that the two readings of the vernier of Figure 3-4 (17°25′ and 342°35′) add to 360°. This should always be true.

The discussion of the particular vernier of Figure 3-4 has been rather thorough here because this vernier scale is one of the most common ones. There are, however, other vernier scales available that allow greater precision. It is hoped that the reader, upon understanding the vernier of Figure 3-4, will be able to use any other vernier scale encountered, since the principle is the same for all cases.

Transits are expensive and delicate pieces of equipment, and it should go without saying that extreme care should be exercised in handling and using one. They should never be allowed to fall or be dropped. When set up, the tripod legs should be well apart and firmly "planted" in the ground. A transit (on tripod) is normally carried on the shoulder, but when crossing any obstacle, such as a fence or stream, it should be removed from the shoulder and passed safely across the obstacle. It is good practice to carry a waterproof cover to place over a transit in case one gets caught in a rainshower. Lacking that, the transit should be protected from precipitation however possible. If it does get wet, it should be allowed to dry, preferably in sunlight. When not in use, the transit should be stored in its carrying case.

3-3 The Level

As noted in the preceding section, a transit can be used to measure vertical distances by running levels. Most surveyors would, however, prefer to use an instrument specifically made for running levels. This instrument is called, appropriately, a "level." It is used almost entirely for this purpose.

A level is essentially a telescope mounted on a leveling head. A Dumpy level is shown in Figure 3-5. When looking through the telescope, a vertical and a horizontal cross hair should be visible. These cross hairs are brought into focus in the same manner as explained for the transit. Just as with the transit, the level has four leveling screws. There is, however, only one leveling bubble, which is attached to the telescope. The telescope must be rotated first

Figure 3-5 Dumpy level. (Courtesy of Keuffel & Esser Co.)

so that it is positioned directly over two (opposite) leveling screws. By manipulating these two leveling screws, it is leveled in this direction. The telescope is then rotated 90 degrees so that it is positioned directly over the other two (opposite) leveling screws. By manipulating these two leveling screws, it is leveled in this direction. The telescope should now be level; however, it is good practice to rotate it an additional 90 degrees and check to see if it remains level in the initial direction. The procedure is repeated, if necessary, until the instrument is exactly level. The level is then ready for running levels, a procedure which will be discussed in Chapter 5.

Although very common and widely used, the level described in the preceding paragraph takes some time to set up and level precisely. If frequent setups are required, leveling work can be quite slow—particularly with a novice running the level. For faster work, several self-leveling, or automatic leveling, levels are available. One of these is shown in Figure 3-6. In general, these have a circular spirit level available for quick, rough leveling. Once the level is leveled roughly, prisms, pendulums, and/or gravity actuated compensators maintain a horizontal line of sight, even though the barrel of the telescope may not be exactly horizontal. Obviously, some saving of time can be achieved by using one of these self-leveling levels.

Like the transit, the level is both expensive and delicate, and it should be handled and used with great care. Virtually everything that was said in the previous section about the care of a transit is applicable to the level.

One additional piece of equipment needed in running levels is the level rod.

Figure 3-6 Automatic level. (Courtesy of Keuffel & Esser Co.)

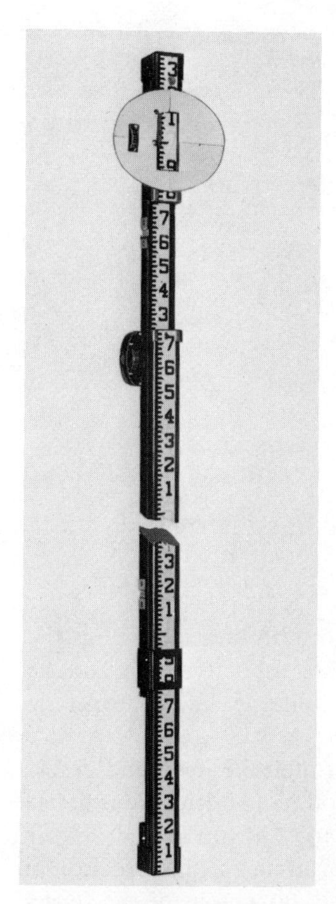

Figure 3-7 Philadelphia leveling rod. (Courtesy of Keuffel & Esser Co.)

Simply speaking, a level rod is a calibrated stick. Level rods are standardized for surveying purposes, however, Some consist of two or three sections that are telescoping or hinged to allow for maximum length while in use but smaller length while being transported or stored. One of the most common rods is the Philadelphia rod, shown in Figure 3-7. It consists of two sections—one for readings between zero and 7 ft and one for readings between 7 and 13 ft. When making a rod reading between 7 and 13 ft, the rod must be fully extended ("high rod"). The rod is calibrated linearly in hundredths of a foot. The calibrations are indicated by alternating black and white spaces on the rod, as shown in Figure 3-8. In reading the rod, the top of a black space should have a reading ending in an even hundredth of a foot (i.e., the second digit to the right of the decimal should be even), while the bottom of a black

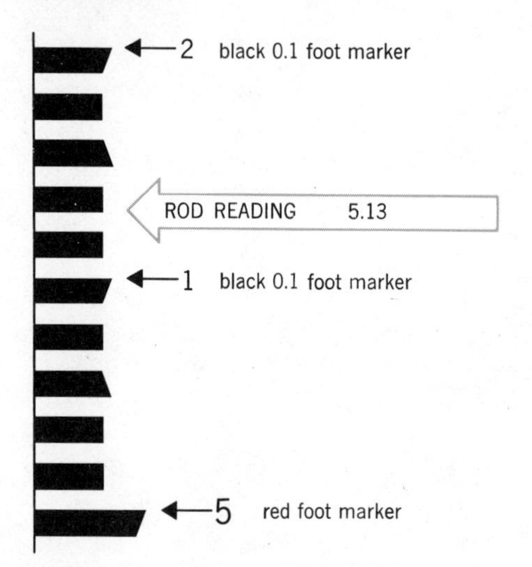

Figure 3-8 Level rod calibration.

space should have a reading ending in an odd hundredth of a foot. Thus the reading indicated in Figure 3-8 is 5.13 ft. There are other types of rods in addition to the Philadelphia rod, but all are similar in nature and perform the same function.

The level rod can be read directly by looking through the level and determining where the horizontal cross hair crosses the rod. An alternative is to use a target on the rod. A target is oval shaped (see target at top of rod, Figure 3-7) and can be moved up and down the rod by the rodman at the direction of the instrumentman. The instrumentman directs the movement of the target until the horizontal line on the target coincides with the horizontal cross hair. The rodman can then determine the rod reading—to the nearest thousandth of a foot, if necessary, by means of a vernier on the target. The target is also very useful in taking long shots, where the numbers and calibration marks may be difficult to read.

Like all equipment the level rod should be handled with care. Common sense indicates that the level rod should not be dragged along the ground or hit against trees, buildings, and the like.

3-4 Miscellaneous Equipment

The preceding sections discuss what might be called the mainstays of surveying—the tape, the transit, and the level (including the associated accessories). These have been used to do the bulk of surveying in the United States for many, many years, and they likely will continue to be so used.

There have developed, however, a number of pieces of equipment that perform surveying measurements. Many of these have been developed during recent years. They are sophisticated, giving very accurate measurements, and they are relatively expensive.

It is felt that most users of this book will be using the more traditional surveying equipment; hence, rather thorough discussion of these individual items was presented in the last three sections of this chapter. Some, however, will likely use the more sophisticated equipment, and therefore a cursory discussion follows for two of these—electronic distance measuring devices and the theodolite. The individual needing additional information is referred to the references at the end of this book or to manufacturers' literature.

Since about the middle of the twentieth century, various types of electronic distance measuring devices have been developed and perfected for use in determining linear distances in surveying. These devices are based on the fact that, if the time required for anything (a car, a light wave, a bullet, etc.) to move from one point to another at constant, known velocity can be determined, the distance covered in traveling from one point to the other can be computed by multiplying the time by the velocity.

The first electronic distance measuring device was the Geodimeter, which was put on the market in the early 1950s. It worked on the principle of determining the length of time for a light wave to travel from the device to a reflector at the distant point and back to the device. Later devices were based on the characteristics of electromagnetic waves instead of merely timing light wave pulses. The latest devices utilize infrared radiation and laser beams.

These devices are extremely beneficial in measuring very long distances (miles) very quickly and very accurately. For measuring distances across lakes and rivers, from one mountain to another, or any long distance (including the distance to a reflector on the moon), there is just no comparison with respect to time required and accuracy attainable between taping and using electronic devices.

As an example of one of the many electronic distance measuring devices on the market, a Hewlett-Packard 3820 Electronic Total Station is shown in Figure 3-9. (This device also measures angles.) Details regarding the use of this device are presented in Chapter 4. Attainable accuracy using electronic distance measuring devices is $\frac{1}{25,000}$ and can be much greater.

With regard to angular measurement, an instrument used to measure angles in addition to the ordinary transit is the theodolite (actually a transit of high accuracy). One particular make of theodolite is shown in Figure 3-10. Although similar to the transit in general appearance, there are several features quite different. The horizontal and vertical circles for measuring angles are made of glass instead of metal and are completely enclosed. By means of prisms, light is reflected from the circles to a reading eyepiece so that circle readings can be read by looking through the reading eyepiece adjacent to the telescope eyepiece. Thus scale readings can be made without

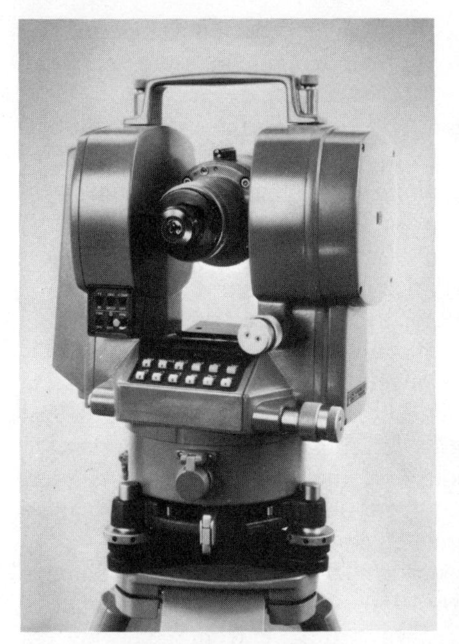

Figure 3-9 Hewlett-Packard 3820 electronic total station for distance and angle measurements. (Courtesy of Hewlett-Packard Co.)

having to walk around the instrument as is sometimes required when using a transit. The scales are lighted for night use. Optical systems are used for centering the instrument over the reference point as opposed to the plumb bob method for the transit.

Theodolites are light weight and easy to use, but their principal advantage is their increased accuracy in measuring angles as compared to the transit. Figure 3-11 shows a scale reading for a theodolite. Attainable accuracy with theodolites varies from one make to another, with possible accuracy to the nearest second or even the nearest tenth of a second.

3-5 Care and Adjustment of Equipment

It has been emphasized throughout this chapter that surveying equipment should be handled with utmost care. This will be emphasized one more time. It is presumed that the surveyor will take care of his or her own equipment. He or she must also see that assistants handle it with care, and that bystanders are not allowed to damage equipment.

Levels and transits of all types have certain adjustments that can be made to be sure the equipment is in top working order. Information on such adjustments can be found in some of the references at the end of this book.

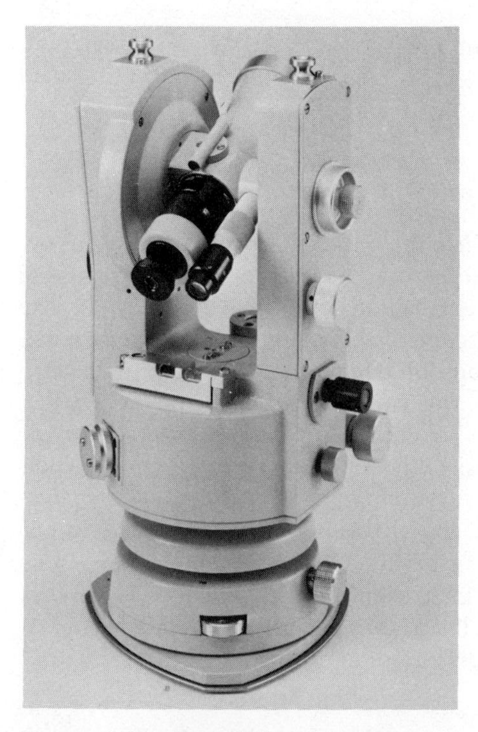

Figure 3-10 Theodolite. (Courtesy of Keuffel & Esser Co.)

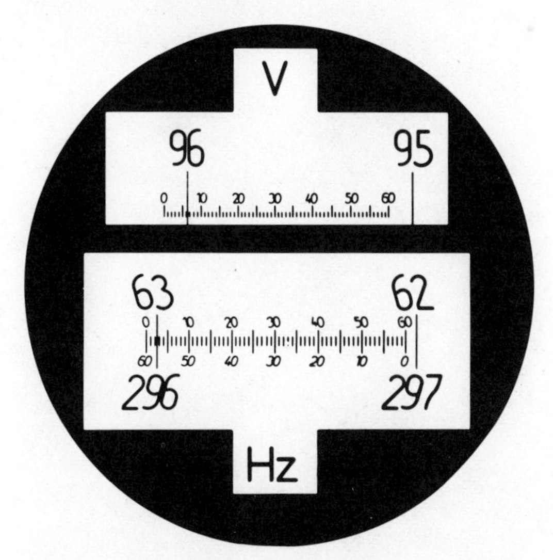

Figure 3-11 Scale reading for theodolite. (Courtesy of Keuffel & Esser Co.)

Although some of these adjustments are simple, it is recommended that, unless one is very good mechanically, adjustments of equipment be entrusted to a professional repairer.

3-6 Summary

This chapter has attempted to introduce the basic equipment used in surveying. Specific pieces of equipment discussed and presented in photographs in this chapter were intended to be representative of what is available. The author feels that it is not within the scope of this book to discuss in detail all available surveying equipment. In addition, technology changes rapidly; and new, more sophisticated equipment will continue to come on the market. Anyone planning to purchase surveying equipment is advised to survey the market and compare performance, cost, and so on, prior to purchase. Additional information concerning surveying equipment can be obtained by referring to some of the references at the end of this book and to manufacturers' literature. Some manufacturers of surveying related equipment are: Wild Heerbrugg Instruments, Inc.; Kern Instruments, Inc.; Keuffel & Esser Co.; Lietz Co.; W. & L. E. Gurley; Hewlett-Packard; Zena Co.; Lenker Manufacturing Co.; C. L. Berger & Sons; and Tellurometer, Inc.

MEASUREMENT OF HORIZONTAL DISTANCES

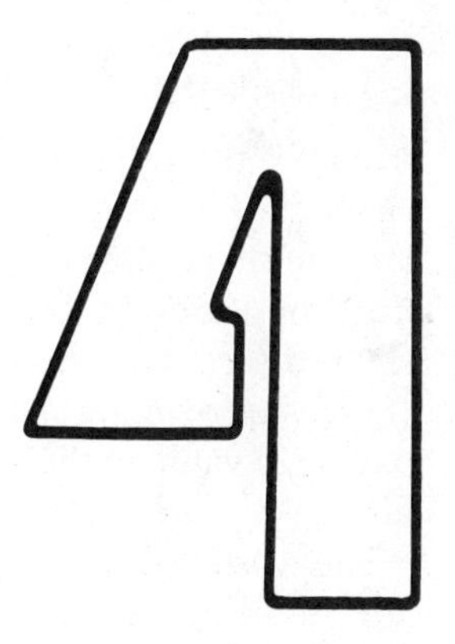

Probably the most basic operation performed in surveying and perhaps the most difficult to do *well* is the measurement of horizontal distances. Give a surveyor a 100-ft tape, and he or she can do a lot of surveying with it. In addition to the obvious measurement of linear distances, angles can also be "measured" with a tape (as will be explained subsequently in this chapter).

It is most important to emphasize the word "horizontal" in the chapter title. In plane surveying, whenever the length of a side of a piece of property or the linear distance between two points on the surface of the earth is given, it is understood that the distance is the horizontal distance regardless of the elevations of the end points of the line. In general, therefore, linear measurements in surveying are made either by measuring along a horizontal line or by measuring indirectly along an inclined line and computing the corresponding horizontal distance. In the latter case the distance to be reported is, of course, the (computed) horizontal distance. This will be covered in greater detail later in this chapter.

There are a number of ways in which horizontal distances may be measured. Listed in order of increasing accuracy (and also in the order in which they are covered herein), the most common methods are as follows: (1) pacing, (2) stadia, (3) taping, and (4) electronic measurement. Other means of measuring horizontal distances (not further discussed in this book) include the subtense bar, the odometer on the automobile, and the measuring wheel.

4-1 Pacing

Probably the crudest method of determining horizontal distances, yet one which can be quite useful, is pacing. Pacing is usually not covered in great detail in surveying books— probably because it is not accurate enough for the precision generally required in professional surveying. Pacing can, however, be used to determine approximate distances relatively quickly in situations where approximate measurements are sufficient. Since it is felt that pacing may be of value to the reader of this book, a rather detailed treatment is given here.

The word "pace," as a noun, means "the length of a step in walking." As a verb, it means "to move with slow or measured steps." Hence in a surveying context, pacing means to move with measured steps; and if the steps are counted and if the length of a step is known, the distance paced can be determined.

There are two methods that can be used to calibrate one's pace. One method is to determine the average length of an individual's normal pace (or step). The other method is to adjust one's pace to some predetermined length, such as 3 ft or 1 m. The former method has the advantage that it utilizes one's normal pace; thus the individual does not have to learn a "new" pace, and one's normal pace is probably more constant from day to day. It has the disadvantage that the average length of a normal pace may be a number that is not easy to use in mentally computing a paced distance. The latter method has the advantage that the length of pace can be forced to be a number that is easy to work with, such as 3 ft, so that mental computations can be done easily. It has the disadvantage that the pace is generally unnatural, which may cause the pace to vary somewhat from day to day. Also a person may tend to tire more easily in pacing long distances while using an unnatural pace.

Once a person has calibrated his or her pace, it is a relatively simple matter to pace a distance. One simply walks at the designated pace from one point to the other and counts the number of paces required. Care should be taken to walk in as near a straight line as possible, while maintaining the designated pace. Probably the most difficult aspect of pacing is counting accurately the number of paces. That may sound somewhat facetious; but the fact of the matter is that pacing can be rather boring, and it is easy for the mind to wander. The result can be that the person ends up with an incorrect number of paces and a corresponding incorrect value for the paced distance. A common mistake would be to record 432 paces instead of 532 paces— the result of not concentrating on the count. Anyone doing surveying must be constantly alert to avoid making mistakes of this type.

After the number of paces has been determined, the corresponding distance is determined by multiplying the number of paces by the length of the pace. It would be wise to pace the distance several times and take an average value. This would help identify large mistakes (such as missing the count by 100) and should also give a more accurate value, since the average of several measurements should be more accurate than a single measurement.

With considerable practice and experience, a person should be able to pace distances with an accuracy of $\frac{1}{50}$, if conditions are favorable. The terminology "accuracy of $\frac{1}{50}$" means that in measuring (by pacing, in this instance) a true distance of 50 ft, the measured (paced) distance should not vary from 50 ft by more than 1 ft either way. In measuring a true distance of 500 ft, the measured distance should not vary from 500 ft by more than 10 ft either way.

Care must be taken in expressing distances by pacing so as not to imply that a distance is more accurate than it really is. Under no circumstances should a distance determined by pacing be recorded indicating fractions of a foot (in other words, the recorded distance should be rounded off to—at least—the nearest whole foot). For example, suppose a total of 289 paces has been recorded by a person whose pace is calibrated to be 2.65 ft. Multiplying 289 paces by 2.65 ft per pace yields 765.85 ft. It would be highly misleading, however, to record the distance so determined as 765.85 ft, because to do so implies that the measured value is accurate to within ±0.005 ft. This accuracy is just not possible in pacing. Probably the best one could hope for would be to be within a foot or so of the true distance, in which case the recorded distance should be rounded off to the nearest foot (i.e., 766 ft, in this example). A paced distance of this magnitude might be considered to be within only 10 ft of the true distance, in which case the recorded distance should be rounded off to the nearest 10 ft (770 ft, in this example).

If paced distances are placed on a sketch or map, it should be indicated that the distances were determined by pacing. They might also be designated as "plus or minus" (as 770± ft).

4-2 Stadia

The stadia method provides a rapid means of determining horizontal distances, although attainable accuracy may not be sufficient for some purposes.

The general principle of stadia measurement involves the reading of values on a leveling rod intersected by two horizontal cross hairs that are always in the telescope of a transit. These two horizontal cross hairs set a fixed angle subtended by difference between the readings taken from the leveling rod. The distance from the transit to the level rod is in direct known proportion to the measured value.

In utilizing the stadia method a knowledge of the use of both the transit and the level rod is required. Also, vertical distances can be determined simultaneously with horizontal distances while using stadia. For these reasons, the complete discussion of stadia is delayed until Chapter 7 after the use of the transit (Chapter 6) and the level rod (Chapter 5) and the measurement of vertical distances (Chapter 5) have been covered.

4-3 Taping

Taping is probably the most common surveying method of measuring horizontal distances. In simple terms, taping consists of stretching a calibrated tape between two points and reading the distance off the tape. Sometimes taping is referred to as "chaining," because originally distances were measured with a surveyor's chain (which was literally a chain). As related in Chapter 1, a "chain" is 66 ft in length and subdivided into 100 links.

The only piece of equipment that is absolutely required for taping is a tape. In measuring distances longer than a tape length, however, taping pins are useful in marking tape lengths. In measuring distances over terrain that is not level, plumb bobs (one for each end of the tape) are necessary to project accurately a point on the tape to the ground below or vice versa. Range poles are helpful in maintaining alignment, and a hand level is helpful in keeping the tape horizontal. These basic equipment items needed to measure by taping were shown in Figure 3-1.

To begin the discussion of the correct taping procedure, let us assume that a measurement of less than one tape length is to be made between two points at the same elevation with level ground between them. Obviously, "the" distance between two points is the straight line distance; hence, the first rule of taping is to keep the tape along a straight line between the two points. That does not sound too difficult, but it can be a matter of concern if the taping is through thick forest, heavy underbrush, or the like. To begin the measurement, both tapemen would proceed to one of the points. The head tapeman would walk the zero end of the tape to the other point while the rear tapeman remained at the initial point. The tape must then be positioned so that the rear tapeman is holding a whole foot mark at his point while the head tapeman is holding some mark within the one foot of tape that is calibrated in tenths and hundredths of a foot.

Sometimes this requires a bit of maneuvering. Suppose a subtracting tape is being used and is placed so that the rear tapeman is holding the 48-ft mark at his point but when the head tapeman pulls the tape tightly, he notes that the point to which measurement is desired falls between the zero mark and the end of the tape. Since a subtracting tape is not calibrated in this part of the tape, the head tapeman would tell the rear tapeman to "give me a foot," meaning loosen the grip and allow the head tapeman to pull the tape a foot toward him. The rear tapeman would now hold the 49-ft mark at his point, and when the head tapeman pulls the tape tightly, the point should now fall between zero and the 1-ft mark, which is calibrated on a subtracting tape. At this point the actual measurement can be made. The rear tapeman braces himself to hold the 49-ft mark at the initial point, while the head tapeman pulls the tape tightly and takes a reading of the fractional part of a foot by reading off the calibrated part of the tape. Suppose the head tapeman reads 0.29. Then the correct distance is determined by subtracting 0.29 from 49, giving 48.71 ft. If an adding tape had been used, the rear tapeman would have held the 48-ft mark at the initial point, and the head tapeman would have read 0.71 off the calibrated part of the tape located between the zero mark and the end of the tape. The correct distance would have been determined by adding 0.71 and 48, giving again 48.71 ft. The comparison of making this measurement using these two types of tape can be better made by referring to Figure 4-1.

Several comments should be added to the previous discussion. It should be noted that only the head tapeman actually applies tension to (i.e., pulls) the tape. The rear tapeman's job is to place the appropriate foot marker at his point and to hold it stationary on the point. If the rear tapeman tries to apply tension by pulling the tape, the unwanted result will be that the foot marker would be moved away from the point of measurement. The rear tapeman should simply place the foot marker at his point and brace himself so that he can hold it stationary there. The head tapeman is free to pull the tape and apply the proper tension. He does not want to place a specified foot marker at his point; he wants to pull the tape and determine which marker lines up with his point.

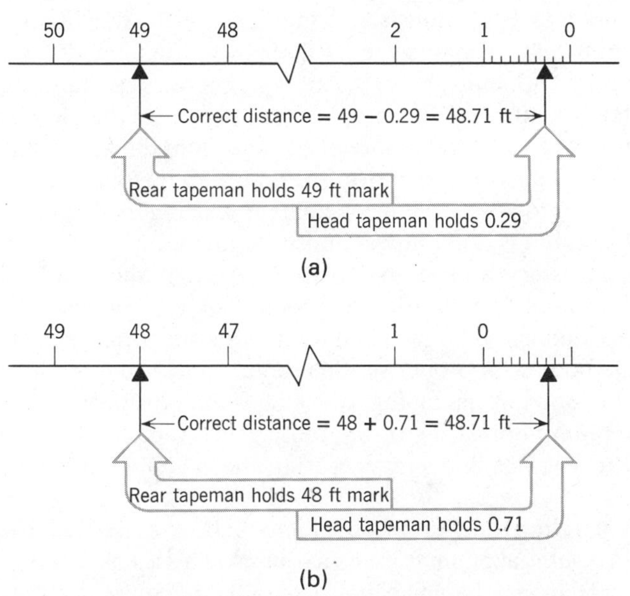

Figure 4-1 Comparison of determining a measurement using (a) a subtracting tape and (b) an adding tape.

Figure 4-2 Measuring distance directly along the ground.

In the example cited previously, since it was stated that the taping was to be done between two points with level ground between them, it was assumed that the tape would be placed directly on the ground, as shown in Figure 4-2. Often, however, taping is done over sloping ground or over level ground with some kind of obstacle between end points. In such cases, either one or two plumb bobs are necessary to allow the tape to be maintained horizontally while a distance is being measured.

The use of the plumb bob is demonstrated in Figure 4-3. When first learning to use the plumb bob, it may seem as if three or four hands are needed, but with practice two hands are sufficient. In general, one hand is used to hold the tape steady while the other hand is used to control the plumb bob. The usual procedure is to stand to the side of the line, hold the tape with one hand, and let the plumb bob string loop over the tape with the plumb bob hanging just above the point to which the measurement is being made. That does not sound too difficult, but it is not an easy position to hold if the tape is being pulled with the necessary appreciable tension. Also, if the wind is blowing, the plumb bob will not remain still. The problem of holding position while the tape is being pulled can be minimized by holding the arms against the body and bracing the body with proper foot placement and body rigidity. The problem of the plumb bob not remaining still can be eased by placing the knee lightly against the plumb bob string or the foot against the plumb bob. The best procedure is to get out and practice using the plumb bob while exercising good common sense.

We have already stated that one or more plumb bobs are needed to maintain the tape in horizontal alignment, but the question arises as to how one knows when the tape is, in fact, being held horizontally. In Figure 4-3, the tape is being held horizontally; however, the beginner generally tends to have the downhill end of the tape too low. One helpful hint is to remember that

Figure 4-3 Using a plumb bob.

when the tape is horizontal, it will make a right angle with the plumb bob string. For very accurate taping, a hand level (Chapter 3) should be used to assure horizontal alignment.

Let us consider now the case of measuring a distance that is longer than a single tape length. Everything that has been presented previously is still applicable, but the problem is complicated by the fact that the tape can no longer be stretched from one point to the other. Hence intermediate points must be marked. Standard procedure is for the head tapeman to begin at a point at one end of the line to be measured, take the zero end of the tape, and walk toward the other point. The rear tapeman, standing at the initial point, watches the tape until the end is at hand, at which time he calls out "chain" or "tape" to the head tapeman. This signals the head tapeman to stop and mark off the first 100 ft (assuming a 100-ft tape is being used). The rear tapeman holds the 100-ft mark at his point and braces himself while the head tapeman pulls the tape. (Either or both tapemen would use plumb bobs when necessary.) When the tape is taut, and the rear tapeman has the 100-ft mark at his point, he calls out "ready" or "O.K." or "mark," and the head tapeman then proceeds to stick a taping pin into the ground opposite the zero mark on the tape. The pin should be stuck into the ground at an angle of 30 to 40 degrees with the ground and leaning away from the line (i.e., in a plane perpendicular to the line). After relaxing the tension in the tape, the procedure should be repeated to verify that the pin has been stuck at the correct location. After verification the head tapeman should call out "100, come ahead" and begin walking and pulling the tape in the direction of the line. The rear tapeman drops the tape, walks to the point just marked by the taping pin, and calls out "tape" (or whatever) when the end of the tape approaches the taping pin. The procedure is repeated to mark off the second hundred feet. When the second pin has been placed and verified, the head tapeman should call out "200, come ahead" and then proceed in the direction of the line. The rear tapeman should drop the tape, remove the pin, and walk to the next pin. The entire procedure is repeated until the other end of the line is reached. The last time the tape is stretched, the measurement will likely be something less than 100 ft.

In measuring a distance longer than one tape length, it is necessary to mark tape lengths at intermediate points, and if the total measurement is to be accurate, it is imperative that these intermediate points be established "on line." This is not too much of a problem if the distance to be measured is not too long and if one end of the line can be seen from the other end. If this is the case, a range pole can be placed at each end of the line. When the first tape length is being marked off, the rear tapeman can sight between the two range poles at each end of the line and signal the head tapeman to move one way or the other until he is at the proper position to stick the pin on the straight line between the two range poles. When the second and succeeding tape lengths are being marked off, both the rear tapeman, by sighting ahead to the range pole at the far end of the line, and the head tapeman, by sighting backward to the range pole at the beginning of the line, can help keep the intermediate points on line. If extremely accurate work is required, a transit can be set up at one end of the line and used to set intermediate points on line.

In the case where one end of the line cannot be seen from the other end, it is likely that the line has been run (established) prior to the time the taping is done, and stakes (possibly with tacks on top) have been set on line at intermediate points along the line. These stakes (with range poles placed next to them, if necessary) may be used to keep the taping on line.

As always, tapemen must remain constantly alert to avoid mistakes. One of the easiest mistakes is to miscount the number of tape lengths and accordingly to record a measured distance incorrectly by a multiple of 100 ft. It would seem that the easiest thing in the world would be to count the number of tape lengths marked off while measuring a distance, but the mind wanders and mistakes creep in. To help avoid this, it was suggested previously that the head tapeman call out the number of feet each time he sets a pin and have the rear tapeman verify it.

As an additional means of avoiding mistakes of this type, it is standard procedure for the rear tapeman to begin a long measurement with 1 pin while the head tapeman has 10. When the first tape length is marked off by the head tapeman placing a pin at the proper point, the rear tapeman comes forward carrying the initial pin. When the second tape length is marked off, the rear tapeman pulls the pin marking the first tape length and carries the two pins forward. If this procedure is carried out faithfully, the rear tapeman should always possess a number of pins equal to the number of tape lengths already laid off. When the tenth tape length is marked off, the head tapeman should be sticking the eleventh pin, and the rear tapeman should be bringing 10 pins forward. He should hand these to the head tapeman, who should count them to verify that the pin sticking does in fact represent 10 tape lengths. If the pin check is not correct, further measurement should not proceed until the reason is determined and any corrections made. It is not uncommon for the rear tapeman to neglect to pull one or more pins. If no explanation can be found for an incorrect pin count, the tapemen should return to the starting point and retape the distance. If the pin check is correct, the head tapeman takes the 10 pins and proceeds to mark off the next 10 tape lengths.

Sometimes the situation arises that the change in elevation is so great from one end of the tape to the other that the tape cannot be held in a horizontal position for the entire tape length. When this situation arises, the logical solution is to drop back and measure something less than a full tape length—a procedure called "breaking the tape." Depending on the topography of the land, the total tape length will have to be measured in two or more shorter segments. For the example illustrated in Figure 4-4, segments of 20 ft, 20 ft, and 60 ft are measured to make up a full tape length of 100 ft. This can be accomplished by making the indicated measurements individually and adding them to determine the total distance. This method has the disadvantage, however, that the individual measurements must be recorded (either mentally or written) and added together, a procedure that is susceptible to mistakes being made.

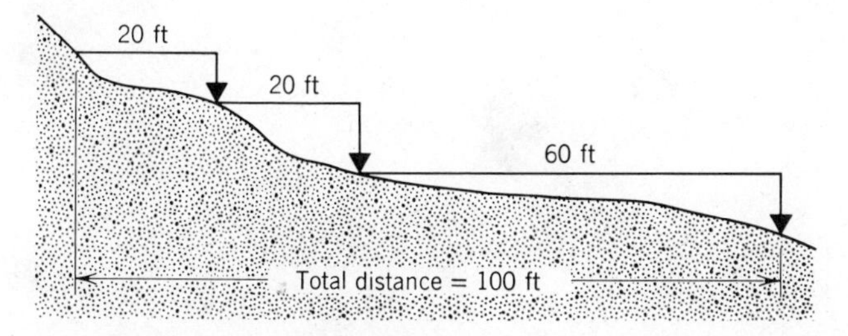

Figure 4-4 Breaking the tape.

A better procedure is to have the head tapeman pull the tape forward a full 100 ft and then return to make the 20-ft measurement. This 20-ft measurement would be made with the rear tapeman holding the 100-ft mark and the head tapeman holding the 80-ft mark. When this point is established and marked by a pin, rather than record the distance measured, the rear tapeman should come forward and take the tape from the head tapeman, grasping it at the 80-ft mark (or whatever mark to which the head tapeman had measured). The head tapeman then moves to the 60-ft mark, and the next 20 ft are laid off and marked with the rear tapeman holding the 80-ft mark and the head tapeman holding the 60-ft mark. The rear tapeman would again go forward and take the tape from the head tapeman, grasping it at the 60-ft mark. The head tapeman would then proceed to the zero mark to lay off the remaining 60 ft.

When this task is accomplished, the net result is that a full 100-ft tape length has been laid off (albeit in three parts), and the tapemen can proceed with the next 100 ft (which may or may not require breaking the tape). Before proceeding, however, the rear tapeman should hand over to the head tapeman any taping pins used to mark intermediate points while breaking the tape so that the pin count can be maintained. This latter procedure has the advantage that no intermediate measurements have to be recorded and added together, thereby eliminating one possibility for making simple mental mistakes.

As stated earlier in this chapter, linear measurements in surveying are usually made either by measuring directly along a horizontal line or by measuring indirectly along an inclined line and computing the corresponding horizontal distance. The discussion to this point has centered on the former. Attention will now focus on the latter.

In some circumstances, it may be easier to measure a slope distance (as AB in Figure 4-5) with the tape and then compute the corresponding horizontal distance (AC). When this is done, it is necessary that either the angle BAC or the deviation from the horizontal (length BC) be measured in order that the horizontal distance (AC) can be computed. If AB and the angle BAC are measured, AC can be computed from the formula

$$AC = AB \cos(BAC) \tag{4-1}$$

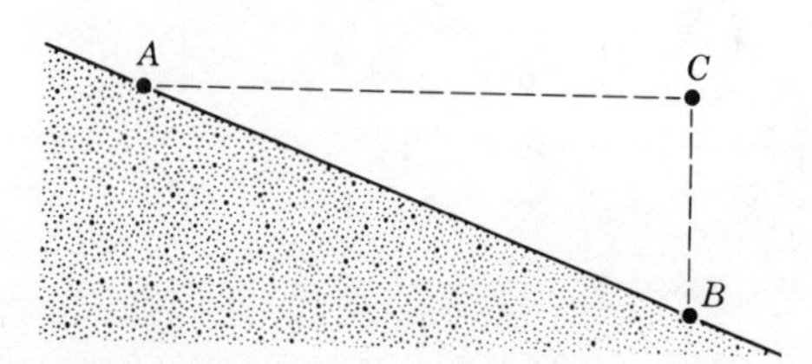

Figure 4-5 Measuring slope distance.

If AB and BC are measured, AC can be computed from the formula

$$AC = \sqrt{AB^2 - BC^2} \tag{4-2}$$

It can be shown that an approximate formula for computing AC when AB and BC are measured is

$$AC = AB - \frac{BC^2}{2AB} \tag{4-3}$$

This formula is accurate to the nearest thousandth of a foot when BC is 10 ft and AB is 100 ft and is accurate to within about 0.02 ft when BC is 20 ft and AB is 100 ft. Equation (4-3) is more easily solved by hand, but if a calculator is available, Eq. (4-2) can be used just as easily and would be preferred since it is exact. Obviously when using this procedure, it is necessary to determine either the angle BAC or the difference in elevation BC. This is somewhat of a nuisance, and in most cases it is more expedient to make measurements with the tape horizontal, even though more error may occur by this procedure.

Attention will now be directed toward the accuracy resulting from taping. If a distance of approximately 10 ft is being measured, it is not unreasonable to expect that the measurement could be read and recorded to the nearest hundredth of a foot (e.g., 10.18 ft). In such a short distance, the tape can be held virtually straight (without sag) and horizontal with no difficulty, and reading the tape should not be difficult.

Suppose now a distance of approximately 90 ft is to be measured. In this case the measurement is more difficult. The tape must be stretched tightly to minimize sag, with the result that it is more difficult for each tapeman to hold the marker at precisely the correct point. Also, in this span of tape, there is bound to be some slight sag, and it is also more difficult to judge when the tape is exactly horizontal. Thus, although the tape could be stretched and a measurement (e.g., 90.18 ft) made and recorded, it is somewhat questionable as to whether the "true" distance is actually that which was recorded. Suppose the tape should have had more tension applied, which would have reduced the sag a bit, resulting in a reading of say 90.16 ft. The result can be that the recorded value is off by a couple hundredths of a foot or perhaps even a tenth of a foot. Experienced tapemen should be able to measure this distance to the nearest hundredth of a foot, and accordingly, it should be recorded to the nearest hundredth of a foot. If however, inexperienced tapemen are measuring the distance and not exercising particular care as to alignment, tension, etc., the measurement is probably not more accurate than the nearest tenth of a foot (maybe not even that accurate) and should be recorded accordingly (i.e., as 90.2 ft instead of 90.18 ft).

Consider one more case—that of measuring a distance of approximately 910 ft. The last measurement to be made would be approximately 10 ft (measuring beyond the pin stuck at the 900-ft point), so the tendency might be to say that this is equivalent to the previous example of measuring approximately 10 ft, and thus the distance measured is accurate to the nearest hundredth of a foot. Such reasoning is faulty, however, because even if the last 10-ft measurement is accurate to the nearest hundredth of a foot, the total measurement is not—unless it is known for certain that the measurement of the first 900 ft is accurate to the nearest hundredth of a foot. For that to be true, each of the nine measurements of a full tape length would have to be done with an accuracy to the nearest hundredth of a foot. While this is not impossible to achieve, it would certainly require extreme care on the part of the tapemen. If such care has not been exercised, the result of a measurement of this magnitude should be reported to the nearest tenth of a foot (e.g., 910.8 ft), or possibly just to the nearest foot.

In the final analysis the surveyor should, for every distance measured, decide how accurate he thinks the measurement is and indicate that accuracy by recording the proper number of places to the right of the decimal, as explained in Section 2-1.

In the discussion of taping to this point, it has been assumed that the tape was exactly 100.00 ft long. This is not always the case, however. Tapes may actually be slightly longer or shorter than 100.00 ft because of imperfect manufacture, stretching, or wear. Corrected distances for measurements made with a tape that is either too long or too short can be computed from the following formula:

$$\frac{l_a}{l_n} = \frac{d_a}{d_m} \tag{4-4}$$

where l_a = actual length of tape
l_n = nominal length of tape
d_a = actual distance
d_m = measured distance, or tape distance to be laid off

Any unit of measure of length—feet or meters, in surveying—may be used in this equation, but the same unit should be used for each term. Equation (4-4) can also be used to compute the tape reading to be used when laying off a specified distance with a tape that is either too long or too short.

Example 4-1

A distance of 268.71 ft was measured with a 100-ft tape that is 0.03 ft too long. What is the correct distance?

Solution

Using Eq. (4-4),

$$\frac{l_a}{l_n} = \frac{d_a}{d_m} \tag{4-4}$$

$l_a = 100.03$ ft

$l_n = 100.00$ ft

$d_m = 268.71$ ft

d_a, the actual distance, is the unknown and is solved for by substituting the above values into the equation.

$$\frac{100.03 \text{ ft}}{100.00 \text{ ft}} = \frac{d_a}{268.71 \text{ ft}}$$

$$d_a = \frac{100.03 \text{ ft}(268.71 \text{ ft})}{100.00 \text{ ft}}$$

$$d_a = 268.79 \text{ ft}$$

In the above computation for d_a, when the numbers are punched into a calculator, the answer is displayed as 268.790613. Since the measured distance (268.71 ft) is expressed as being accurate to the nearest hundredth of a foot, the computed answer (268.79 ft) should be expressed indicating the same accuracy. Hence, the value 268.790613 was rounded off to two places to the right of the decimal.

Example 4-2

We want to lay off a distance of 160.00 ft with a 100-ft tape that is 0.03 ft too short. What tape reading should be used?

Solution

Using Eq. (4-4),

$$\frac{l_a}{l_n} = \frac{d_a}{d_m} \tag{4-4}$$

$l_a = 99.97$ ft

$l_n = 100.00$ ft

$d_a = 160.00$ ft

d_m, the tape distance to be laid off, is the unknown and is solved for by substituting the above values into the equation.

$$\frac{99.97 \text{ ft}}{100.00 \text{ ft}} = \frac{160.00 \text{ ft}}{d_m}$$

$$d_m = \frac{100.00 \text{ ft}(160.00 \text{ ft})}{99.97 \text{ ft}}$$

$$d_m = 160.05 \text{ ft}$$

The difference between these two example problems should be noted carefully. In the first case a distance is measured, and the correct distance is computed. In this case d_a is the unknown when using Eq. (4-4). In the second case it is desired to lay off a specific distance, and the tape reading to be used is computed. In this case d_m is the unknown when using Eq. (4-4).

Observation of Examples 4-1 and 4-2 indicates that multiplication and division of several numbers containing as many as five significant figures may be required in using Eq. (4-4). If a calculator is available, this is no problem. If one is not readily available (not an unlikely prospect when working in the field), these computations can be tedious. An alternative method is to calculate the correction and then either to add or subtract the correction. The correction is easily calculated by multiplying the number of tape lengths by the amount of error per tape length. The real problem arises, however, in deciding whether the correction should be added or subtracted. This decision can be made by referring to Table 4-1.

Table 4-1

	Tape Too Long	*Tape Too Short*
If measuring a distance	add	subtract
If laying off a distance	subtract	add

Referring back to Example 4-1, the correction is (268.71/100)×(0.03), or 0.08 ft. In this case a distance was measured with a tape that is too long, and from Table 4-1 the correction should be added. Hence the correct distance is 268.79 ft. In Example 4-2, the correction is (160.00/100)×(0.03), or 0.05 ft. In this case a distance is to be laid off with a tape that is too short, and from Table 4-1 the correction should be added. Hence the correct tape reading is 160.05 ft. It can be seen that the actual computations required in the latter method are easier; however, the possibility arises that the correction will be added when it should be subtracted, or vice versa. If Eq. (4-4) is used, no decision of whether to add or subtract is necessary.

The last several paragraphs have dealt with corrections to be applied to distances measured with a tape that is either too long or too short. Previously in this section corrections to be applied to distances measured on a slope have been discussed [Eq. (4-1), (4-2), and (4-3)]. There are three additional variable conditions that can cause errors in taping and for which corrections can be calculated and applied. These are temperature variations, sag, and incorrect tension. No discussion of taping would be complete without considering these.

Steel tapes, like other materials, expand with increasing temperature and contract with decreasing temperature. Over a temperature range of 100F°, a 100-ft steel tape changes in length by 0.065 ft, not an insignificant amount. Tapes are standardized at 68°F. Corrections to be applied to measurements made with a standardized tape at temperatures other than 68°F can be computed from the following equation:

$$C_t = 0.0000065(T_t - T_s)L \tag{4-5}$$

where C_t = correction due to temperature to be applied to a distance measured
T_t = temperature of tape at time of measurement, in °F
T_s = temperature of tape when standardized (normally 68°F)
L = total measured length of the line

In Eq. (4-5) the two temperature terms must be in degrees Fahrenheit; but any unit for measuring length may be used for L, and C_t will be in the same unit of measure. The sign of C_t will be either plus or minus, and the correction (including its sign) is applied to the measured length. Thus one does not have to worry about whether the correction should be added or subtracted; the computed sign will take care of that. It should be emphasized that T_t is the temperature of the *tape* at the time of measurement. This is not necessarily the same as the air temperature. Certainly a tape resting on an asphalt surface may be considerably hotter than the air temperature. Finally, the term 0.0000065 in Eq. (4-5) is the coefficient of thermal expansion of steel, and thus the equation, as written, is only applicable to a *steel* tape.

Sag of a tape occurs when a steel tape is supported only at its ends or only at the two points of measurement (i.e., if a distance of about 70 ft is being measured with the tape supported only at the zero and 70-ft mark, there would be sag between the zero and 70-ft mark). The distance between two points is the straight line distance. If a tape being held between two points sags, the actual distance between the two end points is something less than the tape reading. The correction for sag can be computed from the following formula:

$$C_s = -\frac{W_t^2 L}{24P^2} \tag{4-6}$$

where C_s = correction due to sag between points where tape is supported, in feet
W_t = total weight of tape between the two supports, in pounds
L = length of tape between points where tape is supported, in feet
P = total tension applied to tape, in pounds

The value of W_t may be obtained by multiplying unit weight of the tape in pounds per foot by the length of tape between supports. Hence, Eq. (4-6) may be rewritten in the form

$$C_s = -\frac{w^2L^3}{24P^2} \tag{4-7}$$

where w = unit weight of tape in pounds per foot

With regard to error due to incorrect tension, a tape is standardized at a certain pull. Since a tape stretches when it is pulled, if it is pulled with a tension greater than the standard pull, the tape will elongate, resulting in an error of measurement. If a tape is pulled with a tension less than the standard pull, the tape will not elongate as much as it did when it was being standardized. In effect, the tape is too short, resulting in an error of measurement. The correction for tension can be computed by the following formula:

$$C_p = (P_t - P_s)\frac{L}{AE} \tag{4-8}$$

where C_p = correction due to incorrect tension to be applied to measured distance, in feet
P_t = total pull (tension) applied to tape, in pounds
P_s = standard pull for the particular tape, in pounds
L = total measured length of line, in feet
A = cross-sectional area of tape, in square inches
E = modulus of elasticity of tape material, in pounds per square inch (about 29,000,000 lb/sq in. for steel)

Error due to tension can be eliminated by applying the correct pull—that is, the pull that was used when the tape was standardized. Pull can be determined by using a spring balance.

Corrections for temperature, sag, and tension are all demonstrated in Example 4-3, which follows. It will be noted that each correction is computed separately and independently, and the individual corrections are combined to obtain a total correction to be applied to the measured distance.

Example 4-3

A steel tape was standardized at 68°F under a tension of 15 lb with the tape supported throughout its entire length, and the distance between the zero mark and the 100-ft mark was found to be 100.00 ft. The tape weighs 2.60 lb and has a cross-sectional area of 0.0090 sq in. This tape was used in the field to measure a distance that was determined to be 228.48 ft. At the time measurement was made, the temperature of the tape was 20°F, the pull was 22 lb, and the tape was supported only at the end points. What is the correct distance?

Solution

Correction due to temperature:

$$C_t = 0.0000065(T_t - T_s)L \tag{4-5}$$

$$C_t = 0.0000065(20-68)(228.48)$$

$$C_t = -0.071 \text{ ft}$$

Correction due to tension:

$$C_p = (P_t - P_s)\frac{L}{AE} \tag{4-8}$$

$$C_p = (22-15)\frac{228.48}{(0.0090)(29{,}000{,}000)}$$

$$C_p = +0.006 \text{ ft}$$

Correction due to sag:

$$C_s = -\frac{W_t^2 L}{24P^2} \tag{4-6}$$

In the previous computations (for temperature and tension), each correction was made for the entire measurement in one computation. This cannot be done for the sag correction, however, because the sag correction has to be computed in separate parts from one point of tape support to the next. For each 100-ft span, the correction due to sag is

$$C_s = -\frac{(2.60)^2(100.00)}{(24)(22)^2}$$

$$C_s = -0.058 \text{ ft}$$

For the 28.48-ft span, the correction due to sag may be computed from Eq. (4-7) with w (unit weight of tape) = (2.60/100), or 0.0260 lb/ft.

$$C_s = -\frac{w^2 L^3}{24P^2} \tag{4-7}$$

$$C_s = -\frac{(0.0260)^2(28.48)^3}{(24)(22)^2}$$

$$C_s = -0.001 \text{ ft}$$

Since there are two 100.00 ft spans and one 28.48 ft span in measuring a total distance of 228.48 ft, the total correction due to sag is

$$C_s = 2(-0.058) - 0.001$$

$$C_s = -0.117 \text{ ft}$$

Combining the corrections due to temperature, tension, and sag into one (total) correction gives

$$C_{\text{total}} = -0.071 + 0.006 - 0.117$$

$$C_{\text{total}} = -0.182 \text{ ft}$$

Thus the correct distance is

$$228.48 - 0.182 = 228.298 \quad \text{or} \quad 228.30 \text{ ft}$$

In Example 4-3, the error caused by sag is considerably larger than either of the others. Observation of Eq. (4-6) indicates that the error due to sag can be reduced by using a lighter tape, using shorter measuring spans, or increasing the pull. For example, increasing the pull by 3 lb (to 25 lb) in Example 4-3 would decrease the absolute value of the correction due to sag from 0.117 ft to 0.091 ft.

Before ending this discussion on taping, a few additional remarks on mistakes and how to avoid them will be given. Most of these may seem trivial and not worth mentioning here, but the fact is that these mistakes are made every day—even by experienced surveyors. In most cases the real problem is lack of concentration. The problem of keeping an accurate count of the number of tape lengths measured in a long line has already been discussed in some detail.

Another common mistake is to misread the tape. One holds the nearest whole foot mark at his point, reads the foot mark, and calls out (to the head tapeman or recorder) "sixty eight." Is this the correct value? Well, it may be, and then again it may not be. Sixty eight may be the correct value, or the person may be holding the tape upside down, in which case 68 is erroneous, and the correct value is actually 89. To avoid making a mistake of this type, it is good practice to check the foot mark on each side of the mark that is actually read. The really alert tapeman should realize that if the reading is 89, the end of the tape would be about 10 ft away; whereas if the reading is 68, the end of the tape would be about 30 ft away.

Another mistake possibility that has been mentioned previously concerns the opportunity for confusion between using an adding tape and a subtracting tape. It is very tempting to want to use the end of the tape or the last mark on the tape as the zero mark on an adding tape. This can really be a problem when one occasionally has to use inexperienced help. Great care should be used to explain to such help exactly which mark he or she is supposed to use.

Another possible mistake related to misreading the tape is that of transposing digits. A funny thing happens sometimes between the time a distance is called out, perceived by the ear, and written down by the hand. Suppose a distance of 668.23 ft is measured and called out for recording. It is entirely possible that what gets written down may be either 686.23 or 668.32. How can this happen? Well, again, it is lack of concentration. One way to avoid mistakes of this type is to call out a number in the following form: "Six, six, eight, point, two, three."

An accuracy of $\frac{1}{5000}$ or higher can be attained in taping distances with care.

4-4 Electronic Distance Measurement

Electronic distance measuring devices were introduced in Chapter 3, with a Hewlett-Packard 3820 Electronic Total Station shown in Figure 3-9. The operation of this particular device is quite simple. It is set up (on tripod) over one point and the telescope aimed at a prism held on the other point (to which measurement is being made). Then by pressing the appropriate key on the control panel, the measured distance is displayed either in feet or meters. It should be emphasized that this is only one such device out of many on the market, and it is intended to be representative. Anyone using any device is, of course, advised both to be instructed in its use and to study the operating instructions accompanying the device.

4-5 Miscellaneous Taping Problems

As noted at the beginning of this chapter, it is possible to determine the size of an angle using the tape. Suppose it is desired to measure the angle *ABC* in Figure 4-6*a*. One way to do it is to measure off any convenient distance (100 ft is a good value for computational purposes) along both lines *BC* and *BA*, establishing points *D* and *E* (Figure 4-6*b*). If the distance *DE* is measured, the angle *ABC* can be computed from the formula

$$\text{angle } ABC = 2 \arcsin \frac{DE}{2BE} \tag{4-9}$$

Equation (4-9) is valid only if the distances *BD* and *BE* are equal. If unequal distances are laid out, as *BF* and *BG* in Figure 4-6*c*, then upon measuring the distance *FG*, the angle *ABC* can be computed using the law of cosines, as was explained in Chapter 2.

$$\text{angle } ABC = \arccos \frac{BF^2 + BG^2 - FG^2}{2(BF)(BG)} \tag{4-10}$$

Sometimes it might be necessary to lay out a specified angle using the tape.

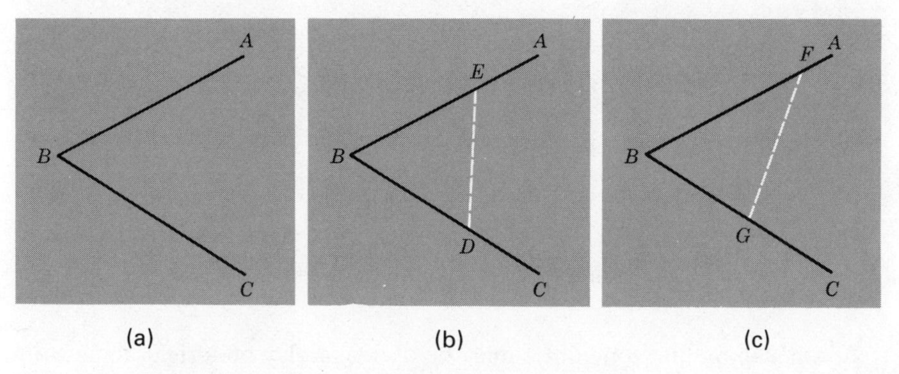

Figure 4-6 Determining the size of an angle using the tape.

To do this, it is convenient to rearrange Eq. (4-9) to the following form:

$$DE = 2(BE)\sin\frac{ABC}{2} \tag{4-11}$$

Then by substituting any convenient distance BE (equal to BD in Figure 4-6b) and the specified angle ABC, the value of DE can be computed and the required angle laid out.

Example 4-4

Describe the steps necessary to lay off an angle of 32°11′ using the tape.

Solution

Use Eq. (4-11) with $BE = 100.00$ ft.

$$DE = 2(BE)\sin\frac{ABC}{2} \tag{4-11}$$

$$DE = 2(100.00)\sin\frac{32°11'}{2}$$

$$DE = 55.43 \text{ ft}$$

In Figure 4-7a, the first step would be to lay off the distance $BE = 100.00$ ft. Then with one person holding the zero mark at B, another person can mark off an arc 100.00 ft from B at a point approximately 55 ft from E (Figure 4-7b). The person holding the zero end of the tape can then move to E, and the other person can mark off an arc 55.43 ft from E (Figure 4-7c). The intersection of these arcs establishes D, and DBE is the required angle.

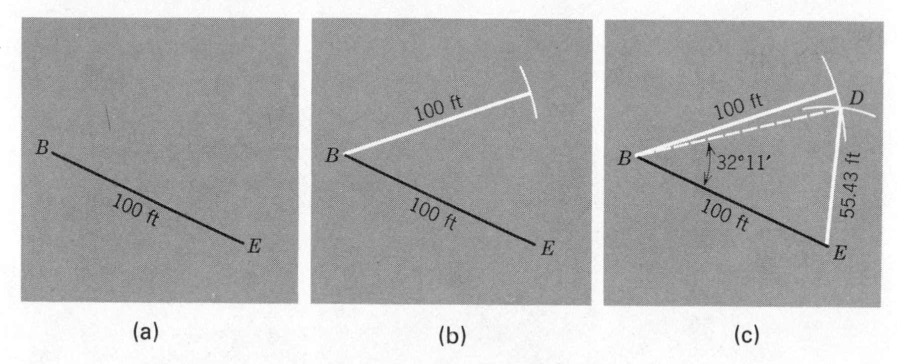

Figure 4-7 Laying off an angle of 32°11′ with the tape.

Another thing that sometimes must be done is to lay off a right angle with the tape. There are several ways to do this. One way is to follow the

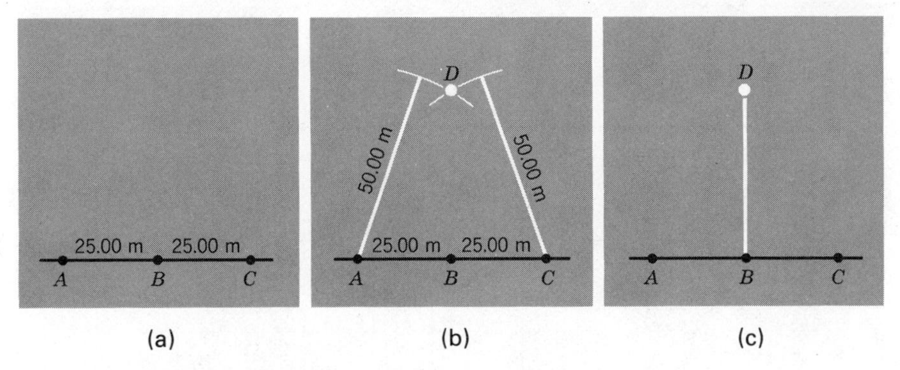

Figure 4-8 Laying off a right angle using a tape.

procedure described in the last paragraph, using a value of 90° for angle *ABC* in Eq. (4-11). If *BE* and *BD* are specified to be 100.00 ft, the distance *DE* to be laid out (Figure 4-7*c*) in order that angle *DBE* be a right angle would be 141.42 ft. This value is obtained by solving for *DE* in Eq. (4-11) when *BE* equals 100.00 and angle *ABC* is 90°.

Another way to lay off a right angle with the tape is to lay off a convenient distance in each direction from a point along a line. This could be done by laying off *AB* and *BC* equal to 25.00 m in Figure 4-8*a*. The next step would be to lay off two arcs of equal radius (e.g., 50.00 m)—one measured from *A* and the other from *C*, with both crossing at *D* (Figure 4-8*b*). The line from *D* to *B* (Figure 4-8*c*) would then form a right angle with line *AC*.

One additional method to lay off a right angle with the tape is based on the fact that a triangle whose sides are in a 3:4:5 ratio is a right triangle. Thus a distance of 50.00 ft could be laid off (*AB* in Figure 4-9*a*). An arc could be marked off 40.00 ft from *A* (Figure 4-9*b*) and another, 30.00 ft from *B*. The intersection of these two arcs would form *C*, and the angle *ACB* would be a right angle.

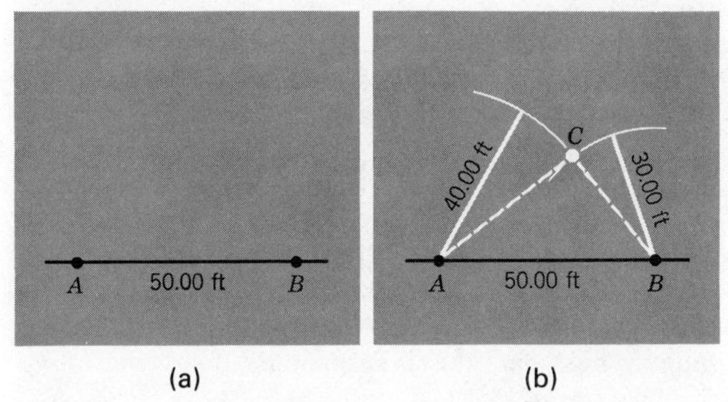

Figure 4-9 Laying off a right angle using a tape and the 3:4:5 ratio.

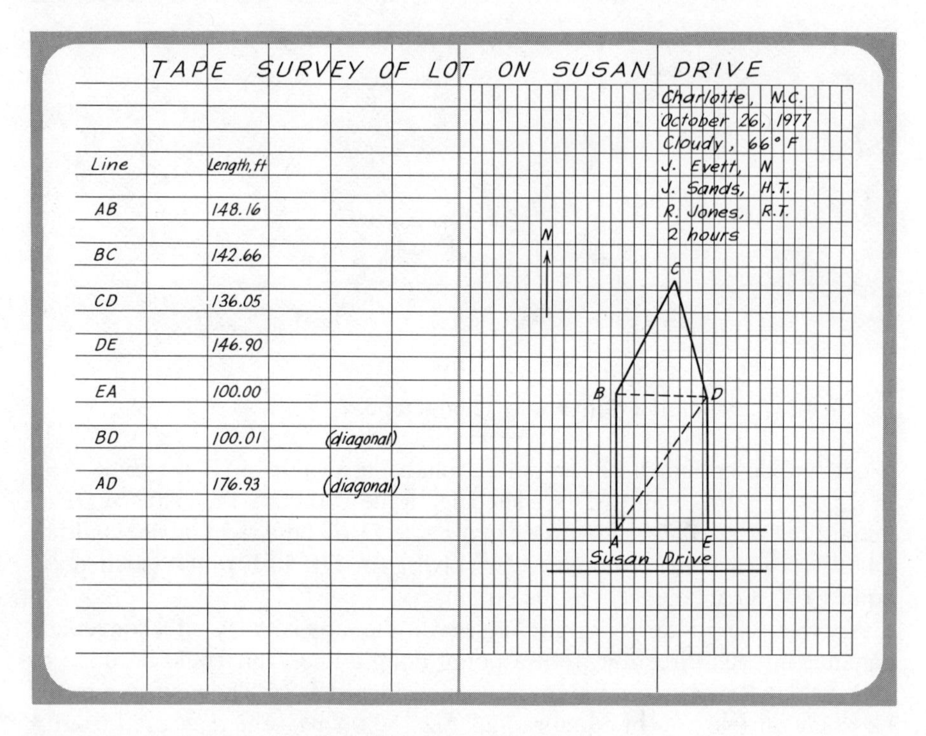

Figure 4-10 Sample field notes for tape survey.

4-6 Illustration of Field Notes

In Figure 4-10, a set of field notes is shown for a tape survey.

4-7 Problems

4-1 A surveyor whose pace is 2.5 ft paced a distance and recorded 209 paces. What should be the recorded distance?

4-2 A surveyor whose pace is 2.75 ft wishes to pace off a distance of 427 ft. How many paces should he step off?

4-3 A person wishing to calibrate her pace laid off a course of 400.0 ft (using a tape). She then paced the distance four times, recording in order the following number of paces: 162, 164, 165, 168. What would be the length of her pace? Is there anything in these data that is suspicious? If so, what?

4-4 What must you think if someone indicated a paced distance as being 298.73 ft?

4-5 On some subtracting tapes, the foot between the 99 and 100 ft marks is calibrated (in addition to the foot between the zero and 1-ft marks). Referring to Figure 4-1*a*, suppose the head tapeman does not realize that he is actually holding the 100 end of the tape instead of the zero end. What would be the correct distance if the rear tapeman reads 49 and the head tapeman reads 0.29 as shown?

4-6 Referring to Figure 4-5, what would be "the" distance from point *A* to point *B* if a slope distance (*AB*) of 48.82 m and a difference in elevation (*BC*) of 1.867 m are measured? Solve using both the exact formula and the approximate formula.

4-7 A distance of 489.27 ft was measured with a 100-ft tape that is 0.02 ft too short. What is the correct distance?

4-8 A distance of 413.66 ft is to be laid off with a 100-ft tape that is 0.03 ft too long. What tape reading should be used?

4-9 Referring to Figure 4-6*c*, what would be the size of angle *ABC* if the following distances are determined: $BF = 29.42$ m, $BG = 26.52$ m, and $FG = 12.41$ m?

4-10 Referring to Figure 4-7*c*, suppose we want to lay off an angle of 40°00′. If *BE* and *BD* are laid off as 17.00 m, what should *DE* be?

4-11 Scott needed to know the distance along the center line of a highway from *A* to *B*. He got in his car and drove along the center line from *A* to *B* and noted from his car's odometer that the distance driven was 1.17 miles. Point *A* is at the top of a mountain, and a sign there indicates an elevation of 8040 ft. Point *B* is at the bottom of the mountain, where a sign indicates an elevation of 7022 ft. What is the horizontal distance (in feet) from *A* to *B*? What is your opinion concerning the accuracy of this value?

4-12 Lennie built an electronic measuring device that depends on determining how long it takes light to travel from the device to a target and back to the device. If Lennie's device indicates that it takes 0.0001056 second for light to travel from device to target to device, what is the distance between the device and the target? Assume Lennie looked in a reference book and noted that the speed of light is 186,218 miles per second for existing atmospheric conditions.

4-13 If a distance of 812.4 ft is measured (recorded) when the tape temperature is 108°F, what measurement (tape reading) would be expected if the measurement is repeated with the same tape at a temperature of 19°F? What would be the "true" distance if the tape is 100.00 ft long at 68°F?

4-14 Careless Carlos was measuring the distance between two points. Pulling the tape between the two points, he read a distance of 82.00 ft. However he had carelessly pulled the tape on the wrong side of a tree, causing the tape to be 1.66 ft off line at the 25-ft mark. What is the correct distance between the points? Would it have made any difference if the tape had been 1.66 ft off line at the 41-ft mark (instead of at the 25-ft mark)? If so, what?

4-15 In taping the distance from point *A* to point *D*, Sarah encountered a cliff that was too steep to "break the tape." She established points *B* and *C* (see Figure 4-11), set her transit over *B*, and determined a vertical angle of 82°16′. She and her assistant then managed to pull the tape tightly between *B*′ and *C*, and determined that *B*′*C* was 71.1 ft. If *AB* and *CD* were measured as 302.6 ft and 407.9 ft, respectively, what is the distance from *A* to *D*? What would be your opinion if she did everything as related above except that the distance *BC* was measured instead of *B*′*C*?

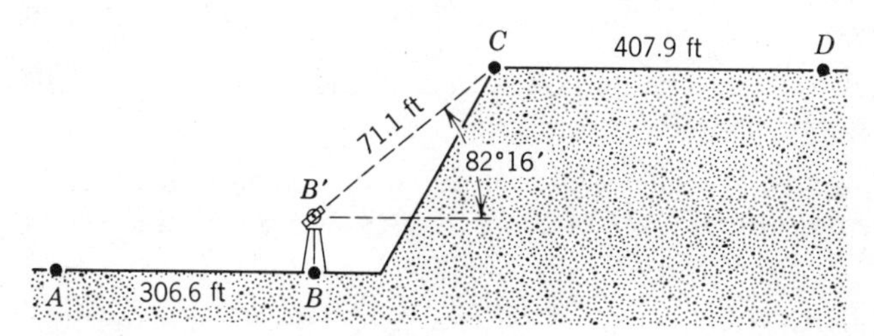

Figure 4-11

4-16 A tape, weighing 0.026 lb/ft with cross-sectional area of 0.0095 sq in., was calibrated at 68°F with a 20-lb pull while the tape was supported throughout. At calibration the 100-ft tape was 100.00 ft long. This tape is used in the field at 68°F with a pull of 10 lb, and a distance is measured and recorded as 478.39 ft. If the tape is supported only at the ends, what is the correct distance?

4-17 Repeat Problem 4-16 assuming the 100-ft tape was found to be 100.04 ft long at calibration.

4-18 Repeat Problem 4-16 assuming the temperature of the tape is 104°F when the measurement is made in the field.

4-19 What would the correction for sag be in Problem 4-16 if the tape is supported throughout when the measurement is made in the field?

4-20 Repeat Problem 4-16 assuming the tape is always supported at the 50 ft mark when the measurement is made in the field.

4-21 What would be the correction for tension (pull) in Problem 4-16 if the pull is 20 lb when the measurement is made in the field? What would be the correction for sag?

A 100-ft steel tape weighs 1.2 lb and has a cross-sectional area of 0.0035 sq in. It is calibrated at 68°F with a 15-lb pull, while the tape is supported throughout. At calibration the actual length of the tape is 99.95 ft.

In Problems 4-22 through 4-26, determine the true length of the measured line. Itemize each individual correction.

Problem	*Recorded Distance*	*Tape Temperature*	*Pull*	*Diff. in Elevation per 100 ft*	*Means of Support*
4-22	222.25 ft	68°F	10 lb	2.5 ft	ends only
4-23	468.00 ft	94°F	25 lb	0	throughout
4-24	188.84 ft	4°F	20 lb	2.8 ft	ends and 50 ft mark
4-25	199.98 ft	32°F	10 lb	4.8 ft	ends only
4-26	644.18 ft	25°F	10 lb	1.9 ft	ends only

4-27 A distance of 389.00 ft was measured with a 100-ft tape. Which of the following would cause the largest error?

- **(a)** When the pin is stuck to mark the 300-ft point, it was set 1.11 ft off line.
- **(b)** The measurement is downhill and each time a tape length (or partial tape length) is marked off, the head tapeman holds his end of the tape 2.0 ft too low.
- **(c)** The tape temperature is 30F° below the temperature at which it was calibrated.

4-28 A tape is exactly 100.00 ft long when calibrated at 68°F. It is desired to lay off a building that is 110.00 ft by 88.00 ft. What measurements should be made if the tape temperature is 9°F? Repeat for 95°F.

4-29 The sides of a rectangular parcel of property were measured and recorded as 250.0 m and 495.9 m. It was determined, however, that the 30-m tape was actually 30.04 m long. What is the correct area of the rectangle? Give answer in hectares.

4-30 It is required to lay off a rectangle of area 1.000 ha, with the length being twice as long as the width. If the tape described in Problem 4-29 is used, what measurements should be made?

4-31 If the tape described in Problem 4-29 is used to measure a distance, which is recorded as 409.8 m, what is the correct distance? What is the percentage error in the length of the line if no correction is made for incorrect tape length?

4-32 Suppose there is 0.275 m between the 30 m mark and the end of the tape. Using inexperienced help as rear tapeman, Richard, acting as head tapeman, measured a distance as 102.73 m. Later, Richard discovered that the rear tapeman had been holding the end of the tape instead of the 30-m mark each time a tape length was marked off. What was the correct length?

MEASUREMENT OF VERTICAL DISTANCES

Like the measurement of horizontal distances, the measurement of vertical distances is an important facet of surveying. Vertical distances are required in order to determine the differences in elevation among different locations. Determination of differences in elevation is important for many purposes, such as topographic mapping, designing drainage systems, designing and laying out highways and airports, setting grade stakes for sewer lines, delineating the shoreline for a proposed reservoir, and many others.

In some cases it is possible to measure vertical distances in the same manner that many horizontal distances are determined—by stretching a tape between end points. For example, the height of a power pole could be determined by extending a tape from the top of the pole to the ground below. In most instances, however, the points in question are not directly above and below each other. For example, it may be necessary to determine the difference in elevation between a point at one location and another point two miles down the road. In a case such as this, it is obviously not feasible to determine the difference in elevation by stretching a tape vertically from one point to the other. In this case the difference in elevation can be determined by leveling. Most of the remainder of this chapter is devoted to leveling.

5-1 Preliminary Considerations

Before discussing leveling in detail, a few preliminary remarks will be made. The terms "vertical distance" and "elevation" were used in the last two

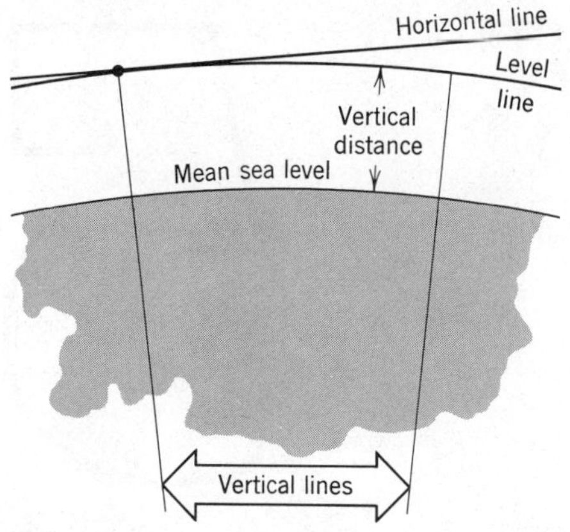

Figure 5-1

paragraphs. A vertical distance refers to a distance along a vertical line, which is the direction of gravity. Thus a vertical line at any given point is the direction of the string supporting a suspended plumb bob passing through the point. Actually, it is the line passing through the given point and the center of the earth. A straight line perpendicular to a vertical line is a "horizontal line." "Elevation" refers to the vertical distance above (or below) a reference level, usually mean sea level (msl). A "level line" is a curved line having all points at the same elevation and perpendicular to the vertical line. These terms may be better understood by referring to Figure 5-1. It should be noted that, while Figure 5-1 is two dimensional, the real world situation is three dimensional; thus a level line is actually a level surface and a horizontal line is actually a horizontal plane.

5-2 Differential Leveling—Procedure

Suppose G in Figure 5-2 is a bench mark (BM) of known elevation 322.08 ft, msl. (A bench mark is a relatively permanent point whose elevation is known. It may be a permanent monument established by the government, or it may be a nail in a tree, a mark on a concrete sidewalk, etc.) We want to determine the elevation of point H on the opposite side of a hill, utilizing differential leveling.

Two people are necessary to run levels—one to carry and hold the level rod (the rodman) and another to carry and set up the level and to make rod readings (the instrumentman). The instrumentman can record the data in a field book, or a third person may act as recorder. Both the characteristics and the manipulation of the level and the level rod were discussed in Chapter 3.

The notes for leveling are recorded in a standard form, as shown in Figure 5-3. The six columns on the left side of the field book are labeled as shown. The remarks indicated in the sixth column may extend over to the right page.

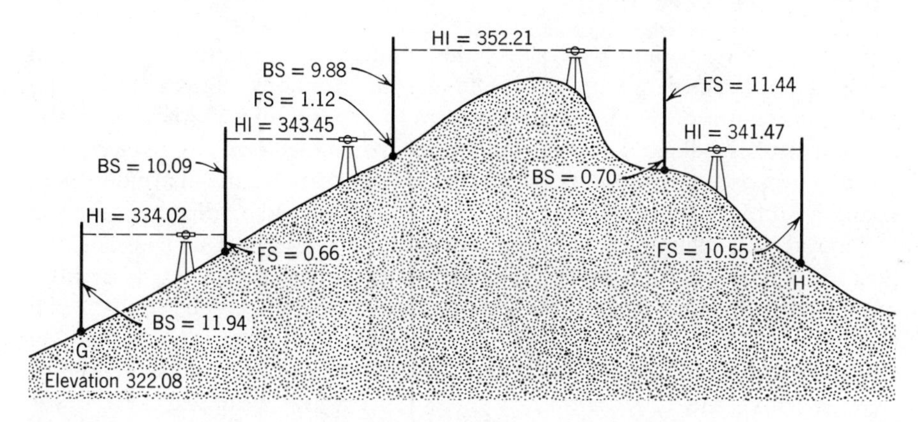

Figure 5-2

For the present illustration the point identification (G) and the elevation of the starting point (322.08 ft) would be entered under the columns headed "Point" and "Elev," respectively. Also, an entry was made on the same line on the right page of the field book to describe the location of the starting point. The remaining column headings are explained in the following discussion.

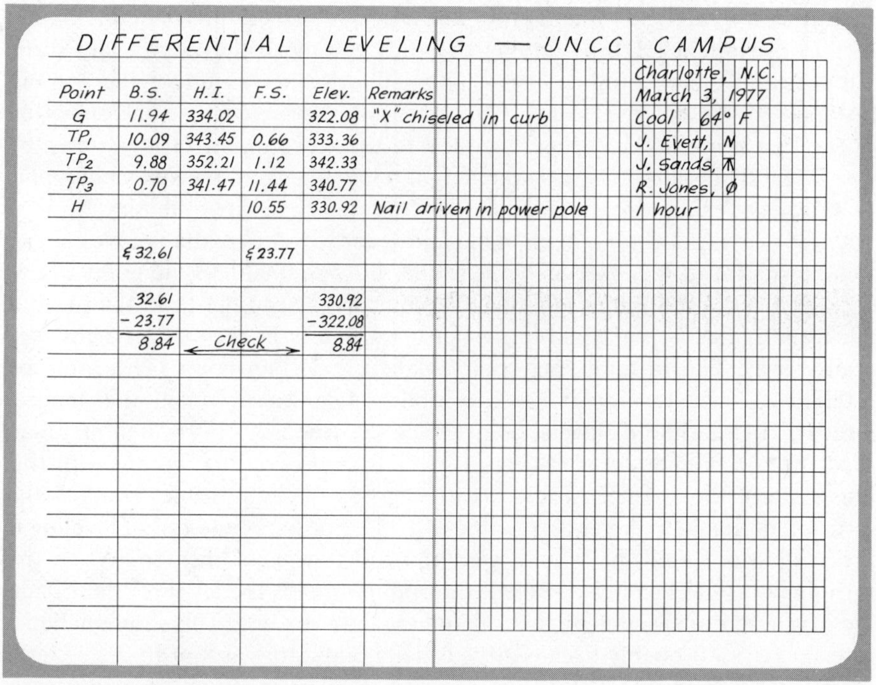

DIFFERENTIAL LEVELING — UNCC CAMPUS

Point	B.S.	H.I.	F.S.	Elev.	Remarks	
						Charlotte, N.C.
						March 3, 1977
G	11.94	334.02		322.08	"X" chiseled in curb	Cool, 64° F
TP_1	10.09	343.45	0.66	333.36		J. Evett, N
TP_2	9.88	352.21	1.12	342.33		J. Sands, ⩚
TP_3	0.70	341.47	11.44	340.77		R. Jones, ϕ
H			10.55	330.92	Nail driven in power pole	1 hour
	Σ 32.61		Σ 23.77			
	32.61			330.92		
	−23.77			−322.08		
	8.84	← Check →		8.84		

Figure 5-3 Standard form for recording level notes.

To begin the leveling procedure, the rodman would hold the level rod vertically on the bench mark at *G*. The instrumentman would proceed in the general direction of *H* and set up the level at some point so that when the instrument is leveled, a rod reading can be made. Generally speaking, if leveling is proceeding uphill (as at first in this illustration), the instrumentman would want to go as far as possible in order to make a rod reading near the top of the rod; if leveling is proceeding downhill, he would like to read near the bottom of the rod. The exact point at which the level is set up is of little importance. What is important is that the level be set up such that a rod reading can be made. Sometimes the level will be set up at too high an elevation when going uphill and when a rod reading is attempted, the line of sight is above the rod. In this case the level must be moved to a lower elevation. Similarly, sometimes the level will be set up at too low an elevation when going downhill and when a rod reading is attempted, the line of sight is below the rod. In this case the level must be moved to a higher elevation. When the instrumentman sets up the level and levels it properly, he sights on the rod held on *G*; and, noting where the horizontal cross hair intersects the rod, he "reads" the rod. This shot is called a backsight (BS), probably because it is usually taken backward, or away from the general direction in which the leveling is proceeding. The meaning of a backsight, however, is a sight (and reading) on a rod held on a point of known elevation to determine the height of the instrument (HI), by adding the rod reading to the known elevation. Suppose this BS reading is 11.94. This value is entered under the column headed "BS" in the level notes (Figure 5-3) on the same line as the elevation of the bench mark. This value is added to the elevation of *G* (known to be 322.08 ft). The sum (334.02) is the height (elevation) of the instrument and is recorded under the column headed "HI" on the same line in the level notes.

When the backsight has been made and recorded, the rodman moves ahead to a new point. If it is possible to read the rod when held on the point whose elevation is desired (point *H*), the rod would be placed there. In the present illustration, this is not possible, because the hill is in the line of sight (see Figure 5-2). Thus a temporary point, called a turning point (TP), must be established. The turning point is established as far away (in the general direction of the line of levels) as practical provided the instrumentman can read the rod. Generally speaking, if leveling is proceeding uphill (as in this illustration), the rodman would want to go as far as possible to allow the instrumentman to make a rod reading near the bottom of the rod; if leveling is proceeding downhill, he would want the rod read near the top. When the turning point has been established, the rodman holds the level rod on it, and the instrumentman sights on the rod and makes a reading. This shot is called a foresight (FS), probably because it is usually taken forward, or in the general direction in which the leveling is proceeding. The meaning of a foresight, however, is a sight (and reading) on a rod held on a point of unknown elevation to determine the elevation of that point by subtracting the rod reading from the height of the instrument. Suppose this FS reading is 0.66. This foresight (0.66) is recorded in the column headed "FS" in the level notes; but since it is used to determine the elevation of the next point (i.e., the turning point), it is placed on the next line (see Figure 5-3). This foresight is subtracted from the height of the instrument (334.02) to determine the elevation of this turning point (333.36), which is recorded on the same line as the foresight in the column headed "Elev." Note also that an entry "TP_1" was placed in the first column on that line to identify the point whose elevation is indicated on that line in the fifth column.

At this stage with the elevation of the first turning point determined, the rodman remains at the turning point while the instrumentman picks up the level and moves ahead to set up at another point in the general direction of point *H*. The procedure of reading a backsight, the rodman moving ahead to establish another turning point, and reading a foresight is repeated, and the elevation of the second turning point is determined. This procedure is repeated as many times as is necessary until the level is in such a position that the (final) foresight can be taken on the point whose elevation is desired (point *H* in this illustration). When this (final) foresight is read and subtracted from the current height of the instrument, the result is the elevation that was desired.

Before proceeding further with this chapter, the reader should study carefully the foregoing procedure for running a line of levels. It is recommended that the reader study the manner in which the indicated readings of backsights and foresights shown in Figure 5-2 are recorded in the field book (Figure 5-3). Finally, it is recommended that the reader take a blank sheet of paper, rule off six columns, look at the readings in Figure 5-2, and record the readings and do the arithmetic in the order it would be done in the field. This is an extremely important procedure, and it must be learned thoroughly. The author has seen many students who thought they understood the method but who, when they went into the field, could not remember where to record rod readings, when to add and subtract, and so on.

5-3 Differential Leveling—Admonitions

The leveling procedure of the preceding section was presented in as elementary a form as possible, in the hope that the reader could grasp the technique without being annoyed with a lot of details. There are, however, a number of items on which elaboration is needed.

To begin with, it is always desirable in surveying to verify work in every possible way. In the case of a set of level notes such as that of Figure 5-3, the difference between the sum of the backsights and the sum of the foresights should be equal to the difference between the ending elevation and the beginning elevation. Thus in Figure 5-3, the difference between the sum of the backsights and the sum of the foresights is 8.84 as is the difference between ending and beginning elevations. This is, however, only an arithmetic check. It tends to verify that the arithmetic was done correctly, but it does not verify that the elevation of point H is correct. For example, if one or more of the rod readings had been read incorrectly, the elevation computed for point H would be incorrect, although the arithmetic check above would be valid.

As a more rigorous means of verifying a set of levels, it is standard practice to continue levels so that they end on the starting point. This is referred to as a level loop. If the last foresight in a level loop is taken with the level rod on the starting point, and that foresight is subtracted from the current height of the instrument, the result, which is in effect the computed elevation of the starting point, should be the same as the initially used elevation of the starting point. Thus in the levels of Figures 5-2 and 5-3, after determining the elevation of point H, the levels should be continued back to the starting point, G. If the computed elevation of G is the same as the initially used elevation (322.08 ft), this would be an excellent verification of both the field readings and the arithmetic. An alternative to a complete loop is to continue the line of levels to a reliable, established bench mark, the elevation of which is known. If the elevation of the bench mark as computed from the line of levels is equal to the known elevation, this would be a good verification also. Neither method is completely foolproof, however, since the possibility always exists that compensating mistakes were made.

In the "closed loop check" it was stated that if the computed elevation of the starting point is the same as the initially used elevation, the work is verified. Suppose, however, these two values differ by 0.01 ft (this difference is referred to as the "closure"). Does this mean that the work is not verified and therefore useless? Actually, if these values differed by only 0.01 ft, the work would almost surely be of acceptable accuracy. Because of errors beyond human control, it is likely that the closure will frequently be 0.02 or 0.03 ft or more but still be of acceptable accuracy. On the other hand, should these differ by a foot or so, it would appear that a mistake (blunder) was probably made in reading the level rod. In this case the work would be unacceptable and should be done again. Allowable closure for ordinary leveling is proportional to the length of line of levels or to the number of times the level is set up. It can be computed from the following formula:

$$C = \pm 0.10\sqrt{M} \tag{5-1}$$

where C = allowable closure, in feet
M = length of the line of levels, in miles

Example 5-1

A line of levels was run in the form of a loop 2040 ft long. The initial elevation of the starting point was 98.33 ft, msl. When the last foresight reading of 4.16 was made on the starting point, the height of the instrument was 102.54. Compute the closure. Is it tolerable?

Solution

The final (computed) elevation of the starting point is obtained by subtracting the foresight from the height of the instrument.

$$\text{Final elevation} = 102.54 - 4.16$$
$$\text{Final elevation} = 98.38 \text{ ft}$$

The closure is 98.38 − 98.33, or 0.05 ft.
Allowable closure for ordinary leveling is computed from Eq. (5-1).

$$C = \pm 0.10\sqrt{M} \tag{5-1}$$

$$C = \pm 0.10\sqrt{\frac{2040}{5280}}$$

$$C = \pm 0.062 \text{ ft}$$

Since the actual closure (0.05 ft) is less than the allowable closure (0.06 ft), the levels are acceptable for ordinary leveling.

Equation (5-1) gives allowable closure for ordinary leveling. For more accurate work, classification of leveling into first order, second order, and third order is done by using the numbers 0.017, 0.035, and 0.050, respectively, in place of the number 0.10 in Eq. (5-1).

Example 5-2

Rework Example 5-1, assuming second-order leveling is required.

Solution

The closure is still 0.05 ft. Allowable closure for second-order leveling is

$$C = \pm 0.035\sqrt{\frac{2040}{5280}}$$

$$C = \pm 0.022 \text{ ft}$$

Since the closure (0.05 ft) is greater than the allowable closure (0.02 ft), the levels are not acceptable for second-order leveling.

In the previous discussion of the procedure for running levels, the term bench mark was defined and used. Bench marks, points of known elevation, are necessary in order to have a starting elevation. (While it is possible to use an assumed starting elevation in some instances, it is preferred, and in many cases required, that elevations be given with reference to mean sea level.) Standard bench marks, consisting of brass plates set in concrete, have been established at locations throughout the United States by the National Geodetic Survey and the U.S. Geological Survey. Many cities have established local bench marks throughout the city so that their survey crews will have a bench mark nearby wherever they are working. These might consist of a mark chiseled on the curb or sidewalk, the top of a fire hydrant, and the like. Information concerning location and elevation of bench marks can be obtained by contacting the appropriate government agency.

Other means of determining starting elevations may have to be devised. On one occasion the author began (and ended) a line of levels by taking a backsight (and subsequently a foresight) on the water surface of a nearby lake. The elevation of the water surface (starting elevation for the levels) was determined by looking in the local newspaper which gave daily lake elevations above mean sea level. Although not extremely accurate, this method was acceptable for the particular job.

Sometimes the surveyor may set his or her own bench marks. For example, if a surveyor is going to be involved over a period of time in obtaining cross-section data referenced to mean sea level at specified intervals along a creek (such data are required in delineating flood profiles), it would be helpful to have a number of bench marks along the creek. If these are available, each day's work can begin at a nearby bench mark instead of having to begin at one available bench mark that might be miles away. The latter method would waste time and introduce considerable opportunity for mistakes and errors. These local, or temporary, bench marks could be established by setting them at various points or marking certain locations (driving a nail into a tree, chiseling an "×" in concrete, etc.) and then running a line of levels, starting at an established bench mark of known elevation, to determine the elevation of each local bench mark. Certainly, the levels run to establish the elevations of such local bench marks must be very accurate and verified by closing the loop or closing into a second bench mark. Also, written descriptions giving the location and elevation of each bench mark should be prepared for future reference.

It was stated previously that the level rod is held on a point while the instrumentman makes a rod reading. Since leveling is a means of determining vertical distances, it is apparent that the level rod should be held *vertically* on a point. A careless rodman may allow the rod to lean backward or forward, or to one side. This should be avoided. The instrumentman can tell if the rod is leaning to one side or the other by observing the rod in comparison to the vertical cross hair, but he cannot tell whether the rod is leaning toward him or away from him. A rod level is available that attaches to the rod and indicates by bubble when the rod is vertical. Another procedure is for the rodman to wave (tilt) the rod slowly backward and forward (away from and toward the instrumentman) while holding it on the point of observation. To the instrumentman it appears that the cross hair is moving up and down the rod. The instrumentman should read the rod at the *lowest* point on the rod where it is intersected by the cross hair. (This works best when the rod is on a rounded point. If the rod rests on a flat or concave surface, it does not work too well because the rod tends to balance on the front and back edge as it is tilted back and forth. Use care.) If this is done properly and if the instrumentman ensures (by observing the vertical cross hair) that the rod is not leaning to one side, a sufficiently accurate reading can be made.

In the level notes of Figure 5-3, all readings and elevations were recorded to the nearest hundredth of a foot. This is typical precision for leveling. In some cases, such as determining elevations of ground points to use for drawing contour lines on a topographic map, determination of elevations to the nearest tenth of a foot could be sufficient. On the other hand, determination of elevations to the nearest thousandth of a foot may sometimes be necessary. This would require extreme care in every phase of the leveling procedure. A target on the level rod, with vernier, would be needed in order to read the rod to the nearest thousandth of a foot.

It should, of course, go without saying that the instrument should be level when rod readings are made. Although the instrument is always leveled when set up and prior to making any rod readings, sometimes rotation of the telescope in reading backsights and foresights or focusing the telescope causes it to become slightly out of level. Accordingly, prior to every rod reading, the level bubble should be checked. If the telescope is not level, it should be properly leveled by turning one leveling screw slightly. This should be only a very slight adjustment.

Thus far in all the discussion of leveling, it has been assumed that when the instrument is properly leveled, the line of sight is along a level line. This, however, is not true. Since light is assumed to travel in a straight line, the line of sight must be a horizontal line rather than a level line, which is desired. Thus, in Figure 5-4, while the desired rod reading is BC (i.e., along the level line), because the line of sight departs from a level line, the reading must be BE. Thus the rod reading would be incorrect by the amount EC. This is not the complete story, however. Light passing through the atmosphere is refracted (bent) toward the earth's surface. Thus the refracted line of sight is as indicated in Figure 5-4, and the rod reading is actually incorrect by the amount DC. This amount DC is the error due to the combined effects of curvature of the earth and atmospheric refraction. It can be computed from

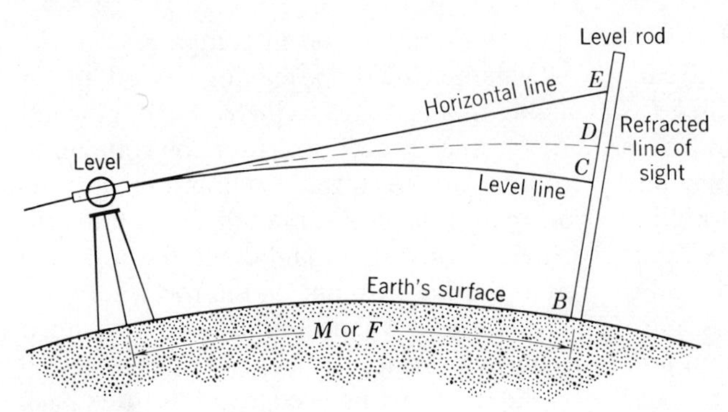

Figure 5-4

the approximate formula

$$e = 0.574M^2 = 0.206 \times 10^{-7}F^2 \tag{5-2}$$

where e = error in rod reading, in feet
M = distance from level to rod, in miles
F = distance from level to rod, in feet

Utilizing Eq. (5-2), it can be determined that in order for the error due to the combined effects of curvature and refraction to be as much as 0.01 ft, the distance from level to rod must be at least 697 ft. This is a very long shot in ordinary leveling; thus the effect is ordinarily slight. Also, since the correction would be subtracted from all rod readings (both backsights and foresights) and since backsights are added and foresights are subtracted, these errors tend to cancel. As a matter of fact, they would exactly cancel if successive backsights and foresights were made at the same distance from level to rod. In view of the preceding, it should be obvious that the actual effect of curvature and refraction is generally negligible, except when extremely accurate leveling is being done.

In Section 5-2 the establishing of a turning point was encountered. A turning point, it will be recalled, is a point the elevation of which is determined temporarily for the purpose of continuing a line of levels. A turning point should be firm; it should not settle during the time a foresight is read on it, the level is moved and set up, and a backsight is read on it. A paved surface, a rock, and very hard soil can make good turning points. If the area is marshy or the soil sandy, the rodman might carry a small wooden peg to drive into the ground for use as a turning point.

Also on the general topic of turning points, it should be emphasized that during the time a foresight is read on a turning point, the level is moved and set up, and a backsight is read on the turning point, the rodman should keep the level rod on the turning point. If the rodman is careless and moves the rod while the instrumentman is setting up the level, the entire line of levels will have to be rerun, unless the location of the turning point can be determined with certainty.

One final consideration in this section is the nature of errors and mistakes in running levels. Some of the most prevalent sources of error are: level rod not held vertically, settling of the level rod or of the level, leveling bubble not perfectly centered when rod reading is made, certain weather conditions (wind, heat waves, etc.) that cause error in rod readings, improper instrument adjustment, failure to equalize lengths of backsights and foresights, and incorrect rod length. Some of these have been discussed previously; most can be dealt with by exercising care by both the rodman and the instrumentman.

Some of the most common sources of mistakes are: reading the rod incorrectly, failing to extend the rod properly when using the "high rod," incorrectly recording field notes (such as recording a backsight in the foresight column), and allowing the turning point to change before its purpose is satisfied. To elaborate a bit on the first of these, reading the rod incorrectly is a very real possibility with serious consequences. Assume the cross hair intersects the rod at 8.98 ft. When the instrumentman is making this reading, the big red "nine" indicating the 9-ft mark on the rod is plainly visible; and it is very easy to record 9.98 instead of 8.98. The result of such a mistake is to throw the levels off by one foot, although the mistake should be detected when the loop is completed. The other mistakes listed above are more or less self-explanatory. In leveling, as in all phases of surveying, the surveyor must be constantly vigilant to avoid mistakes and to minimize errors.

5-4 Leveling Involving Intermediate Foresights

In all the discussion to this point, the levels have been described with all level rod readings alternating between backsights and foresights. This is not unusual if the levels are simply being run from one point to another distant point. Frequently, however, it is also necessary to determine elevations of various points along the way (intermediate points). For example, it may be necessary to determine accurately the elevations of the center of a paved road at 10-ft intervals.

Once the level has been set up and leveled, a backsight taken, and the height of the instrument determined, any number of foresights may be read for the purpose of determining elevations of various points without moving the level. Each foresight is recorded on a separate line in the notes and is subtracted from the current height of instrument to determine the elevations of the respective points. The foresights that do not establish turning points are called intermediate foresights (IFS). When all intermediate foresights have been read for a given set-up of the level, a turning point can be established in order to move the level to a higher or lower elevation so that additional intermediate foresights can be made or just to continue the line of levels.

Example 5-3

A surveyor has been assigned the task of determining accurately the elevations of the center line of a paved highway at 25-ft intervals starting at point A and proceeding to point B, which is 200 ft from A. A bench mark (elevation 608.22 ft, msl), a nail in an oak tree, is available near point A. The surveyor and his assistant first used his tape and marked off the 25-ft intervals from A to B. (He marked an "×" on the highway at each 25-ft interval.) He then proceeded to run a line of levels, making the following rod readings in the order indicated.

BS on BM:	0.64
FS on TP_1:	10.16
BS on TP_1:	2.49
IFS on A:	6.41
IFS on $A+25$:	8.00
FS on $A+50$:	10.99
BS on $A+50$:	1.12
IFS on $A+75$:	5.05
IFS on $A+100$:	10.20
FS on TP_2:	12.01
BS on TP_2:	3.16
IFS on $A+125$:	4.43
IFS on $A+150$:	1.17
FS on $A+175$:	8.84
BS on $A+175$:	1.29
FS on $A+200$ (B):	4.19
BS on $A+200$ (B):	11.80
FS on TP_3:	0.04
BS on TP_3:	11.97
FS on TP_4:	0.33
BS on TP_4:	10.62
FS on TP_5:	1.50
BS on TP_5:	8.66
FS on BM:	3.71

Prepare the level notes in proper form to record these readings. Is the closure tolerable?

Solution

See Figure 5-5 for level notes. Closure = 0.02 ft. Assuming ordinary leveling, since the length of levels run is approximately 400 ft,

$$C = \pm 0.10\sqrt{M} \tag{5-1}$$

$$C = \pm 0.10\sqrt{\frac{400}{5280}}$$

$C = \pm 0.028$ ft

Since the actual closure (0.02 ft) is less than the allowable closure (0.03 ft), these levels are satisfactory for ordinary leveling.

Several comments should be made regarding Example 5-3. When the usual arithmetic check (difference between sum of backsights and sum of foresights compared to difference between starting and ending elevations) is made, the intermediate foresights are not included in the sum of the foresights. When the elevation of point $A+50$ was determined, it was obvious to the surveyor in the field that a turning point was needed. To expedite matters, the point $A+50$ was used as a turning point. The same is true for points $A+175$ and $A+200$ (which is point B).

It will be noted that, in Figure 5-5, all rod readings and elevations were recorded to the nearest hundredth of a foot. In many situations, elevations to the nearest tenth of a foot are quite adequate. In example 5-3, if the road had been unpaved, elevations to the nearest tenth of a foot would have been satisfactory. Since it is much easier and quicker to read the level rod to the

PROFILE LEVELS

Point	B.S.	H.I.	F.S.	Elev.	Remarks
BM	0.64	608.86		608.22	BM, nail in oak tree
TP_1	2.49	601.19	10.16	598.70	
A			6.41	594.78	Pt. A
A + 25			8.00	593.19	
A + 50	1.12	591.32	10.99	590.20	
A + 75			5.05	586.27	
A + 100			10.20	581.12	
TP_2	3.16	582.47	12.01	579.31	
A + 125			4.43	578.04	
A + 150			1.17	581.30	
A + 175	1.29	574.92	8.84	573.63	
A + 200	11.80	582.53	4.19	570.73	Pt. B
TP_3	11.97	594.46	0.04	582.49	
TP_4	10.62	604.75	0.33	594.13	
TP_5	8.66	611.91	1.50	603.25	
BM			3.71	608.20	
	Σ 51.75		Σ 51.77		
	51.75			608.20	
	−51.77			−608.22	
	−0.02	← Check →		−0.02	

Scott Street
Charlotte, N.C.
July 1, 1977
Hot, 94°F
J. Evett, N
J. Sands, ⊼
R. Jones, Ø
3 hours 20 minutes

Figure 5-5

			PROFILE			LEVELS	Scott Street
Point	B.S.	H.I.	F.S.	I.F.S.	Elev.		Charlotte, N.C.
BM	0.64	608.86			608.22	Nail in oak tree	July 1, 1977
TP_1	2.49	601.19	10.16		598.70		Hot, 94° F
A				6.4	594.8	Pt. A	J. Evett, N
A + 25				8.0	593.2		J. Sands, ⊼
A + 50	1.12	591.32	10.99		590.20		R. Jones φ
A + 75				5.0	586.3		3 hours 20 minutes
A + 100				10.2	581.1		
TP_2	3.16	582.47	12.01		579.31		
A + 125				4.4	578.1		
A + 150				1.2	581.3		
A + 175	1.29	574.92	8.84		573.63		
A + 200	11.80	582.53	4.19		570.73	Pt. B	
TP_3	11.97	594.46	0.04		582.49		
TP_4	10.62	604.75	0.33		594.13		
TP_5	8.66	611.91	1.50		603.25		
BM			3.71		608.20		
	Σ51.75		Σ51.77				
	51.75				608.20		
	−51.77				−608.22		
	−0.02	←	Check	→	−0.02		

Figure 5-6

nearest tenth of a foot than to the nearest hundredth, it is common practice to record intermediate foresights and elevations computed therefrom to the nearest tenth of a foot. (This should be done only if elevations to the nearest tenth of a foot are satisfactory for the specific situation.) If this is done, in order to ensure proper overall leveling control, backsights and foresights used in determining the height of instrument and elevations of turning points should still be read to the nearest hundredth of a foot.

One final observation is that some surveyors prefer to list intermediate foresights in a column separate from the one in which foresights are recorded. This probably makes the notes easier to read and interpret. To demonstrate the use of a separate column for intermediate foresights and the recording of intermediate foresights to the nearest tenth of a foot, the level notes of Figure 5-5 have been redone to reflect these changes (see Figure 5-6).

5-5 Adjustment of Level Loops

As discussed previously in this chapter, it is not unusual that there will be some error of closure in running a level loop, no matter how carefully the work is done. If a line of levels is run for the purpose of establishing the elevations of several bench marks and there is a significant error of closure on completing the loop, the elevations of the bench marks should be adjusted to reflect the error of closure. It is reasonable to assume that the propensity to incur error in the determination of the elevation of a bench mark is proportional to the ratio of the length of the line of levels from starting point to the bench mark in question to the total length of the level loop. This procedure is illustrated by Example 5-4. It should be emphasized that the adjustment considered here is for error of closure, which should not, in general, exceed the allowable error of closure [Eq. (5-1)]. If the closure is unacceptable, the levels should be rerun.

Example 5-4

A surveyor ran a line of levels from A, a bench mark of known elevation 301.52 ft msl, to B, to C, to D, and back to A for the purpose of determining the elevations of bench marks B, C, and D. The length of each segment of the loop and the elevations determined are given below.

	Distance, ft		*Elevation from Levels, ft*
A to B	5600	A	301.52 (known)
B to C	4200	B	288.01
C to D	3800	C	169.45
D to A	5300	D	250.08
Total	18,900	A	301.37

What adjusted elevations should be reported for these bench marks?

Solution

The error of closure is (301.37 − 301.52), or −0.15 ft. The allowable error of closure for ordinary leveling is computed from Eq. (5-1).

$$C = \pm 0.10\sqrt{M} \tag{5-1}$$

$$C = \pm 0.10\sqrt{\frac{18{,}900}{5280}}$$

$$C = \pm 0.19 \text{ ft}$$

Since the actual error of closure (0.15 ft) is less than the allowable (0.19 ft), the levels are acceptable, and the elevations of bench marks B, C, and D should be adjusted. These adjustments are shown below.

Bench Mark	*Distance from Starting Point, ft*	*Adjustment, ft*	*Adjusted Elevation, ft*
A	—	—	301.52
B	5600	$+\frac{5600}{18{,}900}(0.15)=0.04$	288.05
C	9800	$+\frac{9800}{18{,}900}(0.15)=0.08$	169.53
D	13,600	$+\frac{13{,}600}{18{,}900}(0.15)=0.11$	250.19
A	18,900	+0.15	301.52

In the solution of Example 5-4, the second column gives the distance from the starting point to each bench mark. Since the correction applied to the last (computed) elevation for A to make it agree with the initial (known) elevation of A is +0.15 ft, the correction for each other bench mark is determined by multiplying +0.15 ft by the ratio of distance each specific bench mark is from the starting point to the total length of the loop. For example, the distance from A to C is 9800 ft and the total length of the loop is 18,900 ft. Hence the correction to be applied to C is (9800/18,900)(+0.15), or +0.08 ft.

5-6 Other Methods of Determining Vertical Distances

Most of this chapter has been devoted to differential leveling, as this is the most common method of determining vertical distances. There are other methods available, some of which are mentioned briefly here.

One method was mentioned briefly in the introduction to this chapter. That method is to stretch a tape vertically between the two points.

Another method is to use a barometer (or altimeter). An airline pilot uses this method to determine, approximately, the vertical distance between his plane and the ground below. The principle is that atmospheric pressure, which is measured by a barometer, varies inversely with elevation. With proper correlation, a recorded difference in barometer readings between two ground points can be converted to a difference in elevation. Atmospheric pressure is influenced by weather conditions, and determination of elevations by barometer is generally not accurate enough for ordinary surveying. It may be satisfactory for some purposes, such as reconnaissance surveys.

One additional method of determining elevations is by trigonometric leveling. This involves measuring horizontal and inclined distances and vertical angles and using these data to compute vertical distances. The distances can be measured by taping and the vertical angles with a transit. The stadia method (Chapter 7) can also be used as a form of trigonometric leveling.

5-7 Problems

In Problems 5-1 through 5-3 complete the level notes.

5-1

Pt	*BS*	*HI*	*FS*	*Elev (ft)*	*Remarks*
BM_1	6.22			94.47	
TP_1	8.04		2.65		
TP_2	0.27		3.65		
TP_3	0.96		4.02		
BM_2			9.81		

5-2

Pt	*BS*	*HI*	*FS*	*Elev (m)*	*Remarks*
BM_1	0.275			82.140	
TP_1	1.025		1.223		
TP_2	2.407		3.281		
BM_2	3.101		3.134		
TP_3	1.079		2.177		
BM_3			1.887		

5-3

Pt	*BS*	*HI*	*IFS*	*FS*	*Elev (ft)*
BM_1	8.19				152.22
TP_1	9.23			2.02	
A			6.6		
B			7.5		
C			0.4		
TP_2	4.43			8.12	
BM_2				2.27	

In Problems 5-4 through 5-6, set up level notes in proper form for the levels indicated in the illustrations. Make the appropriate check.

5-4 See Figure 5-7.

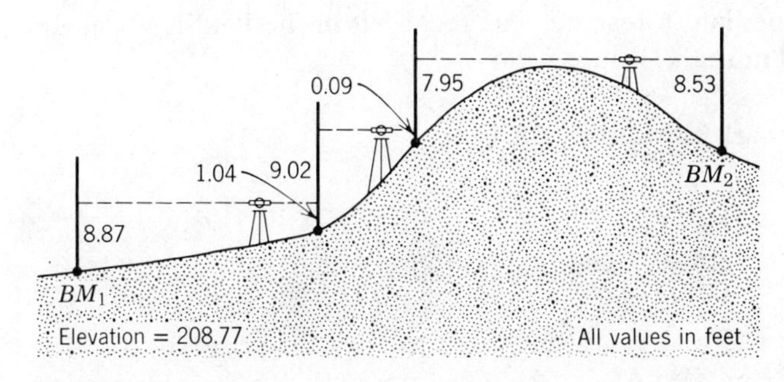

Figure 5-7

5-5 See Figure 5-8.

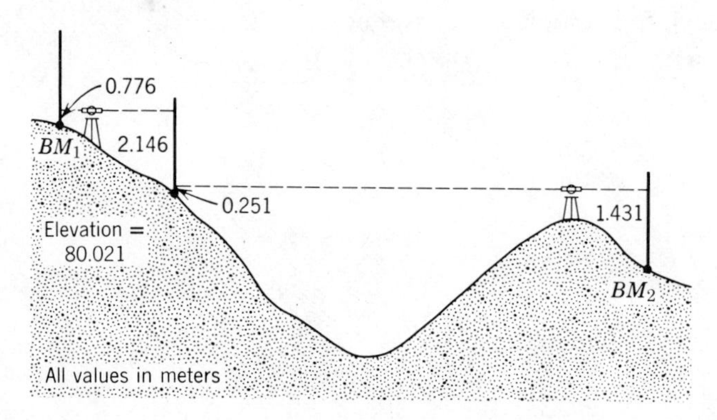

Figure 5-8

5-6 See Figure 5-9.

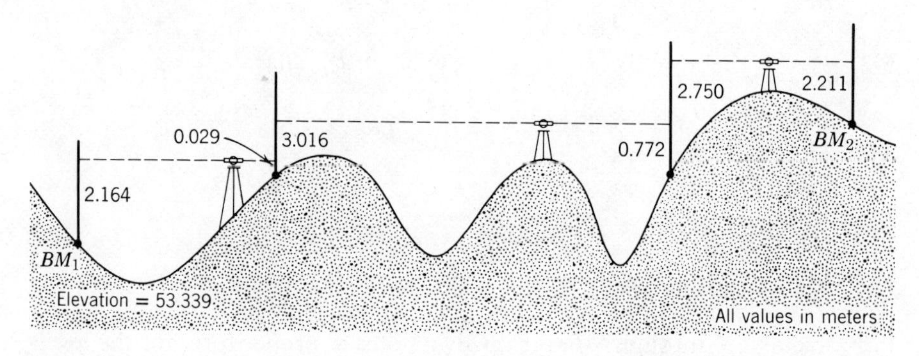

Figure 5-9

In Problems 5-7 through 5-9, the order in which a series of backsights, foresights, and intermediate foresights was recorded in the field is given. In each case set up level notes in proper form.

5-7 Starting bench mark elevation: 602.14 ft

BS on BM_1: 4.33
FS on TP_1: 6.28
BS on TP_1: 9.03 (Rod readings in feet)
IFS on A: 6.2
IFS on B: 0.5
FS on BM_2: 10.01

5-8 Starting bench mark elevation: 80.33 ft

BS on BM_1:	10.06	
FS on TP_1:	0.93	
BS on TP_1:	8.03	
FS on TP_2:	1.23	
BS on TP_2:	8.08	(Rod readings in feet)
IFS on R:	2.25	
IFS on S:	1.09	
FS on TP_3:	3.33	
BS on TP_3:	10.07	
FS on BM_2:	2.21	

5-9 Starting bench mark elevation: 33.095 m

BS on BM_1:	2.088	
IFS on *A*:	3.117	
IFS on *B*:	1.199	(Rod readings in meters)
FS on TP_1:	0.082	
BS on TP_1:	1.109	
FS on BM_2:	1.002	

5-10 A line of levels about 8000 ft long was run and the closure was 0.11 ft. Is this acceptable for ordinary leveling? Is this acceptable for first-order leveling?

5-11 How long a line of levels must be run in order for the allowable closure to be 1.00 ft? (a) Assume ordinary leveling. (b) Assume first-order leveling.

5-12 Suppose you send a survey crew out to run a line of levels approximately 1000 ft in length, and they report a closure of 0.99 ft. As boss of the crew, what would be your opinion and recommendation?

5-13 Dick was trying to determine very accurately the elevation of point *B* by running a line of levels from point *A*, the elevation of which is 227.408 ft. Using a target on the rod, he made the following readings:

BS (200 ft shot) on *A*:	2.438	
FS (525 ft shot) on TP:	9.030	(Rod readings in feet)
BS (226 ft shot) on TP:	1.003	
FS (612 ft shot) on *B*:	10.039	

Based solely on the indicated rod readings, what would be the elevation of *B*? What would be the computed elevation of *B* if curvature of the earth and refraction are considered?

5-14 How long must a level shot be in order for the effect of curvature of the earth and refraction to be 0.001 ft?

5-15 A level loop was run to establish the elevation of several bench marks. The following data were obtained.

	Distance, ft		*Elevations from Levels, ft*
E to F	4200	E	612.22 (known)
F to G	6000	F	643.18
G to H	2300	G	653.25
H to I	3800	H	599.02
I to E	7000	I	582.16
		E	612.39

Is the closure acceptable for ordinary leveling? If so, adjust the elevation of all bench marks.

5-16 Repeat Problem 5-15 for the following data.

	Distance, ft		*Elevations from Levels, ft*
R to S	1100	R	23.16 (known)
S to T	1200	S	32.04
T to U	800	T	36.36
U to R	1900	U	20.05
		R	23.07

5-17 In Figure 5-10 Dorothy was trying to determine the elevation of E based on the known elevation of 209.62 ft for D. She set up her transit and, with level telescope, read the rod being held on D. This rod reading was 8.62 ft. She then rotated the transit and read the (vertical) angle α. Finally, she stretched the tape from the center of the telescope on the transit to point E, obtaining a reading of 62.25 ft. If the angle α was 31°11′, what is the elevation of E?

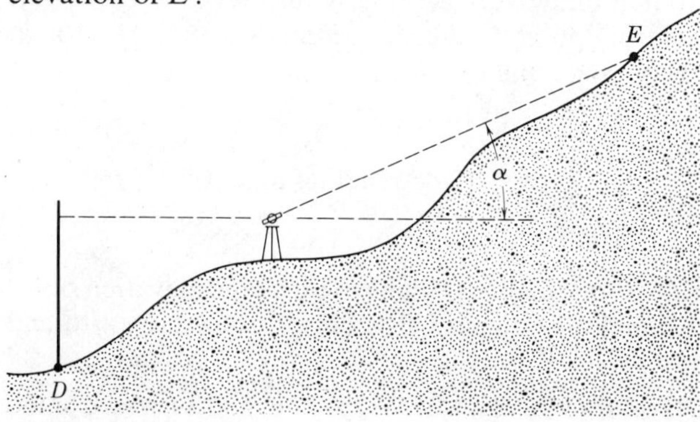

Figure 5-10

5-18 A line of levels was run from a point of known elevation 62.87 ft to another point, the elevation of which was determined (from the levels) to be 193.27 ft. It was later determined that the level rod (which was 13 ft long) was 0.03 ft too long because of imperfect manufacture. What is the correct elevation of the second point?

5-19 For the data given in Problem 5-18, what would be the answer if someone sawed 0.03 ft off the bottom of the rod? (Assume the rod was exactly 13.00 ft long before someone sawed off the bottom.)

5-20 Dean wanted to determine the height of a building some distance away. Using an electronic distance measuring device, he determined that the distance from him to the bottom of the building was 2128.733 m and the distance from him to the top of the building was 2136.330 m. The angle formed by the line from him to the bottom of the building and the line from him to the top of the building was 3°32′30″. What is the height of the building? See Figure 5-11.

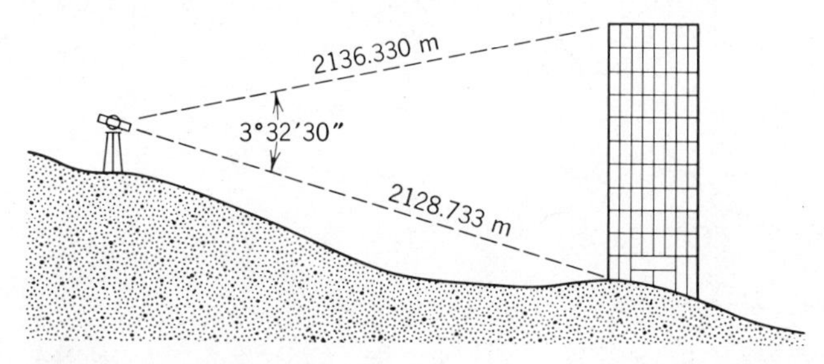

Figure 5-11

5-21 A lifeguard is sitting in a chair at the edge of the sea such that her eye is 10.0 ft above sea level. If a small ball starts floating out to sea, how far out could it go before, in theory, she could no longer see it? Assume a calm sea and that she has binoculars to view the ball.

5-22 For the same lifeguard in the same chair (Problem 5-21), how far away, in theory, could she see the top of a man's head if he is sitting in a raft and his head is 3.2 ft above sea level?

ANGULAR AND DIRECTIONAL MEASUREMENTS

Previous chapters have dealt with the measurement of distances—both horizontal and vertical. Another basic type of measurement required in surveying is the measurement of direction. In general, surveying in the field answers two questions: "How far?" and "Which way?" The answers are determined by measuring distances and directions.

The term "direction" is used in a general sense here. In surveying, direction is indicated by the measurement of horizontal angles, vertical angles, bearings, and azimuths. These are discussed in detail in the remainder of this chapter.

6-1 Horizontal Angles

It is assumed that the reader remembers the high school geometry definition of an angle as being two rays (lines) drawn from the same point (vertex). The size of an angle is a measure of the amount of opening between the two lines expressed, in effect, as a percentage of one complete rotation. As related in Section 1-1, angles are generally measured in degrees (and minutes and seconds), where a degree is $\frac{1}{360}$ of one complete rotation.

In surveying, horizontal angles are measured with the transit. To begin with, the transit is set up (and leveled) exactly over (using plumb bob) the vertex of the angle to be measured. The next step is to line up the zero mark on the circular scale on the lower plate with the zero mark on the vernier scale on the upper plate. This is referred to as "zeroing the vernier." It is accomplished by loosening the upper plate clamp and rotating the vernier scale on the upper plate until the two zero marks line up approximately. The upper plate clamp is then tightened, and the upper plate tangent screw is used to rotate the upper plate slightly until the two zero marks line up exactly. A magnifying glass that comes with the transit is helpful in doing this and in reading the vernier later. The next step is to loosen the lower plate clamp, rotate the telescope, and sight on a point on one of the lines of the angle. The lower plate tangent screw is used to zero in on the target. All of this must be done with the upper plate clamp tightened. The next step is to loosen the upper plate clamp, rotate the upper plate, and sight on a point on the other line of the angle. The upper plate tangent screw is used to zero in on this target. All of this must be done with the lower plate clamp tightened. The final step is to read the vernier (i.e., to read the position of the vernier scale on the upper plate with regard to the circular scale on the lower plate) to determine the size of the angle. (Reading the vernier was discussed thoroughly in Section 3-2.) Since the circular scale is marked off from zero to 360° in both directions (clockwise and counterclockwise), care must be exercised in reading the vernier. It must be read in the same direction the transit was rotated. Using the preceding technique, angles are often read to the nearest minute.

In the preceding discussion of measuring an angle, it is not absolutely necessary to zero the vernier initially. Instead, the upper plate clamp may be tightened wherever it happens to be at the start of the measurement, and the vernier may be read initially. After sighting on one point, rotating the transit, and sighting on the other point, the vernier is read again. The difference between the two vernier readings is the size of the angle.

The discussion thus far has referred to *the* vernier, but actually there are two verniers on a transit located on opposite sides. It is good practice to read both of these verniers each time a reading is made. The results should be the same, but they may not be because of imperfection or improper adjustment of the transit. If the results of reading both verniers are not the same, an average of the two should be used.

Errors in measuring horizontal angles may result from improper adjustment of the transit, from not setting up exactly over the vertex, and from not sighting exactly on the points on the lines. Mistakes may result from turning the wrong tangent screws, from reading the vernier incorrectly—especially, failing to add 30 minutes to the vernier reading when necessary to do so—and from recording the value incorrectly.

6-2 Vertical Angles

A vertical angle is an angle in a vertical plane, one side of the angle being horizontal and the other being inclined—either upward or downward. If the inclined line is upward, the vertical angle is considered to be positive; if inclined downward, it is considered to be negative.

To measure a vertical angle, the transit is set up and leveled carefully. If the telescope is leveled by using the leveling bubble on the telescope, the reading on the vertical circle and vernier should be 0°00′. If it is not zero, either the transit was not leveled properly or it is out of adjustment. The vertical motion clamp is loosened and the telescope is rotated to sight on the point defining the inclined line. The vertical motion tangent screw may be used to set the (horizontal) cross hair on target. The value of the vertical angle is read using the vertical circle and vernier. It should be indicated as either plus (if upward) or minus (if downward). It should be emphasized that the actual angle measured has its vertex at the center of the telescope of the transit (the intersection of the center line of the telescope and the plumb bob line). One side of the angle is the horizontal line passing through the center of the telescope, and the other side is an inclined line extending from the center of the telescope to a reference point upward or downward.

Errors and mistakes in measuring vertical angles are the same as those for measuring horizontal angles, except that there should be no mistake caused by turning the wrong tangent screw, since there is only one such screw to turn.

6-3 Bearings

A bearing is an indicator of the direction of a line. It is indicated by giving the angle between the line and a reference line, or reference meridian. The usual reference meridian for bearings is the north-south meridian (more about that later). Specifically, the bearing of a line is indicated by stating (1) the angle between the line and the north-south meridian and (2) the quadrant in which the line lies (either NE, SE, SW, or NW). For example, in Figure 6-1, the bearing of line *PQ* is N 62° E. At first glance, one might decide that the bearing of line *PR* would be N 93° E. This is incorrect, however, because *PR* actually lies in the SE quadrant and not in the NE quadrant. Since the bearing angle is the angle between the line and the north-south meridian, the correct bearing of *PR* is S 87° E. Obviously, bearing angles are acute angles.

As a final observation, if the bearing of *PQ* is N 62° E, what is the bearing of *QP* (the opposite direction)? If the line *QP* is extended into the SW quadrant as line *PT*, it can be observed that the bearing of *PT*, which is the same as that of *QP*, is S 62° W. Thus the bearing of a line in one direction has the same angular value as the bearing of the line in the opposite direction but the quadrant indicators are opposite. If the direction of a line is the same as one of the reference meridians, the bearing is indicated by the word "due" in front of the direction, as due north or due west.

The measurement of bearings is among the simplest measurements made in surveying work. It is made using a device that has been used by navigators,

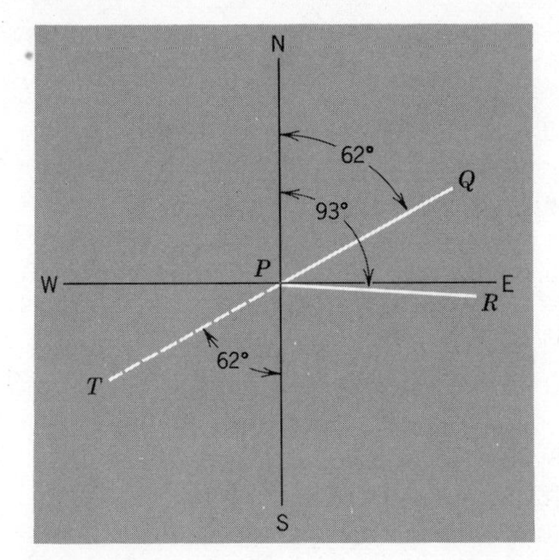

Figure 6-1

surveyors, and others for many centuries—the compass. As if by magic, but really by attraction of the earth's magnetic field for certain metallic material, a freely floating compass needle lines up in the earth's magnetic north-south magnetic field.

As related in Chapter 3, a transit is usually equipped with a compass needle that rests on a pivot in a recessed well in the horizontal plate of the transit. The well has a glass cover. When the needle is unclamped, it oscillates back and forth until it comes to rest in the earth's north-south magnetic field. Readings are made by means of a circular scale located on the upper plate of the transit adjacent to the ends of the needle (see Figure 6-2). This scale is

Figure 6-2 Circular scale for measuring bearings. (Courtesy of Keuffel & Esser Co.)

calibrated with north-south and east-west axes. Each quadrant is graduated in degrees from zero to 90 degrees from north to east and to west, and from south to east and to west. Each degree may be divided into half degrees or quarter degrees.

In order to determine the bearing of a line *AB*, the transit is set up either over point *A* or point *B*. (This must be done carefully using a plumb bob, of course.) After the transit is leveled, the telescope is directed toward the other point, and with the aid of the vertical cross hair, the telescope is "set" on the far point. The sight on the far point should be made by aligning the vertical cross hair with the center of the far point. This can be done if the far point is a tack in a stake and is visible from the transit. Often, however, the far point itself cannot be seen—because the point is too far away, or the point is concealed by underbrush, or the point is on the other side of a small hill—in which case a range pole can be placed directly behind the point and the sighting can be made on the range pole. Obviously, the range pole should be held plumb. When the telescope is sighted approximately on the far point, both the lower plate clamp and the upper plate clamp are tightened. One of the tangent screws can then be used to zero in on the far point. The needle is then unclamped and allowed to float freely until it comes to rest. When it comes to rest, the magnetic bearing can be read by noting where the end of the needle intersects the circular scale. Theoretically, either end of the needle can be read to ascertain the angular part of the bearing, but for consistency it is good practice always to read the north end of the needle. The quadrant in which the bearing lies is determined by observing the bearing scale.

As explained previously, ordinary transits have circular scales for reading bearings. The scales are marked to half degrees or quarter degrees. Thus bearings read from ordinary transits can generally not be read more precisely than to the nearest 15 minutes or perhaps 10 minutes. However, recorded bearings indicating precision to the nearest minute or even to the nearest second are often encountered. Rather than using compass readings, such bearings likely have been computed, based on angle readings (interior angles, deflection angles, etc.).

Most anyone who has ever played with a toy compass knows that a magnet placed near a compass will cause the needle to deviate from the north. While surveyors do not normally carry magnets as such that might have an effect on the compass needle, they do carry or come in contact with certain metallic and other items that are magnetized sufficiently locally to disturb the earth's magnetic field and cause the needle to deviate from its proper position. These are referred to as "local attraction." Some examples include wristwatches, fences, automobiles, electric power lines, local ore deposits, and even the surveyor's tape, taping pins, and ax. Some of these can be controlled; others cannot. Some wristwatches will cause the needle to deviate. This is potentially a real problem because it is normal to have the hands near the transit for

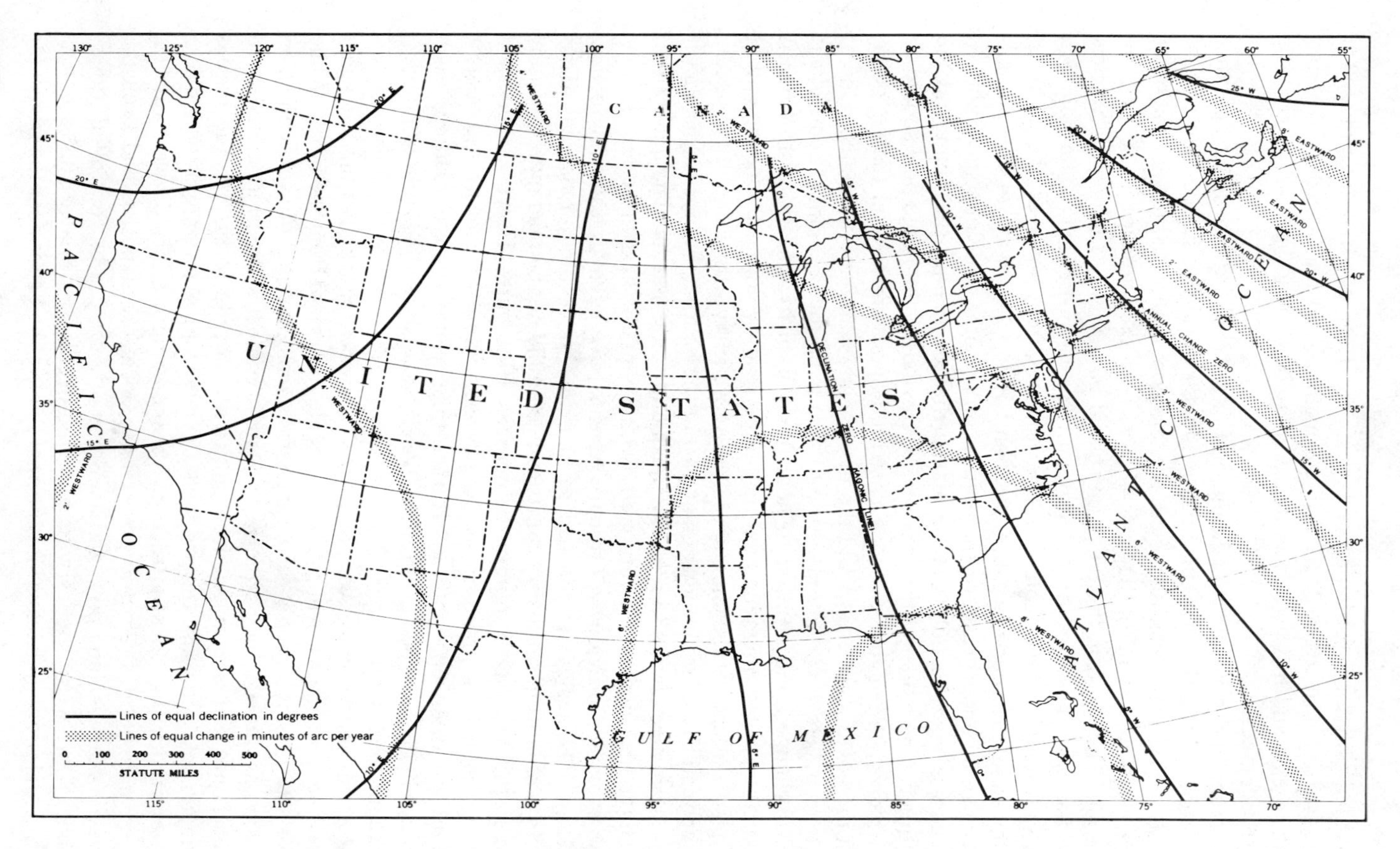

Figure 6-3 Magnetic declination in the United States for 1975. (Courtesy of U.S. Geological Survey.)

manipulating clamps and screws or perhaps shading the direct sunlight in reading a bearing. Obviously, the instrumentman should not wear a watch while using the transit if the watch affects the needle. Fences and power lines are potential problems, because it is not unusual for property lines (which are being surveyed using the transit) to run along fences or beneath power lines. Problems of this nature can sometimes be avoided by moving the transit to another location where the same bearing can be read. For example, if one end of a line is near a fence or power line and the other end is not, the transit could be set up over the end that is not near the fence or power line and the sight could be made on the other end. In the case of automobiles, usually the interfering automobile can be moved.

The treatment of bearings thus far has referred to the north-south meridian as the reference from which bearing angles are measured. The true north-south meridian at any point on the earth's surface is a curved line passing through the geographic north and south poles of the earth and the point of observation. This true north-south meridian can be determined by sighting on the North Star (Polaris), details of which can be obtained from references 1, 3, and 5 listed at the end of this book.

The true north-south meridian is not, however, the reference meridian along which a compass needle points. (At certain points the compass needle will point in the direction of the true north-south meridian, but this is mostly coincidental.) The entire earth acts as a gigantic magnet with the magnetic poles near the geographic poles. The surveyor's compass needle lines itself up in the earth's magnetic field. Thus the surveyor takes the magnetic north-south meridian as the line from the observation point to the magnetic north pole.

The difference between the true north-south meridian and the magnetic north-south meridian is called the magnetic declination. If the declination is known at a specific location, a magnetic bearing (bearing referenced to the magnetic north-south meridian) may be converted to the true bearing (bearing referenced to the true north-south meridian). The entire situation is complicated, however, by the fact that the declination itself changes at all points with time. Hence the declination at a point today will almost certainly be somewhat different from what it was 50 years ago. Maps are available (Figure 6-3) which give approximate declinations for specific points in time. The surveyor must be careful in making the conversion from magnetic bearing to true bearing and vice versa. Sometimes the declination is added; sometimes it is subtracted. Two examples follow.

Example 6-1

A magnetic bearing of S 56°30′ E was read in the field. If the declination is 6°15′ E, what is the true bearing?

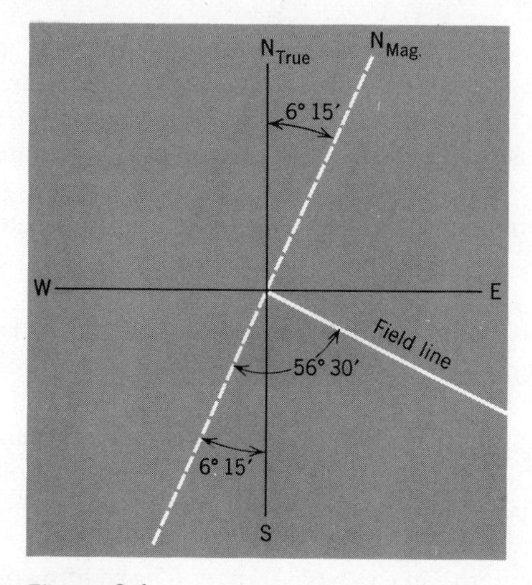

Figure 6-4

Solution

This type of problem is easily solved if an accurate sketch is drawn. The true north-south meridian should be drawn vertically, and then the magnetic north-south meridian should be drawn relative to the true. See Figure 6-4.

The next step is to lay off the line on the sketch. In this case the magnetic bearing is S 56°30′ E; hence the line makes an angle of 56°30′ with magnetic south, and since the line is in the southeast quadrant, the angle is marked off from south toward east. The true bearing is indicated by the angle between the line and true south. From the sketch, it should be obvious that this angle is determined by *subtracting* the declination from the magnetic bearing angle. Hence, the true bearing angle is 56°30′ − 6°15′, or 50°15′, and the true bearing is S 50°15′ E.

Example 6-2

In 1916, when the declination was 7°15′ W, the magnetic bearing of a line was determined in the field to be N 44°45′ W. What magnetic bearing should be used to retrace that line today, when the declination is 2°30′ E?

Solution

Since the true bearing of a line does not change, the true bearing can be computed from the 1916 data. From the true bearing, the present magnetic bearing can be computed based on the present declination. To determine the true bearing from the 1916 data, a sketch (Figure 6-5*a*) was prepared. From the sketch it should be obvious that in this case, the true bearing angle is determined

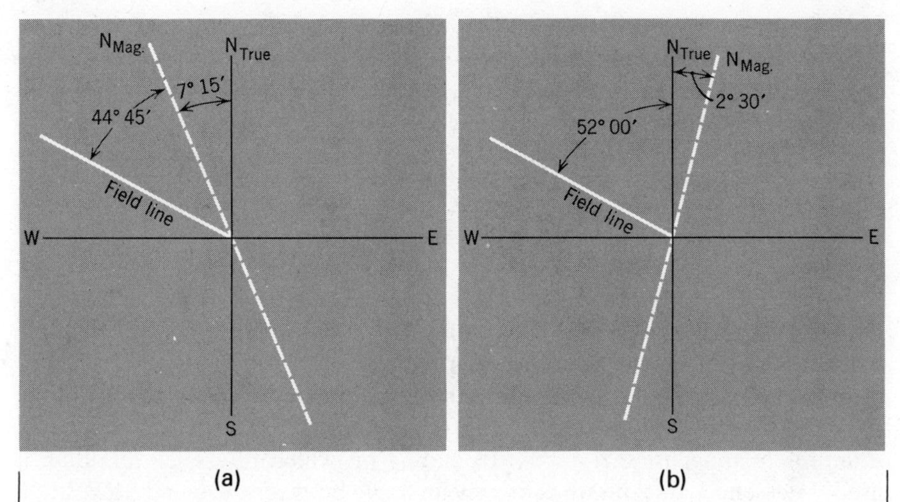

Figure 6-5

by *adding* the declination to the magnetic bearing angle. Hence the true bearing is N 52°00′ W. Since the true bearing never changes, the problem now is to find the magnetic bearing if the true bearing is N 52°00′ W and the present magnetic declination is 2°30′ E. From the sketch (Figure 6-5*b*) it can be seen that the magnetic bearing (today) is N 54°30′ W.

From the previous discussion it should be apparent that the determination and use of bearings must be done somewhat carefully. To begin with, the usual scale for reading bearings generally permits readings to perhaps the nearest 10 or 15 minutes. The effects of declination and local attraction subtract from the reliability of bearing readings. Add to this the fact that the bearing readings of the same line observed with different transits may differ from each other because of instrumental errors (mentioned in the next paragraph). From this discussion one begins to get the impression that bearings may not be accurate enough for precise surveying. This may be true, but bearings are still read and recorded. They can be used for indication of general direction and for rough checking of measured angles of a traverse. They are usually recorded on plats, although they may be computed from angle readings rather than the values read from the compass in the field.

Errors in reading bearings may result from the instrument being out of adjustment (needle bent, sluggish needle, for example), personal errors (not setting the instrument up exactly over the reference point or not sighting exactly on the far point), and natural errors (undetected local attraction). Mistakes may result from reading the needle incorrectly or recording the bearing incorrectly. With regard to reading the needle, one must carefully note the direction in which the degrees are marked off on the scale. In the northeast quadrant on the scale, the angles are marked off counterclockwise; whereas in the northwest quadrant they are marked off clockwise. It is easy, for example, to read a bearing angle of 49° when it should be 51° by reading in the wrong direction.

6-4 Azimuths

Like a bearing, an azimuth is an indicator of the direction of a line. It is indicated by giving the clockwise angle measured from either the north or south end of the reference meridian (the north-south meridian) to the line. In this book azimuths will be measured from the north. Azimuths are either true azimuths or magnetic azimuths, depending on which reference meridian is used. In essence, an azimuth is merely an angle between zero and 360°, but it should be clear as to whether a given azimuth is a true azimuth or a magnetic azimuth and whether it is measured from the north or the south. In Figure 6-1 the azimuths of *PQ*, *PR*, and *PT* are 62°, 93°, and 242°, respectively. Note that the azimuth of *PQ* and that of *PT* (where *PQ* and *PT* are really along the same line but opposite in direction) differ by 180°.

If it is necessary to measure the azimuths to a number of points all measured from the same point, the transit should be set up and leveled at the common point. After zeroing the vernier, the telescope should be oriented in the north direction, this being done with the upper plate clamp tightened and the lower plate clamp loosened. When the approximate north direction has been achieved, the lower plate clamp should be tightened and final adjustment made with the lower plate tangent screw. At this point both upper and lower plate clamps should be tight, the telescope should be oriented in a north direction, and the reading on the circular scale should be 0°00′. By manipulating the upper plate clamp and the upper plate tangent screw, the transit may be oriented in any direction and the azimuth can be read off the circular scale. Since the circular scale goes from zero to 360°, the azimuths can be read directly off the scale.

6-5 Miscellaneous

Horizontal angles are classified in several ways, primarily for use in traversing (Chapter 8). At the point of intersection of two sides of a closed figure, two angles are formed. The angle within the figure is called an *interior angle*; the angle outside the figure is called an *exterior angle.* In Figure 6-6*a*, angles *A* and *C* are interior angles, and angles *B* and *D* are exterior angles. An *angle to the right* is used in connection with going from one line in one direction to the

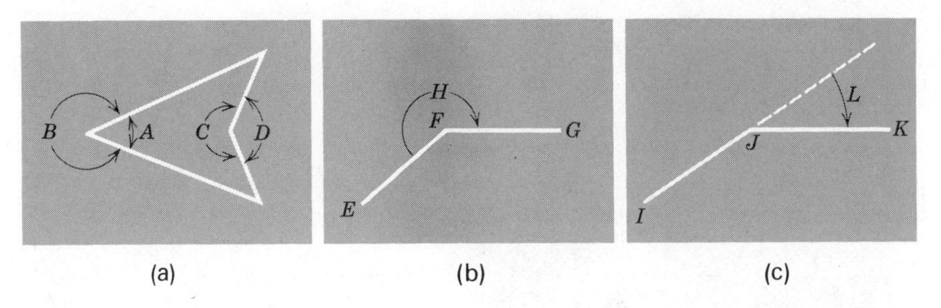

(a) (b) (c)

Figure 6-6

next line in another direction. It is the clockwise angle measured from the preceding line to the succeeding line. In Figure 6-6*b*, in proceeding from *EF* to *FG* angle *H* is an angle to the right. (*Angles to the left* can be used in the same context except that the angle is turned in the counterclockwise direction.) A *deflection angle* is also used in connection with going from one line in one direction to the next line in another direction. It is the angle measured from the extension of the preceding line to the succeeding line. In Figure 6-6*c*, in proceeding from *IJ* to *JK* angle *L* is a deflection angle. Since a deflection angle may be to the right or to the left of the extension of the preceding line, an indication of "*R*" or "*L*" must be placed after the value of each deflection angle to indicate right or left. Thus if angle *L* in Figure 6-6*c* has the value 37°42′, the deflection angle would be recorded as 37°42′ *R*.

The surveyor must be able to make computations involving bearings, azimuths, interior angles, and the like. For example, the survey of a closed figure may be done by measuring the interior angles, from which it is desired to compute the bearing of each line. There are no hard and fast rules for making such computations. The best procedure is to prepare a good sketch showing the information given and then visualize what computation is needed. Two examples follow.

Example 6-3

Line *AB* has a bearing of N 16°30′ E. At *B* a deflection angle of 29°25′ *L* is measured, defining the direction of line *BC*. Determine the bearing and azimuth of *BC*.

Solution

With the information given, the sketch shown in Figure 6-7 is prepared. From the sketch, it is apparent that the bearing angle is 29°25′ − 16°30′, or 12°55′; hence the bearing of *BC* is N 12°55′ W. The azimuth is determined by subtracting the bearing angle (12°55′) from 360°. Thus the azimuth is 347°05′.

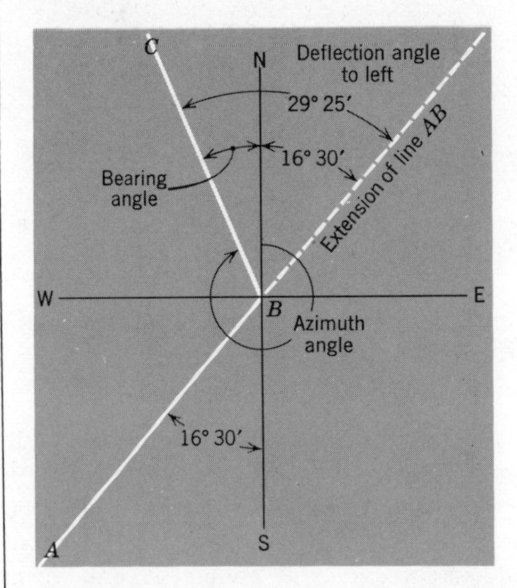

Figure 6-7

Example 6-4

In Figure 6-8, determine the values of the interior angles at Q and R, based on the given bearings.

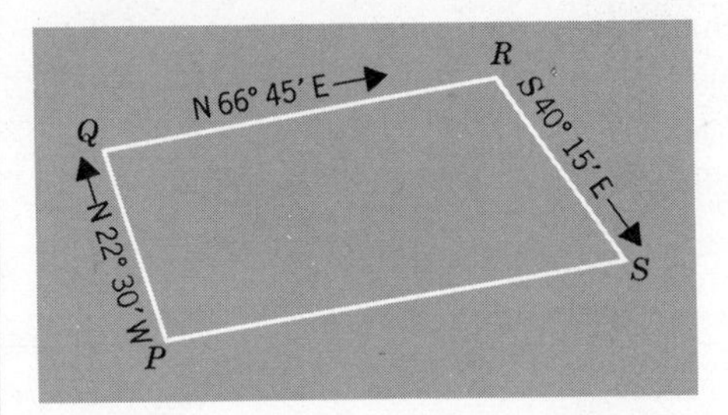

Figure 6-8

Solution

It is helpful to prepare a sketch of point Q, in order to determine the interior angle at Q. This is shown in Figure 6-9a. From the sketch, it is apparent that the interior angle at Q is $180° - 66°45' - 22°30'$, or $90°45'$. In order to determine the interior angle at R, a sketch of point R is prepared (Figure 6-9b). From the sketch, the interior angle at R is $66°45' + 40°15'$, or $107°00'$.

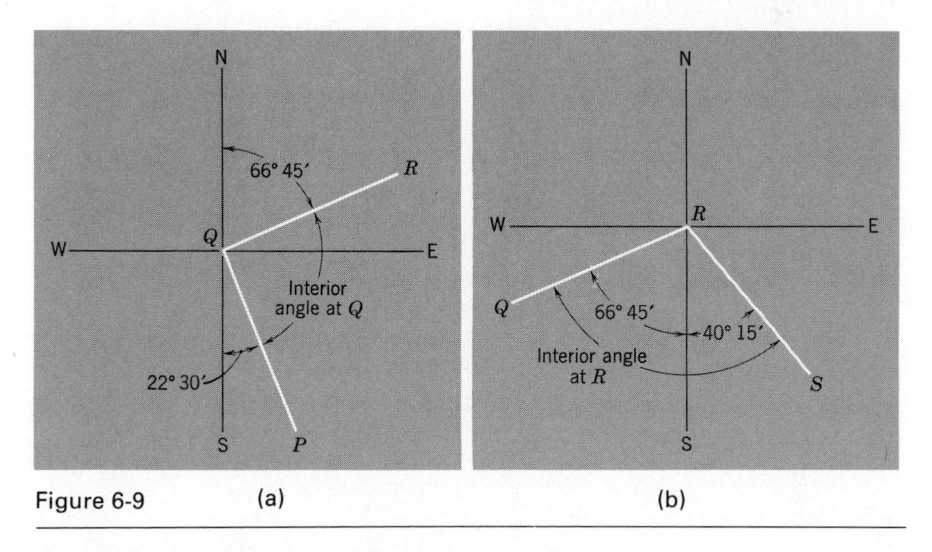

Figure 6-9 (a) (b)

6-6 Illustration of Field Notes

In Figure 6-10, a set of field notes is shown for a survey involving angles, bearings, and distances.

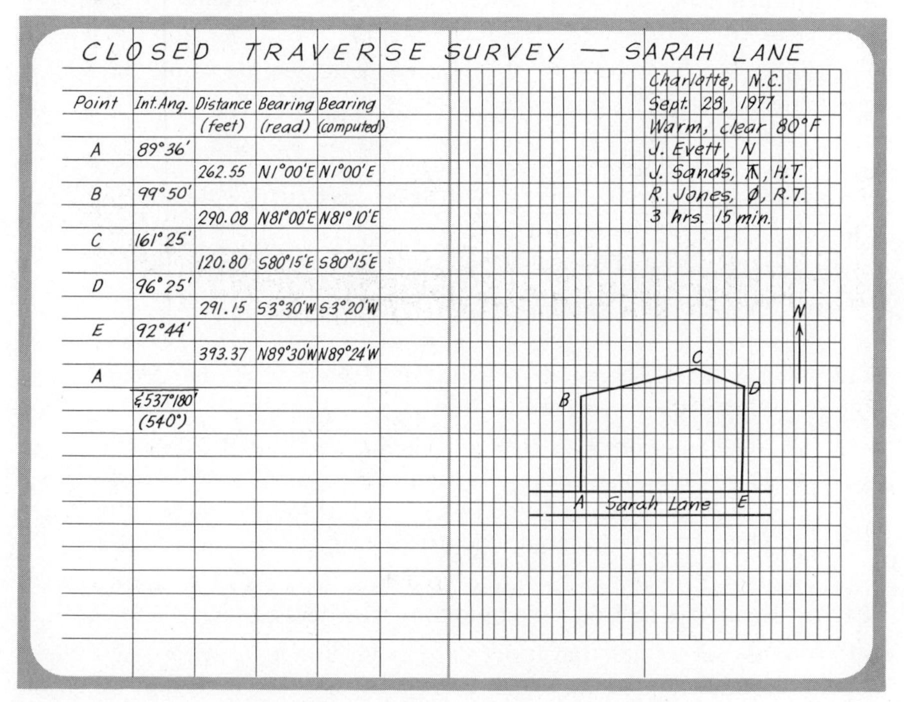

CLOSED TRAVERSE SURVEY — SARAH LANE

Point	Int. Ang.	Distance (feet)	Bearing (read)	Bearing (computed)
A	89° 36′			
		262.55	N1°00′E	N1°00′E
B	99° 50′			
		290.08	N81°00′E	N81°10′E
C	161° 25′			
		120.80	S80°15′E	S80°15′E
D	96° 25′			
		291.15	S3°30′W	S3°20′W
E	92° 44′			
		393.37	N89°30′W	N89°24′W
A				
	Σ537°180′			
	(540°)			

Charlotte, N.C.
Sept. 28, 1977
Warm, clear 80°F
J. Evett, N
J. Sands, ⊼, H.T.
R. Jones, ∅, R.T.
3 hrs. 15 min.

Figure 6-10 Sample field notes for closed traverse survey.

6-7 Problems

6-1 A transit was set up over point *A* and the bearings of lines *AB* and *AC* were read and recorded. From these bearings, the angle *CAB* was computed. Later it was ascertained that there was local attraction at point *A* when the two bearings were read. The local attraction caused the needle to deflect 2°30′ clockwise. What would be the effect of the local attraction on the angle *CAB*?

6-2 A bearing of S 25°30′ W was read when the declination was 4°45′ W. What is the true bearing?

6-3 The true bearing of a line is N 5°00′ W. If the declination is 2°15′ W, what bearing should be used to retrace the line in the field?

6-4 If the magnetic bearing of a line is S 4°00′ E and the declination is 6°30′ E, what is the true bearing?

6-5 Repeat Problem 6-4 for a declination of 6°30′ W.

6-6 The magnetic bearing of a line is S 89°30′ W and the declination is 4°30′ E. What is the true bearing?

6-7 Repeat Problem 6-6 for a declination of 4°30′ W.

6-8 The magnetic bearing of a line is due east, and the declination is 2°30′ E. What is the true bearing of the line?

In Problems 6-9 through 6-11, what magnetic bearing would be used to retrace the line today?

Problem	*Line*	*Magnetic Bearing 1910*	*Declination 1910*	*Declination Today*
6-9	AB	S 25°30′ W	5°30′ E	2°15′ E
6-10	RS	N 16°15′ W	2°45′ W	4°00′ W
6-11	YZ	S 82°00′ E	1°15′ W	2°00′ E

6-12 The magnetic azimuth of a line is 316°25′, and the declination is 6°40′ W. What is the true azimuth?

6-13 A line is run from *A* to *B*, *B* to *C*, and *C* to *D*. At *B*, a deflection angle of 10°06′ *R* is read, defining the direction of *BC*. At *C*, a deflection angle of 94°10′ *L* is read, defining the direction of *CD*. If the true bearing of *AB* is known to be S 10°15′W, compute the true bearing of *BC* and of *CD*.

6-14 Repeat Problem 6-13 if the true bearing of *CD* is known to be N 40°00′ E and it is desired to compute the true bearing of *AB* and of *BC*.

6-15 Annie set up her transit over one point in her front yard and read a bearing to another point by reading the compass needle. The bearing read was S 18°30′ W. Later when it got dark, she set the transit up on the same point and, by sighting on the North Star and carefully following procedures, determined true north. She then measured the clockwise angle from true north to the line (the bearing of which she had determined earlier). The clockwise angle turned was 200°55′. What is the declination?

6-16 The bearing of each of the four sides of a four-sided figure was read, with the values indicated in Figure 6-11. Compute the interior angles.

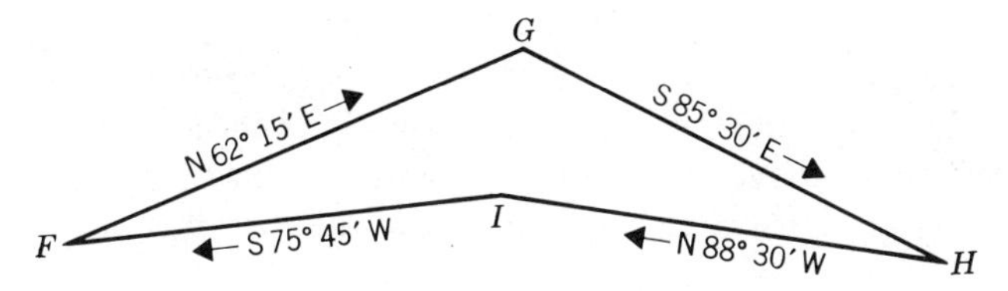

Figure 6-11

6-17 A line is run from *J* to *K*, *K* to *L*, and *L* to *M*. At *K*, an angle to the right of 62°07′ is read, defining the direction of *KL*. At *L*, an angle to the right of 316°11′ is read, defining the direction of *LM*. If the azimuth of *JK* is 95°00′, compute the azimuth of *KL* and of *LM*.

6-18 Repeat Problem 6-17 if the azimuth of *LM* is known to be 98°15′ and it is desired to compute the azimuth of *JK* and of *KL*.

6-19 What is the angle between two intersecting lines if the true bearing of one is N 84°10′ W and the true azimuth of the other is 11°11′?

6-20 Determine the azimuth of each line in Figure 6-11.

6-21 Referring to Figure 6-8, determine the angle to the right at *Q* and at *R* in going from *PQ* to *QR* and from *QR* to *RS*.

6-22 Dan was using the transit to measure a horizontal angle *ABC*. When he set the transit on point *C*, he could see only the top part of the rod and in doing this he had the vertical cross hair 0.051 m off the point. If the angle he read was 28°49′, what is the correct angle? See Figure 6-12. The distance *BC* is 36 m.

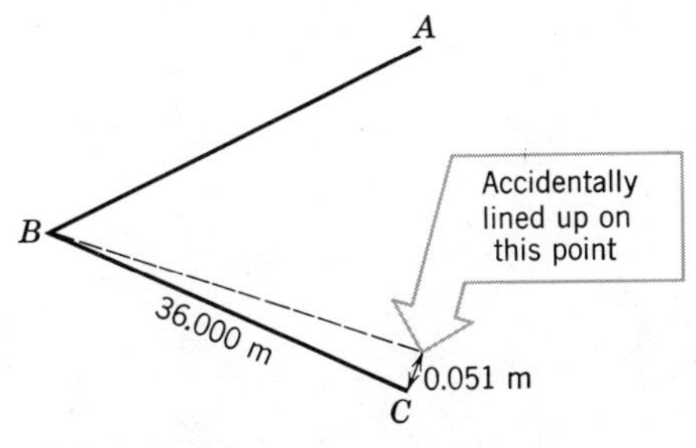

Figure 6-12

6-23 The interior angles of a five-sided figure were measured with the following results:

A	111°33′
B	56°56′
C	88°07′
D	69°20′
E	214°14′

The interior angles of a closed figure should add to $(n-2)\times 180°$, where n is the number of sides. In this case, since $n=5$, the interior angles should add to 540°. Adjust the angles listed above in accordance with the information in Section 2-5. If the bearing of AB is S 10°30′ E, compute the bearings of the remaining lines.

6-24 Repeat Problem 6-23, assuming weights of 1:2:3:3:1 are applied to the interior angles A, B, C, D, and E, respectively.

6-25 Bob sent the 10 students in his surveying class into the field to measure (independently) the magnetic bearing of a particular line, which was marked by two wooden stakes about 200 m apart. The students reported the following results: S 28°45′ W, S 28°30′ W, S 28°15′ W, S 28°40′ W, S 29°00′ W, S 28°15′ W, S 28°50′ W, S 28°15′ W, S 29°00′ W, S 28°30′ W.

What is the most probable value of the bearing of this line and the probable error of the most probable value?

STADIA

In Chapter 4 measurement of horizontal distance was covered in detail, and in Chapter 5 measurement of vertical distance was covered in detail. The most common method for measuring horizontal distance for the type of surveying covered in this book is probably taping, and the most common for vertical distance is probably differential leveling. Both of these methods give good results, and neither would be characterized as being unusually slow procedures.

There are times, however, when faster methods would be very beneficial—even with some sacrifice in precision with which the measurements are made. Such a situation might exist when a large number of data are to be collected, but less than ordinary precision is acceptable. A method of making relatively fast measurements of both horizontal and vertical distances with some sacrifice in precision is "stadia." An example of the use of stadia for faster measurement with less precision is the collection of data needed to draw contour lines on a topographic map. This will be covered in detail in Chapter 10.

7-1 Theory of Stadia

The horizontal distance between two points at the same elevation may be determined by stadia by setting up a transit at one of the points and sighting on a rod held at the other point. (The rod may be a stadia rod, or an ordinary

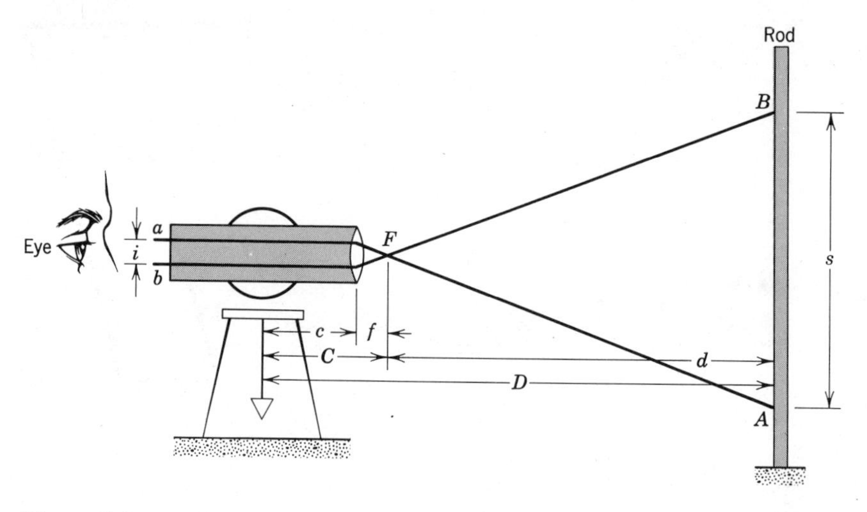

Figure 7-1

level rod will suffice.) Figure 7-1 shows, schematically, a telescope of a transit at the left and a stadia rod, or level rod, at the right. A person at the left looking through the telescope to the right sees the upper and lower cross hairs intersecting the rod at two points, *A* and *B*, on the rod. A ray of light from point *a* (representing the upper cross hair) proceeding parallel to the axis of the telescope will be bent by the lens, will pass through the principal focus (*F*), and will continue to the rod at point *A*. Similarly, one from point *b* (representing the lower cross hair) parallel to the axis will be bent, will pass through *F*, and continue to the rod at *B*. From similar triangles the following relationship is obtained.

$$\frac{f}{i}=\frac{d}{s} \tag{7-1}$$

where f = the distance from the lens to the principal focus (called the focal length)
i = the distance between upper and lower cross hairs
d = the distance from the principal focus to the rod
s = AB, the distance (along the rod) between the points on the rod where the upper and lower cross hairs intersect the rod (this distance is called the "stadia interval")

Solving for d

$$d=\left(\frac{f}{i}\right)s \tag{7-2}$$

Since d is only the distance from the rod to the principal focus, the values of f, the distance from the lens to the principal focus, and c, the distance from the center of the transit to the lens, must be added to the value of d to obtain D, the (total) distance from the center of the transit to the rod. Thus

$$D = \left(\frac{f}{i}\right)s + c + f \tag{7-3}$$

The values of f, i, and c are all constants for a given transit. Thus $\frac{f}{i}$ can be replaced by one constant K, called the stadia interval factor, and $(c+f)$ can be replaced by another constant C, which is the distance from the center of the transit to the principal focus. Thus Eq. (7-3) becomes

$$D = Ks + C \tag{7-4}$$

In Eq. (7-4), since K and C are constant for a given transit, it can be observed that D, the distance from transit to rod, is a function of s, the stadia interval.

If the values of K and C are known (or can be determined), the distance between two points can be determined easily by setting up the transit over one of the points and reading the stadia interval on the rod being held on the other point. The value of the stadia interval s is substituted into Eq. (7-4) along with the known values of K and C, and the distance D is computed.

The values of K and C depend only on the manufacture of the transit (how far apart the cross hairs are placed, how far the lens is placed from the center of the transit, and the focal length of the particular lens used). Thus, when a transit is manufactured, the values of K and C can be determined, and these values are specified for each transit. They are usually written on the transit box. In order to make the computation of a distance using Eq. (7-4) as easy as possible, many transits are manufactured such that K is equal to 100 and C is equal to about 1. In this case a stadia interval can be mentally multiplied by 100 and the result added to 1 to obtain the distance. For example, if a stadia interval of 4.62 is read, the distance is 100(4.62)+1, or 463 ft.

In all of the preceding, it has been assumed that the transit was oriented horizontally. The situation is somewhat more complicated when an inclined line of sight is necessary. Figure 7-2 illustrates such an inclined sight. If the stadia interval was read with the rod inclined so that it is perpendicular to the line of sight, the stadia interval would be read as $A'B'$. The values could be substituted into Eq. (7-4) to determine the value of D, the inclined distance. It is just not practical, however, to expect the rodman to be able to hold the rod perpendicular to the line of sight with sufficient accuracy. Instead, the rod is generally held vertically, and the equation is modified to account for this.

It can be shown that the distance D in Figure 7-2 can be computed from the formula

$$D = Ks \cos \alpha + C \tag{7-5}$$

Equation (7-5) gives the distance D, from transit to rod, in terms of s, the

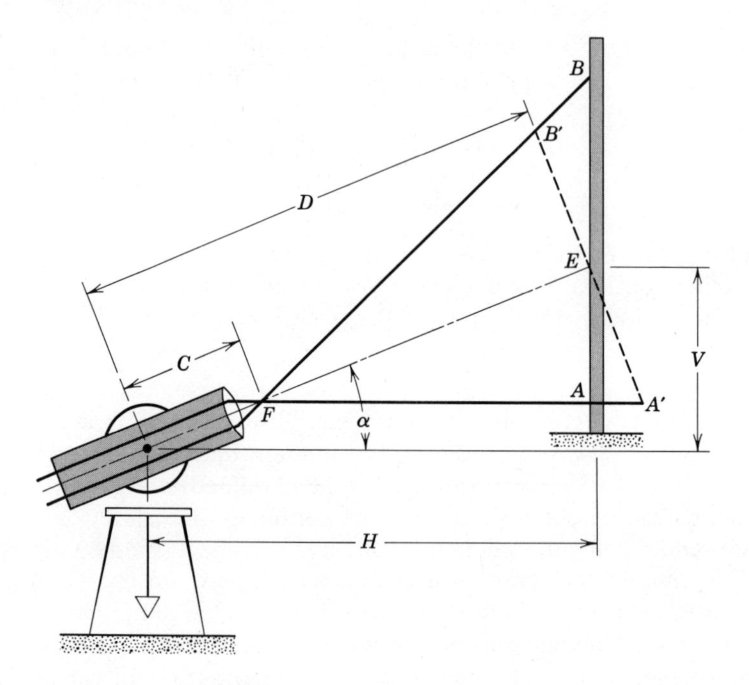

Figure 7-2

stadia interval read on a vertical rod, and the angle of inclination α. α is a vertical angle measured upward (positive angle) or downward (negative angle). If α is zero, Eq. (7-5) becomes the same as Eq. (7-4).

D, computed by Eq. (7-5) is, of course, the slope distance from transit to rod. In surveying one is usually more interested in the horizontal distance and the vertical distance. From Figure 7-2

$$\cos \alpha = \frac{H}{D} \tag{7-6}$$

Substituting from Eq. (7-5) for *D*

$$\cos \alpha = \frac{H}{Ks \cos \alpha + C} \tag{7-7}$$

or

$$H = Ks \cos^2 \alpha + C \cos \alpha \tag{7-8}$$

Also, from Figure 7-2

$$\sin \alpha = \frac{V}{D} \tag{7-9}$$

Substituting from Eq. (7-5) for D

$$\sin \alpha = \frac{V}{Ks \cos \alpha + C} \tag{7-10}$$

or

$$V = Ks \cos \alpha \sin \alpha + C \sin \alpha \tag{7-11}$$

Thus if a stadia interval (s) and angle of inclination (α) are read, the horizontal distance (H) and the vertical distance (V) can be computed, using Eq. (7-8) and Eq. (7-11). It should be emphasized that the vertical distance V is the difference in elevation between the center of the transit (the HI) and the point on the rod where the *middle* cross hair intersects the rod.

In both Eqs. (7-8) and (7-11) the last term is relatively small and is frequently neglected without appreciable loss of accuracy. With regard to precision of stadia, horizontal distances are usually recorded to the nearest foot and vertical distances to the nearest tenth of a foot.

7-2 Application

When using stadia to determine distance between two points at the same (or nearly the same) elevation, the stadia interval is read off the rod and multiplied by K and the product added to C to determine the distance between the points. It does not matter particularly where on the rod the stadia interval is read. What is important is the difference between the readings of the rod at the upper and lower cross hairs. Thus if the lower cross hair is read at 2.47 and the upper one at 5.93, the stadia interval is $(5.93 - 2.47)$, or 3.46. The computation can be made much simpler, however, by deliberately setting the lower cross hair at 1.00 and reading the upper cross hair. If this is done, the subtraction is simpler and can be done mentally. In the previous example, if the lower cross hair had been set at 1.00, the upper cross hair would have been read at 4.46. Thus the stadia interval is $(4.46 - 1.00)$, or 3.46. The answer is the same but the computation is easier. If the value of K is 100 and C is 1, the horizontal distance can be computed mentally once the stadia interval is known. For other values of K and C, Eq. (7-4) may be used. (The reader is cautioned that deliberately setting the lower cross hair at 1.00 on the rod might introduce an appreciable vertical angle (α), in which case Eq. (7-8) must be used to determine the distance between the two points.)

When using stadia to determine horizontal and/or vertical distance between two points at different elevations, the stadia interval is read off the rod and the vertical angle indicating the inclination of the telescope is also determined by reading the vertical circular scale. The horizontal and vertical distances between transit and rod are determined, using Eq. (7-8) and Eq. (7-11). It should be emphasized again that the vertical distance V determined from Eq. (7-11) is the difference in elevation between the center of the transit (the HI) and the point on the rod where the middle cross hair intersects the rod.

Frequently, when the vertical distance V is determined, it is for the purpose of finding the elevation of the point on which the rod is resting, with the elevation of the point over which the transit rests being known. In this case, the elevation of the point on which the rod is resting is determined by (1) adding the elevation of the ground point (or bench mark) at the transit to the distance from the ground point (or bench mark) at the transit to the center of the transit telescope (this gives the HI), (2) adding (or subtracting if the telescope is inclined downward) the vertical distance V, as determined by stadia, and (3) subtracting the rod reading of the middle cross hair on the rod. This can perhaps be better understood by referring to Figure 7-2.

Since the distance from the ground point at the transit to the center of the transit telescope is added, and the rod reading of the middle cross hair on the rod is subtracted, computations would be simplified appreciably if these two distances were equal. This can be easily done by determining the distance from the ground point at the transit to the center of the transit telescope (can be done by standing the rod next to the transit and reading the distance to the center of the transit telescope), and then sighting on the rod on the distant point so that the middle cross hair intersects a rod reading equal to that distance. The vertical angle should be read at this point. Under these circumstances, the difference in elevation between ground points at the transit and at the rod is equal to V. After sighting on the rod at a rod reading equal to the distance from the ground point at the transit to the center of the transit telescope and reading the vertical angle, the telescope can then be raised or lowered as needed in order to set the lower cross hair at 1.00 on the rod and to read the upper cross hair to determine the stadia interval.

The above procedure is very important and is repeated below step by step for future reference in determining elevations by stadia at a distant point based on a known elevation at the transit.

1. Place rod next to transit and determine distance from ground point (or bench mark) at transit to center of transit telescope.
2. Sight on the rod so that the middle cross hair intersects the rod at a rod reading equal to the distance determined in (1).
3. Read and record the vertical angle.
4. Raise or lower telescope until lower cross hair intersects rod at 1.00.
5. Read upper cross hair and compute and record stadia interval.
6. Use Eq. (7-11) to determine the vertical distance V.
7. Compute elevation of distant point by adding (or subtracting) V to the ground (or bench mark) elevation at the transit.
8. Use Eq. (7-8) to determine the horizontal distance H.

These steps provide the best procedure for accomplishing the stated task with a minimum of computation. Deviations from these steps are permissible as long as one keeps up with what he is doing, records the data properly, and makes the correct computations. The most important thing to remember is that the measurement of the vertical angle (α) is associated with the reading of the middle cross hair on the rod; they must both be read with the telescope in the same position.

Several example problems follow, which should help to clarify the use of stadia.

Example 7-1

A surveyor needed to know the distance from point A to point B and the elevation of point B. He knew that the elevation of A was 124.8 ft, msl. He set his transit over point A, set the rod next to it, and determined that the center of the transit telescope was 5.0 ft above point A. He then sighted on the rod held on B and read a vertical angle of $-8°40'$ while the middle cross hair was on 5.0 on the rod. He then rotated the telescope until the lower cross hair was at 1.00 on the rod and the upper cross hair was at 4.16. Based on these data, calculate the distance from A to B and the elevation of B. Assume that $K = 100$ and $C = 1$.

Solution

$$s = 4.16 - 1.00 = 3.16 \text{ ft}$$

$$\alpha = -8°40'$$

The horizontal distance can be computed from Eq. (7-8).

$$H = Ks \cos^2 \alpha + C \cos \alpha \tag{7-8}$$

$$H = 100(3.16) \cos^2 8°40' + 1(\cos 8°40')$$

$$H = 310 \text{ ft}$$

The vertical distance can be computed from Eq. (7-11).

$$V = Ks \cos \alpha \sin \alpha + C \sin \alpha \tag{7-11}$$

$$V = 100\ (3.16) \cos 8°40' \sin 8°40' + 1(\sin 8°40')$$

$$V = -47.2 \text{ ft}$$

V is indicated as negative because the vertical angle was negative. (Note that the horizontal distance H was rounded off to the nearest foot and the vertical distance V, to the nearest tenth of a foot. This is standard for stadia measurement.) Since the vertical angle was read with the middle cross hair at 5.0 on the rod and the center of the transit was 5.0 ft above point A, the elevation of B is equal to the elevation of A plus V.

$$\text{elevation } B = 124.8 + (-47.2)$$

$$\text{elevation } B = 77.6 \text{ ft}$$

Example 7-2
Repeat Example 7-1 except that the middle cross hair was on 6.1 on the rod when the vertical angle was read.

Solution
The values of H and V are not affected. However, since the vertical angle was not read with the middle cross hair on the rod at the same distance up the rod as the telescope is above A, the distance the telescope is above A must be added and the rod reading subtracted to determine the elevation of B.

$$\text{elevation } B = 124.8 + 5.0 + (-47.2) - 6.1$$
$$\text{elevation } B = 76.5 \text{ ft}$$

Example 7-3
Repeat Example 7-1 except that the surveyor set the middle cross hair on 5.0 on the rod but forgot to read the vertical angle. He then rotated the transit until the lower cross hair was at 1.00 and read the stadia interval. Then he remembered that he had to read the vertical angle and, with the lower cross hair still at 1.00, he read the vertical angle indicated in Example 7-1.

Solution
Again, the values of H and V are not affected. Since the measurement of the vertical angle is associated with the reading of the middle cross hair on the rod, it is necessary to know what the reading of the middle cross hair was on the rod at the time the vertical angle was read. Since the lower cross hair was at 1.00 on the rod and the upper cross hair was at 4.16 at the time the vertical angle was read, the middle cross hair must have been at 2.58 on the rod. Hence the elevation of B is

$$\text{elevation } B = 124.8 + 5.0 + (-47.2) - 2.58$$
$$\text{elevation } B = 80.0 \text{ ft}$$

These typical problems should clarify the relationship among the stadia interval, vertical angle, and reading of the middle cross hair on the rod.

7-3 Stadia Tables

In Example 7-1, Eqs. (7-8) and (7-11) were used to compute a horizontal and a vertical distance, using a stadia interval and a vertical angle. These equations are not difficult if a calculator with trig functions is available. Before such calculators were available, trig tables were required to determine the values of the sine and the cosine of the vertical angle. These values, along with the values of K, C, and the stadia interval, had to be substituted into the two equations to determine H and V. This required a fairly tedious arithmetic computation.

In order to simplify the computation of H and V, stadia tables were developed. One is included in Table B of the Appendix. Corresponding to each angle between zero and 30° (in increments of 2 minutes) are two columns—one labeled "Horizontal Distance" and another labeled "Difference in Elevation" (or "Vertical Distance"). At the bottom of each column there are three rows—one labeled "$C = 0.75$," one labeled "$C = 1.00$," and one labeled "$C = 1.25$." To use this table to compute a horizontal distance, find the value of the vertical angle in the table and read the corresponding value in the column labeled "Horizontal Distance." Then at the bottom of the same column, read a second value corresponding to the value of C for the particular transit. The horizontal distance is computed by multiplying the first value determined from the table by the stadia interval and adding to this product the second value read from the table. To compute a vertical distance, the procedure is the same except that the values from the table are taken from the second column.

It should be noted that the stadia table is based on a value of $K = 100$. For any other value of K, the values read out of the table corresponding to the vertical angle must each be multiplied by $(K/100)$.

Example 7-4
Solve Example 7-1 using the stadia table.

Solution
From Example 7-1, $s = 3.16$ ft and $\alpha = -8°40'$.
To determine the horizontal distance, look in the stadia table at 8°40′ and read a corresponding value of 97.73 in the "Horizontal Distance" column. At the bottom of the same column, read 0.99 opposite the "$C = 1.00$" row. Thus the horizontal distance is

$$H = 3.16(97.73) + 0.99$$

$$H = 310 \text{ ft}$$

To determine the vertical distance, look in the stadia table at 8°40′ and read a corresponding value of 14.90 in the "Difference in Elevation" column. At the bottom of the same column, read 0.15 opposite the "$C = 1.00$" row. Thus the vertical distance is

$$V = 3.16(14.90) + 0.15$$

$$V = -47.2 \text{ ft}$$

The remainder of the problem is the same as Example 7-1.

There are other methods of making stadia computation, including the use of diagrams, stadia slide rules, and special stadia arcs on the vertical circle of the transit. For more information on these, the reader is referred to the references listed at the end of this book.

A specific application of stadia, including recording of field data, is given in Chapter 10 in conjunction with collecting data for preparation of a topographic map.

7-4 Errors and Mistakes

Errors in stadia work include instrumental errors (improper spacing of stadia cross hairs, incorrect rod length), personal errors (rod not held plumb, inaccurate rod readings due, for example, to poor focusing), and natural errors (wind, moisture, temperature change).

Typical mistakes in stadia work include incorrect reading of rod, incorrect determination of stadia interval computed from upper and lower cross hair readings, incorrect recording of data, and failing to indicate whether vertical angle is plus or minus.

7-5 Problems

Note: Unless otherwise stated, assume that $K = 100$ and $C = 1$.

7-1 In determining the horizontal distance between two points by stadia, a stadia interval of 6.29 ft and a vertical angle of +9°46′ were read. What is the horizontal distance?

7-2 What would be the percentage error in the horizontal distance computed in Problem 7-1 if the $C \cos \alpha$ term was omitted?

In Problems 7-3 through 7-8, determine the vertical and horizontal distances.

Problem	*Bottom Cross Hair, ft*	*Top Cross Hair, ft*	*Vertical Angle*
7-3	1.00	3.77	+5°56′
7-4	1.00	4.29	+12°12′
7-5	1.66	8.20	−26°22′
7-6	4.37	9.92	−21°11′
7-7	2.00	4.19	+8°00′
7-8	2.50	8.88	−7°59′

In Problems 7-9 through 7-12, the transit is set at A and the rod held at B. Determine the elevation of B.

Problem	*Elevation A, ft*	*Distance Telescope above A, ft*	*Stadia Interval, ft*	*Vertical Angle*	*Middle Cross Hair on Rod*
7-9	193.6	5.0	6.65	+7°08′	5.0
7-10	225.5	4.8	3.89	−11°04′	4.8
7-11	309.2	4.9	7.14	−12°12′	9.2
7-12	90.5	5.2	4.40	+18°00′	4.3

7-13 Bunny set up her transit about midway between two points *Y* and *Z*. Point *Y* was a bench mark of known elevation 339.9 ft. With the rod held on *Y*, she read a vertical angle of +25°28′ with the middle cross hair on 7.2 ft on the rod. She then rotated the transit so that the bottom cross hair was on 1.00 and the top cross hair was at 5.66. With the rod held on *Z*, she read a vertical angle of −16°18′ with the middle cross hair on 5.00 on the rod, the upper cross hair on 6.60, and the lower cross hair on 3.40. What is the elevation of *Z*?

7-14 Repeat Problem 7-13 if $K = 95$.

7-15 Doug had an internal-focusing telescope for which *C* is nearly zero. As a means of determining the *K* value for the transit, he laid off with tape three lines of length 216.2 ft, 488.7 ft, and 326.8 ft. He then used transit and rod to read the stadia interval for each of these lines, obtaining respective values of 2.30 ft, 5.20 ft, and 3.48 ft. What is the value of *K*, assuming *C* to be zero?

7-16 A line was run by stadia from point *R* to point *V*. The elevation of *R* is 866.4 ft. For the data given below, compute the distance between successive points and the elevation of each point. Assume all vertical angles were read with the middle cross hair at rod reading equal to 5.0 ft, and each set-up of transit is 5.0 ft above ground point.

Transit At	*Rod At*	*Stadia Interval, ft*	*Vertical Angle*
S	*R*	8.61	+6°06′
	T	3.18	+16°50′
U	*T*	9.42	−13°20′
	V	5.56	+12°00′

7-17 Laura needed to know the difference in elevation between the two property corners at the front of her piece of land. She set her transit over one of the corners and read a stadia interval of 3.86 ft on the rod held on the other corner. She was in such a hurry that she forgot to determine the vertical angle.

That night she got her notes out to calculate the difference in elevation between the corners and discovered she had forgotten to read the angle. Needing the answer immediately, she looked on the plat of her property and noted that the distance between the property corners was 317.06 ft. She was then able to calculate the required difference in elevation. If $K = 100$ and $C = 0$, what was the difference in elevation between the two property corners?

TRAVERSING

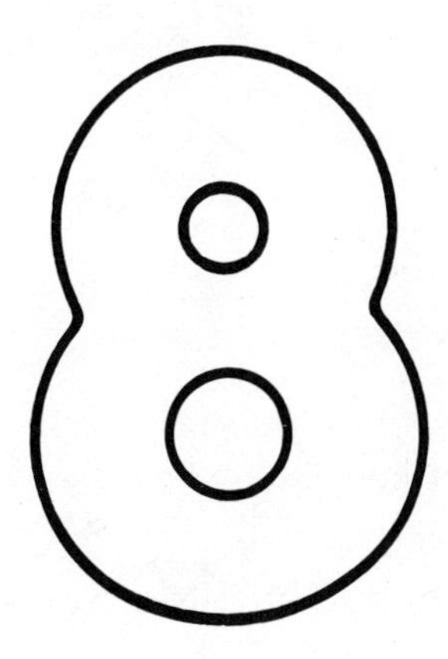

The word traverse means, in general, to pass across. In surveying the word takes on a more specific meaning. As a noun it refers to the survey of any number of consecutive lines by measuring the length and direction of each line. As a verb it means to determine the lengths and directions of consecutive lines.

There are two kinds of traverse in surveying—an open traverse and a closed traverse. In general, an open traverse is one that does not return to (or close on) the starting point. A closed traverse is one that does. An open traverse, shown in Figure 8-1, is employed mostly for "route" surveys—delineating the path of a highway, pipeline, power transmission line, and so on. A closed traverse, shown in Figure 8-2, is employed for surveying property, for example.

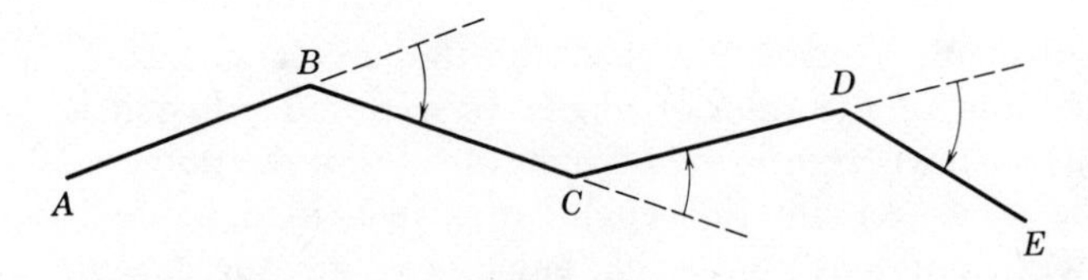

Figure 8-1 An open traverse.

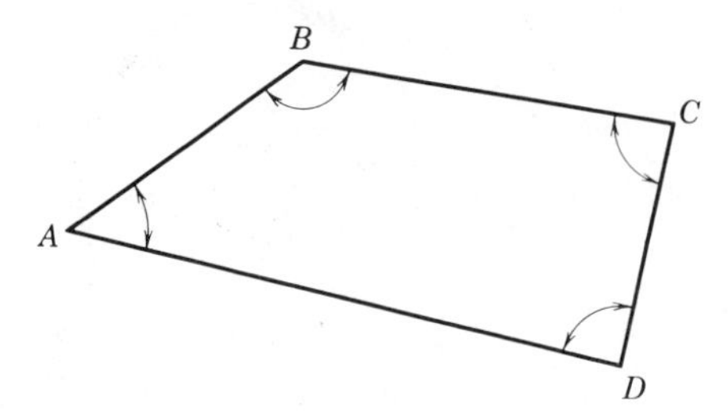

Figure 8-2 A closed traverse.

8-1 Field Methods

Since traversing is defined as the measuring of the lengths and directions of successive lines, it follows that lengths and directions must be measured in the field.

Lengths of lines can be determined by any of the methods discussed in Chapter 4. The method to be used depends on the precision required for a specific job and on the equipment available. If only approximate results are required, pacing may be acceptable. On the other hand, if high precision is required and long lines are involved, electronic distance measuring devices might be preferred. For ordinary surveying, taping is perhaps most common.

Directions of lines can be determined by any of the methods discussed in Chapter 6. These would include (1) determining the deflection angle or angle to the right at each point where adjacent lines connect, (2) determining interior angles, and (3) determining the azimuth or bearing of each line. For open traverses, deflection angles are commonly used (as shown in Figure 8-1). For closed traverses, interior angles are often used (as shown in Figure 8-2). When a traverse is being run using deflection angles, interior angles, or angles to the right, it is wise to record also the bearing of each line. Bearing values can be very useful in finding mistakes, such as recording a deflection angle incorrectly to the right when it should be to the left.

A point where two adjacent lines connect is normally marked by a "hub"—a wooden stake with a tack on it. For low-precision surveying the tack may be omitted.

8-2 Angle Closure

When a closed traverse is run, an excellent check on the angular measurements is available. The sum of the interior angles of a closed polygon is exactly equal to $(n-2)(180°)$, where n is the number of sides of the polygon. Thus if the sum of the measured interior angles of a five-sided figure is 540°00′, one would feel confident that each angle was measured very accurately. (There is always the slight possibility, however, that large, compensating mistakes were made.) If, however, the sum is 539°59′, the closure (difference between computed sum of measured angles and the geometrically correct total) is 01′. As a general rule, allowable closure is equal to the least division readable with the vernier multiplied by the square root of the number of angles. For the common vernier where angles are read to the nearest minute, allowable closure would be $01'\sqrt{n}$. For the five-sided figure, allowable closure would be 2.24′, or 02′. Thus if the sum of the five measured angles falls within the range 539°58′ to 540°02′, the values would be considered acceptable. If the sum is not exactly what it should be but is within the acceptable range, the individual angles should be adjusted as discussed in Section 2-5.

For an open traverse, there is no angle check as there is for a closed traverse. There are, however, several possible means of securing angular check. One is, upon completion of the open traverse, to continue traversing back to the starting point. This would form a closed traverse, which could be checked by the method described above. This continuation of the traverse back to the starting point could possibly be done by an easier route than the open traverse. For example, if an open traverse is run to survey a roadway that roughly follows contours around hills or mountains, the continuation back to the starting point could possibly consist of one or two sights from one hillside or mountainside to another. Another possible means of checking an open traverse would be to tie in both the starting point and the ending point to nearby monuments or bench marks, the coordinates of which are known.

8-3 Latitudes and Departures

As discussed in Section 8-2, an excellent check with regard to angular measurement for a closed traverse is available. Even if the angles should check out perfectly, however, this is no guarantee that the entire survey is correct. The reason is, of course, that there can be considerable error or mistake in the measurement of the lengths of the individual lines. Such errors or mistakes would not show up in the angular check. In order to check the closure of the traverse as a whole, it is necessary to compute the latitude and the departure of each line.

The "latitude" of a line is equal to the length of the line multiplied by the cosine of the bearing angle. The latitude is positive if the bearing is north and negative if it is south. The "departure" of a line is equal to the length of the line multiplied by the sine of the bearing angle. The departure is positive if the bearing is east and negative if it is west. Upon reflection, one will see that the latitude and departure of a line are simply the y- and x-components of the line, respectively, as used in mathematics and engineering mechanics.

The basic premise on which latitudes and departures are used to check closure for a closed traverse is that both the algebraic sum of the latitudes and the algebraic sum of the departures should equal zero. (In computing latitudes and departures, the bearings should be placed on each line in a clockwise direction.) In practice the sums of the latitudes and of the departures will not likely add to exactly zero—certainly not likely if the computations are carried to four or five places beyond the decimal. This is due to the inherent error involved in making measurements. For good work, the sums should be near zero, however.

The error of closure is computed from the equation

$$e_c = \sqrt{(\Sigma L)^2 + (\Sigma D)^2} \tag{8-1}$$

where e_c = error of closure
ΣL = algebraic sum of latitudes
ΣD = algebraic sum of departures

The precision of closure is computed from the equation

$$P_c = \frac{e_c}{p} \tag{8-2}$$

where P_c = precision of closure
e_c = error of closure
p = total perimeter of the traverse

All terms in Eqs. (8-1) and (8-2) except P_c should be expressed in the same unit of linear measure (feet or meters). The precision of closure is expressed as a fraction with a numerator of unity and denominator rounded off usually to the nearest multiple of 100. For rough work a precision of closure of $\frac{1}{1000}$ might be reasonable; for more accurate work, $\frac{1}{4000}$ might be reasonable; for extremely accurate work, $\frac{1}{10,000}$ might be reasonable. In some localities, required precision of closure may be set by law.

Example 8-1

A closed traverse, *EFGHIE*, was run with the results shown below. Compute the latitudes and departures and determine the error of closure and the precision of closure. The bearing of line *FG* is S 82°15′ E and that of line *GH* is S 10°30′ E.

Point	*Length of Line, ft*	*Interior Angle*
E		131°53′
	162.42	
F		114°19′
	233.71	
G		108°15′
	319.19	
H		117°31′
	327.51	
I		68°03′
	425.44	
E		

Solution

The first step is to add the interior angles. That sum is 540°01′. As determined in Section 8-2, this is acceptable closure for a five-sided figure. The angles should, however, be adjusted so that the sum is exactly 540°. Since there is no reason to believe any angle is more (or less) reliable than any other, 01′ will be subtracted arbitrarily from the angle at *E*, giving 131°52′. The sum of the angles is now exactly 540°. The computation of latitudes and departures is dependent on the bearings; therefore the next step is to calculate the bearing of each line, beginning with the known bearing of *FG* and using the interior angles. The procedure for doing this was explained in Section 6-5 (see Examples 6-3 and 6-4). These bearings are listed in the third column below. The fourth column lists the cosine of each bearing angle, from which each latitude is computed and listed in the fifth column. The sixth column lists the sine of each bearing angle, from which each departure is computed and listed in the seventh column. (If a calculator with trig functions is used, it is not necessary that the values of the trig functions be listed here.) It will be noted that the bearings beginning with "S" have negative latitudes and those ending with "W" have negative departures. Some surveyors prefer to list the latitudes in two columns (one for the "N" or positive latitudes and one for the "S" or negative latitudes) and the departures in two columns (one for the "E" or positive departures and one for the "W" or negative departures).

Line	*Length, ft*	*Bearing*	*Cosine*	*Latitude, ft*	*Sine*	*Departure, ft*
EF	162.42	N 32°04′ E	0.84743	137.64	0.53091	86.23
FG	233.71	S 82°15′ E	0.13485	−31.52	0.99087	231.58
GH	319.19	S 10°30′ E	0.98325	−313.85	0.18224	58.17
HI	327.51	S 51°59′ W	0.61589	−201.71	0.78783	−258.02
IE	425.44	N 16°04′ W	0.96094	408.82	0.27676	−117.74
	1468.27			−0.62		0.22

The error of closure is computed from Eq. (8-1).

$$e_c = \sqrt{(\Sigma L)^2 + (\Sigma D)^2} \qquad (8\text{-}1)$$

$$e_c = \sqrt{(-0.62)^2 + (0.22)^2}$$

$$e_c = 0.66 \text{ ft}$$

The precision of closure is computed from Eq. (8-2).

$$P_c = \frac{e_c}{p} \qquad (8\text{-}2)$$

$$P_c = \frac{0.66}{1468.27}$$

$$P_c = \frac{1}{2200}$$

This is acceptable for "rough" work.

8-4 Balancing Latitudes and Departures

If the precision of closure for a closed traverse is not within acceptable limits, one must try to find a mistake that can be corrected to effect acceptable precision of closure, or one must prepare to do the field work again and to do it more accurately. (The first check for mistakes should be of the arithmetic.) If the precision of closure is within acceptable limits, one should go back and adjust (balance) the latitudes and departures so that they will add to zero (assuming they do not add to zero to begin with). Once this is done, the length and bearing of each line can be calculated, if desired.

There are several analytical methods for balancing latitudes and departures, including the compass rule, the transit rule, Crandall method, and least squares method. Perhaps the simplest and most commonly used of these methods is the compass rule, and it is the only one considered in this book. Information on the other methods may be obtained by referring to the references listed at the end of this book.

Before discussing the compass rule, it would be well to discuss what might be called the "arbitrary method." One might choose to adjust the latitudes and departures arbitrarily according to his assessment of the conditions surrounding the survey. For example, if one long line of a survey was through dense swamp, it might be that throwing all or most of the error of closure into that one line would balance the survey satisfactorily. Such a procedure would likely be as good as, if not better than, any of the analytical methods.

The premise of the compass rule is that the error in the latitude (or departure) of each line in a traverse is proportional to the length of the line. In other words, the longer the line, the more error in the line. The compass rule may be stated as follows: The correction in latitude (or departure) for any line is equal to the total error in latitude (or departure) multiplied by the ratio of the length of the line to the perimeter of the traverse. The sign of the correction should be opposite that of the total error in latitude (or departure).

Example 8-2

Adjust the latitudes and departures of Example 8-1 using the compass rule. Then compute the bearing and length of each line, based on adjusted latitudes and departures.

Solution

The correction for the latitude of *EF* is

$$0.62\left(\frac{162.42}{1468.27}\right)=0.07 \text{ ft}$$

The correction for the departure of *EF* is

$$-0.22\left(\frac{162.42}{1468.27}\right) = -0.02 \text{ ft}$$

The corrections for all other lines are computed in a similar manner. These corrections are tabulated below, and the corrections are applied to the respective latitudes and departures to obtain the adjusted values.

		(Computed)		*Corrections*		*(Adjusted)*	
Line	*Length*	*Latitude*	*Departure*	*Latitude*	*Departure*	*Latitude*	*Departure*
EF	162.42	137.64	86.23	+0.07	−0.02	137.71	86.21
FG	233.71	−31.52	231.58	+0.10	−0.04	−31.42	231.54
GH	319.19	−313.85	58.17	+0.13	−0.05	−313.72	58.12
HI	327.51	−201.71	−258.02	+0.14	−0.05	−201.57	−258.07
IE	425.44	408.82	−117.74	+0.18	−0.06	409.00	−117.80
	1468.27	−0.62	0.22			0	0

Once the latitudes and departures have been adjusted, the length and bearing of each line can be computed. For example, the length and bearing of line *EF* can be computed as follows (see Figure 8-3):

$$L_{EF} = \sqrt{(86.21)^2 + (137.71)^2}$$

$$L_{EF} = 162.47 \text{ ft}$$

$$\tan\beta = \frac{86.21}{137.71}$$

$$\tan\beta = 0.62603$$

$$\beta = 32°03'$$

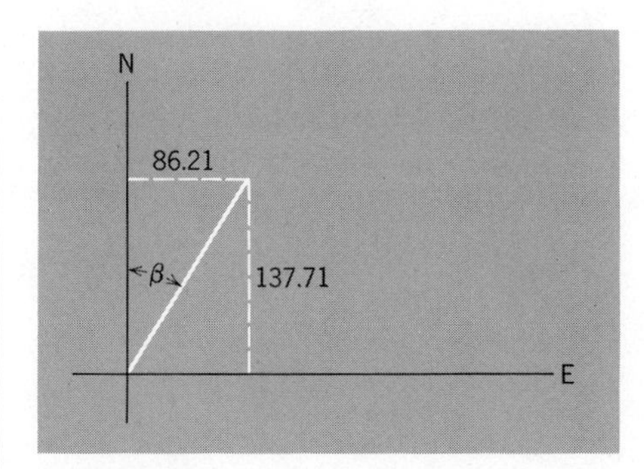

Figure 8-3

β is the bearing angle, but its quadrant must now be determined. This may be done by means of sketch (Figure 8-3) or by recalling that a positive latitude results from an "N" bearing and a positive departure results from an "E" bearing, and vice versa. In this example, both latitude and departure are positive. Thus the bearing of *EF* is N 32°03′ E. The length and bearing of each remaining line are computed in the same manner. The computations are not shown here, but their values are listed below. The reader might wish to verify these results.

Line	*Adjusted* *Length, ft*	*Bearing*
EF	162.47	N 32°03′ E
FG	233.66	S 82°16′ E
GH	319.06	S 10°30′ E
HI	327.46	S 52°00′ W
IE	425.63	N 16°04′ W

8-5 Coordinates

In some cases (e.g., map plotting) it is convenient to have available the coordinates of each point of a traverse. As explained in Chapter 2, the coordinates of a point consist of two numbers (usually separated by a comma and enclosed in parentheses), the first of which specifies distance along the x axis and the second, along the y axis. In some surveying contexts, coordinates are indicated as "north" and "east" (analogous to y and x), with the first number in parentheses referring to north and the second to east.

Assuming latitudes and departures are known for each line, if the coordinates of one point are known, the coordinates of the next point can be calculated fairly easily. In Figure 8-4 it will be noted that the x-coordinate of B is equal to the x-coordinate of A plus the x component (departure) of line AB. Similarly, the y-coordinate of B is equal to the y-coordinate of A plus the y component (latitude) of line AB. Obviously, if latitudes and departures have been computed (and adjusted), and if the coordinates of one point are

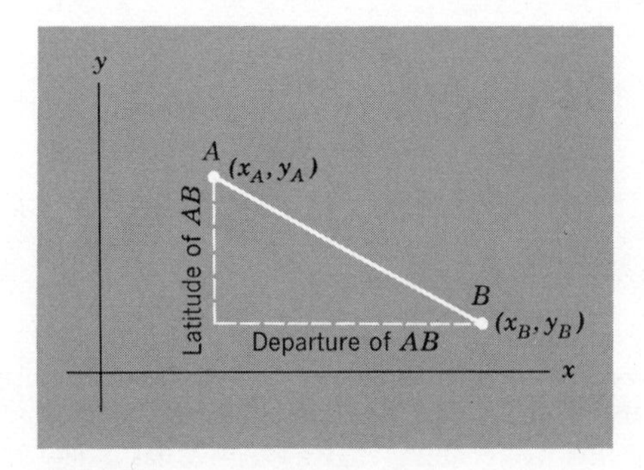

Figure 8-4

known, the coordinates of all points can be determined by adding successive departures to the previous x-coordinates and successive latitudes to the previous y-coordinates.

Example 8-3

For the adjusted latitudes and departures of the traverse of Example 8-1, calculate the coordinates of each point if the coordinates of E are (5000.00, 5000.00) ft.

Solution

The x-coordinate of F is determined by adding the departure of EF to the x-coordinate of E.

$$x\text{-coordinate of } F = 86.21 + 5000.00$$
$$x\text{-coordinate of } F = 5086.21 \text{ ft}$$

The x-coordinate of G is determined by adding the departure of FG to the x-coordinate of F.

$$x\text{-coordinate of } G = 231.54 + 5086.21$$
$$x\text{-coordinate of } G = 5317.75 \text{ ft}$$

The y-coordinate of F is determined by adding the latitude of EF to the y-coordinate of E.

$$y\text{-coordinate of } F = 137.71 + 5000.00$$
$$y\text{-coordinate of } F = 5137.71 \text{ ft}$$

Remaining coordinates are computed in the same manner. The results are presented below.

Line	*Point*	*Latitude, ft*	*Departure, ft*	*x-coordinate*	*y-coordinate*
	E			5000.00	5000.00
EF		137.71	86.21		
	F			5086.21	5137.71
FG		−31.42	231.54		
	G			5317.75	5106.29
GH		−313.72	58.12		
	H			5375.87	4792.57
HI		−201.57	−258.07		
	I			5117.80	4591.00
IE		409.00	−117.80		
	E			5000.00	5000.00

8-6 Missing Data

When the latitudes and departures are calculated for a closed traverse and the sums of the latitudes and of the departures are each set equal to zero, the result is two simultaneous equations. With two simultaneous equations, it is possible to have two unknowns, and the two equations can be solved to determine the values of the unknowns. What this means is that the measurement of any two elements of a closed traverse (length of a line and bearing of a line, lengths of two lines, or bearings of two lines) may be omitted in the field, and the two unknowns may be solved for by calculating latitudes and departures, setting their sums equal to zero, and solving the two resulting simultaneous equations.

Generally speaking, the practice of omitting measurements in the field and solving for them in the manner described above is not a good one. The reason is that this method tends to throw all possible error and possible mistakes into the computation of the unknowns. It eliminates the check on the closure of the traverse. This is not to say, however, that such should never be done. There are circumstances where this technique can be used to advantage. For example, if a traverse of a large parcel of land has been run, the technique above may be used to compute the bearing and length of a proposed line to separate the parcel into parts of specified area. After these data are determined, presumably the division line would be run in the field. Other examples would include computing unknown building dimensions and locating the position of columns in laying out a building.

Example 8-4

A builder was reviewing the plans for construction of an apartment building (Figure 8-5*a*). Because there is a local ordinance stating that no building can be closer than 10.00 ft to any property line, he wondered if the rear corner of the building is at least 10 ft from the property line. Using the method of latitudes and departures, determine the distance from the rear corner to the property line. (It is known that the front line of the building is parallel to the front property line.)

Solution

A closed traverse *ABCDEFA* can be made from the information given (see Figure 8-5*b*). The bearing of *BC* is determined by finding the bearing of a line perpendicular to *AB*. The bearing of *CD* is the same as that of *AB* except that the direction is reversed. The bearing of *DE* is the same as that of *BC*. *EF* is a line from point *E* to line *FA* and is perpendicular to *FA*. Its bearing is determined by finding the bearing of a line perpendicular to *FA*. The length of *EF* is, of course, unknown. The distance *FA* is also unknown. Using these data, one can calculate the following latitudes and departures.

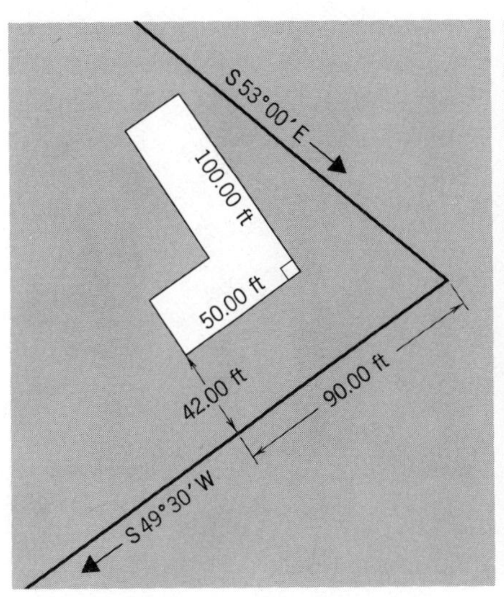

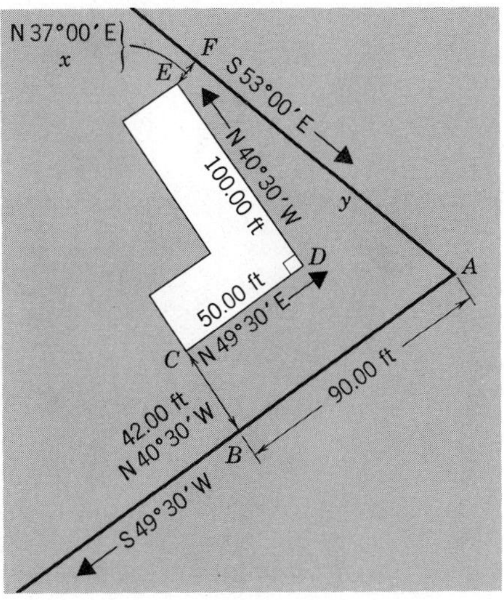

Figure 8-5 (a) (b)

Line	*Length, ft*	*Bearing*	*Latitude*	*Departure*
AB	90.00	S 49°30′ W	−58.45	−68.44
BC	42.00	N 40°30′ W	31.94	−27.28
CD	50.00	N 49°30′ E	32.47	38.02
DE	100.00	N 40°30′ W	76.04	−64.94
EF	x	N 37°00′ E	$0.7986x$	$0.6018x$
FA	y	S 53°00′ E	$-0.6018y$	$0.7986y$

Summing latitudes and setting equal to zero gives

(1) $0.7986x - 0.6018y + 82.00 = 0$

Summing departures and setting equal to zero gives

(2) $0.6018x + 0.7986y - 122.64 = 0$

These two equations must be solved simultaneously. Dividing (1) by 0.6018 and (2) by 0.7986 gives

(3) $1.3270x - y = -136.26$

(4) $0.7536x + y = 153.57$

Adding (3) and (4) gives

(5) $2.0806x = 17.31$

(6) $x = 8.32$ ft

Thus, the building as indicated on the plans is too close to the line (less than 10 ft). Although not required, y (the distance from F to A) is 147.30 ft.

Figure 8-6 Programmable calculators. (Courtesy of Hewlett–Packard Co.)

8-7 Programmed Computations of Latitudes and Departures

The calculation of latitudes and departures and subsequent balancing of surveys are done frequently. Any means of reducing the manual computations involved could amount to a large saving in time. Two such means will be presented in this section—one involving preprogrammed and programmable calculators and one involving the digital computer.

There are a number of preprogrammed and programmable calculators on the market today. Some are designed for desk use; some are hand held and can be battery powered for portable use in the field. When using such a calculator with a preprogrammed surveying package, the procedure is mostly that of following program instructions. For demonstration purposes here, a Hewlett-Packard calculator (Model HP-65) with Surveying Pac 1 has been used. Technology is rapidly changing in the field of calculators; the HP-65 used here (because of its availability to the author) has already been superceded by the HP-97 and HP-67 (Figure 8-6). It must be emphasized that there are a number of such calculators on the market today, and without doubt new ones will come on the market in the future.

Example 8-5

Repeat Examples 8-1, 8-2, and 8-3 using the HP-65 calculator.

Solution

The procedure is to follow the instructions in the Surveying Pac 1.

Line	*Action*	*Calculator Display*	*Interpretation*
1	Enter SURV 1-O2A		
2	Enter 5000. Punch *A*	5000.00	North (*y*) coordinate of *E*
3	Enter 5000. Punch *A*	5000.00	East (*x*) coordinate of *E*
4	Enter 32.04* Punch *B*	32.07	Bearing angle of *EF*
5	Enter 1† Punch *C*	32.07	Quadrant code
6	Enter 162.42 Punch *D*	90.00	
7	Punch *D*	162.42	Length of *EF*
8	Punch *E*	5137.64	North coordinate of *F*
9	Punch R/S	5086.23	East coordinate of *F*
10	Enter 82.15* Punch *B*	82.25	Bearing angle of *FG*
11	Enter 2† Punch *C*	97.75	Quadrant code
12	Enter 233.71 Punch *D*	90.00	
13	Punch *D*	233.71	Length of *FG*
14	Punch *E*	5106.12	North coordinate of *G*
15	Punch *R/S*	5317.80	East coordinate of *G*
16	Enter 10.30* Punch *B*	10.50	Bearing angle of *GH*
17	Enter 2† Punch *C*	169.50	Quadrant code
18	Enter 319.19 Punch *D*	90.00	
19	Punch *D*	319.19	Length of *GH*
20	Punch *E*	4792.28	North coordinate of *H*
21	Punch *R/S*	5375.97	East coordinate of *H*
22	Enter 51.59* Punch *B*	51.98	Bearing angle of *HI*
23	Enter 3† Punch *C*	231.98	Quadrant code
24	Enter 327.51 Punch *D*	90.00	
25	Punch *D*	327.51	Length of *HI*
26	Punch *E*	4590.57	North coordinate of *I*
27	Punch *R/S*	5117.95	East coordinate of *I*
28	Enter 16.04* Punch *B*	16.07	Bearing angle of *IE*
29	Enter 4† Punch *C*	343.93	Quadrant code
30	Enter 425.44 Punch *D*	90.00	
31	Punch *D*	425.44	Length of *IE*
32	Punch *E*	4999.39	North coordinate of *E*
33	Punch *R/S*	5000.21	East coordinate of *E*
34	Punch *R/S*	135,999.06	Area of traverse (sq ft)
35	Punch *R/S*	3.12	Area of traverse (acres)
36	Enter SURV 1-03A		
37	Punch *C*	0.64	Error of closure
38	Punch *R/S*	18.46*	Error bearing
39	Punch *R/S*	2.00†	Quadrant code

40	Punch *D*	1468.27	Total perimeter of traverse
41	Punch *E*	2281.56	Precision of closure
42	Enter SURV 1-07A		
43	Enter 5137.64 Punch *A*	5000.00	North coordinate of *F* (unadjusted)
44	Enter 5086.23 Punch *B*	5086.23	East coordinate of *F* (unadjusted)
45	Punch *C*	5137.71	North coordinate of *F* (adjusted)
46	Punch *D*	5086.21	East coordinate of *F* (adjusted)
47	Enter 5106.12 Punch *A*	5106.12	North coordinate of *G* (unadjusted)
48	Enter 5317.80 Punch *B*	5317.80	East coordinate of *G* (unadjusted)
49	Punch *C*	5106.28	North coordinate of *G* (adjusted)
50	Punch *D*	5317.74	East coordinate of *G* (adjusted)
51	Enter 4792.28 Punch *A*	4792.28	North coordinate of *H* (unadjusted)
52	Enter 5375.97 Punch *B*	5375.97	East coordinate of *H* (unadjusted)
53	Punch *C*	4792.58	North coordinate of *H* (adjusted)
54	Punch *D*	5375.87	East coordinate of *H* (adjusted)
55	Enter 4590.57 Punch *A*	4590.57	North coordinate of *I* (unadjusted)
56	Enter 5117.95 Punch *B*	5117.95	East coordinate of *I* (unadjusted)
57	Punch *C*	4591.00	North coordinate of *I* (adjusted)
58	Punch *D*	5117.80	East coordinate of *I* (adjusted)
59	Enter 4999.39 Punch *A*	4999.39	North coordinate of *E* (unadjusted)
60	Enter 5000.21 Punch *B*	5000.21	East coordinate of *E* (unadjusted)
61	Punch *C*	5000.00	North coordinate of *E* (adjusted)

62	Punch *D*	5000.00	East coordinate of *E* (adjusted)
63	Enter SURV 1-04A		
64	Enter 5000.00 Punch *A*	5000.00	North coordinate of *E*
65	Enter 5000.00 Punch *A*	5000.00	East coordinate of *E*
66	Enter 5137.71 Punch *B*	137.71	North coordinate of *F*
67	Enter 5086.21 Punch *B*	86.21	East coordinate of *F*
68	Punch *C*	162.47	Adjusted length of *EF*
69	Punch *D*	32.03*	Adjusted bearing angle of *EF*
70	Punch *R/S*	1.00†	Quadrant code
71	Enter 5106.28 Punch *B*	−31.43	North coordinate of *G*
72	Enter 5317.74 Punch *B*	231.53	East coordinate of *G*
73	Punch *C*	233.65	Adjusted length of *FG*
74	Punch *D*	82.16*	Adjusted bearing angle of *FG*
75	Punch *R/S*	2.00†	Quadrant code
76	Enter 4792.58 Punch *B*	−313.70	North coordinate of *H*
77	Enter 5375.87 Punch *B*	58.13	East coordinate of *H*
78	Punch *C*	319.04	Adjusted length of *GH*
79	Punch *D*	10.30*	Adjusted bearing angle of *GH*
80	Punch *R/S*	2.00†	Quadrant code
81	Enter 4591.00 Punch *B*	−201.58	North coordinate of *I*
82	Enter 5117.80 Punch *B*	−258.07	East coordinate of *I*
83	Punch *C*	327.47	Adjusted length of *HI*
84	Punch *D*	52.00*	Adjusted bearing angle of *HI*
85	Punch *R/S*	3.00†	Quadrant code
86	Enter 5000.00 Punch *B*	409.00	North coordinate of *E*
87	Enter 5000.00 Punch *B*	−117.80	East coordinate of *E*
88	Punch *C*	425.63	Adjusted length of *IE*
89	Punch *D*	16.04*	Adjusted bearing angle of *IE*
90	Punch *R/S*	4.00†	Quadrant code
91	Punch *E*	134,871.71	Area of traverse (sq ft)
92	Punch *R/S*	3.10	Area of traverse (acres)
93	Punch *R/S*	1468.26	Total perimeter of traverse

* Angle expressed as 32.04 means 32°04′.

† Quadrant code for bearings: NE, 1; SE, 2; SW, 3; NW, 4.

Comments:

1. Lines 1–35 make up one program (SURV 1-02A) that computes the coordinates of each point based on the field measured bearings and distances. *Lines 34 and 35* give the area of the traverse, but this is the *area based on the unadjusted traverse.*
2. Lines 36–41 make up another program (SURV 1-03A) that computes the error of closure, error of bearing, total perimeter, and precision of closure. These differ slightly from the corresponding values computed in Example 8-1, probably because of round-off errors.
3. Lines 42–62 make up another program (SURV 1-07A) that takes the (unadjusted) coordinates computed in SURV 1-02A (lines 1–35) and computes adjusted coordinates. These agree with the coordinates computed in Example 8-3, with occasional difference of 0.01 ft.
4. Lines 63–93 make up another program (SURV 1-04A) that takes the adjusted coordinates computed in SURV 1-07A (lines 42–62) and computes adjusted lengths and bearings of all lines. These agree with corresponding values computed in Example 8-2, with occasional difference as high as 0.02 ft. *Lines 91 and 92* give the *area of the adjusted traverse.*

The reader who first encounters the calculator solution of Example 8-5 may be initially overwhelmed by it. With a small amount of practice, however, one will become proficient at entering the data, punching the right buttons, and recording the pertinent answers in the proper sequence. The surveyor who can do this and who finds himself in the field in the middle of nowhere with no electricity and needing to compute closure on a traverse he has just completed, or needing to compute the bearing (or interior angle) required to complete the last line of a traverse, will appreciate a hand held, battery-powered calculator that can do all of this quickly and accurately. To him, it must surely be "worth its weight in gold."

The other means of reducing manual computations involving latitudes and departures mentioned at the beginning of this section is the digital computer. Most universities and many engineering or surveying companies have access to high-speed digital computers; thus some students or surveyors may find it more convenient to collect their field data, bring them into the office, punch the required data cards, leave them with the computer operator, and pick up the results later. For these students or surveyors, the author has prepared a computer program (written in Fortran IV). This program is listed in Figure 8-7.

To use this program, one data card is prepared for each line of the (closed) traverse, giving the bearing and length of the line. As the program is written, a

```
      DIMENSION DATE(2),TITLE(12),NL(100),NR(100),D(100),BL(100),
     $     NDEG(100),NMIN(100),BR(100),ALAT(100),ADEP(100),XCOOR(100),
     $     YCOOR(100)
   99 FORMAT (2A4,12A6)
   98 FORMAT (A2,1X,A2,F15.2,4X,A1,1X,I2,1X,I2,1X,A1,1X,2F10.2)
   97 FORMAT ('1',12A6,2A4,////,' GIVEN DATA',//,'  LINE       LENGTH
     $   BEARING')
   96 FORMAT (1X,A2,'-',A2,F12.2,6X,A1,2I3,1X,A1)
   95 FORMAT (///,' SURVEY ADJUSTED BY COMPASS RULE ',//,' POINT LINE
     $   LENGTH   BEARING       LATITUDE   DEPARTURE    X-COORDINATE   Y-
     $COORDINATE')
   94 FORMAT (2X,A3,58X,F10.2,5X,F10.2,/,6X,A2,'-',A2,2X,F10.2,2X,A1,
     $     2I3,1X,A1,2X,F10.2,2X,F10.2)
   93 FORMAT (//,' ERROR OF CLOSURE = ',F5.2,//,' PRECISION OF CLOSURE =
     $ 1/',I5,//,' AREA = ',F11.0,' SQ FT        OR',F11.3,' ACRES')
      DSUM = 0.
      SUMD = 0.
      SUML = 0.
      AREA = 0.
      PI = 3.14159265
      READ (1,99) DATE, TITLE
      DO 100  J=1,100
      N = J-1
      IF(J.EQ.1)READ(1,98)NL(J),NR(J),D(J),BL(J),NDEG(J),NMIN(J),BR(J),
     $     XBEG,YBEG
      IF(J.NE.1)READ(1,98)NL(J),NR(J),D(J),BL(J),NDEG(J),NMIN(J),BR(J)
      IF (D(J).LT..01) GO TO 101
      DEG = NDEG(J)
      AMIN = NMIN(J)
      BRAD = (DEG+AMIN/60.)*PI/180.
      ALAT(J) = D(J)*COS(BRAD)
      IF (BL(J).EQ."S") ALAT(J) = -ALAT(J)
      SUML = SUML+ALAT(J)
      ADEP(J) = D(J)*SIN(BRAD)
      IF (BR(J).EQ."W") ADEP(J) = -ADEP(J)
      SUMD = SUMD+ADEP(J)
      DSUM = DSUM+D(J)
  100 CONTINUE
  101 WRITE (3,97) TITLE,DATE
      WRITE (3,96)(NL(J),NR(J),D(J),BL(J),NDEG(J),NMIN(J),BR(J),J=1,N)
      WRITE (3,95)
      EC = SQRT(SUML*SUML+SUMD*SUMD)
      PC = DSUM/EC
      JPC = PC
      DO 102  J=1,N
      ALAT(J) = ALAT(J)-SUML*D(J)/DSUM
      ADEP(J) = ADEP(J)-SUMD*D(J)/DSUM
      D(J) = SQRT(ALAT(J)**2+ADEP(J)**2)
      ANG = ATAN(ABS(ADEP(J)/ALAT(J)))
      ADEG = ANG*180./PI
      NDEG(J) = ADEG
      T = NDEG(J)
      U = (ADEG-T)*60.+.5
      NMIN(J) = U
      BL(J) = "N"
      IF(ALAT(J).LT.0.) BL(J) = "S"
      BR(J) = "E"
      IF (ADEP(J).LT.0.) BR(J) = "W"
      IF (J.EQ.1) XCOOR(J) = XBEG
      IF (J.EQ.1) YCOOR(J) = YBEG
      IF (J.NE.1) XCOOR(J) = XCOOR(J-1)+ADEP(J-1)
      IF (J.NE.1) YCOOR(J) = YCOOR(J-1)+ALAT(J-1)
  102 CONTINUE
      WRITE (3,94)(NL(J),XCOOR(J),YCOOR(J),NL(J),NR(J),D(J),BL(J),
     $     NDEG(J),NMIN(J),BR(J),ALAT(J),ADEP(J),J=1,N)
      DO 104  J=1,N
      IF (J.NE.N) AREA = XCOOR(J)*YCOOR(J+1)-XCOOR(J+1)*YCOOR(J)+AREA
      IF (J.EQ.N) AREA = XCOOR(J)*YCOOR(1)-XCOOR(1)*YCOOR(J)+AREA
  104 CONTINUE
      AREA = ABS(AREA/2.)
      ACRE = AREA/43560.
      WRITE (3,93) EC,JPC,AREA,ACRE
      STOP
      END
```

Figure 8-7 Computer program for computing traverse closure.

blank data card should be placed after the card giving the data for the last line of the traverse. The data for each line must be presented as follows.

Columns 1–5	Identification of end points of line, with identification of one end of line in columns 1 and 2 and identification of other end of line in columns 4 and 5. (Identifications must proceed in order—i.e., *A* to *B* for first line, *B* to *C* for second line, etc.)
Columns 6–20	Length of line with decimal in column 18.
Columns 25–33	Bearing of line with "N" or "S" in column 25, number of degrees in columns 27 and 28 (if a one digit number, it must go in column 28), number of minutes in columns 30 and 31 (if a one digit number, it must go in column 31), and "E" or "W" in column 33.

There must be one data card for each line of the traverse, and the cards must be arranged in the order they occur in the traverse. In addition to these data, on the *first* card (*only*) for the first line of the traverse, the east and north coordinates of the starting point must be given. The east coordinate must be given in columns 35–44 with decimal in column 42; the north coordinate must be given in columns 45–54 with decimal in column 52. One additional data card must be placed at the beginning of the data deck. This card allows for the date to be placed in columns 1–8 and a title to be placed in columns 9–80.

To use this program data cards must be prepared as described above. These cards are placed after the program (Figure 8-7) and run through the computer. The output gives the measured length and bearing of each line, the adjusted length and bearing of each line, the adjusted latitude and departure of each line, the adjusted coordinates of each point, the error of closure, precision of closure, and area in square feet and acres. The program has been written with angles indicated to the nearest minute and distances to the nearest 0.01 ft. If one wishes greater precision, the program could easily be modified to reflect this.

Example 8-6

Repeat Examples 8-1, 8-2, and 8-3 using the computer program of Figure 8-7.

Solution

The data cards are prepared as shown in Figure 8-8. These data cards were appended to the program of Figure 8-7 and processed by a computer. The results are shown in Figure 8-9. It will be observed that these results compare favorably with the results of Examples 8-1, 8-2, and 8-3 (manual computations) and Example 8-5 (programmed calculator).

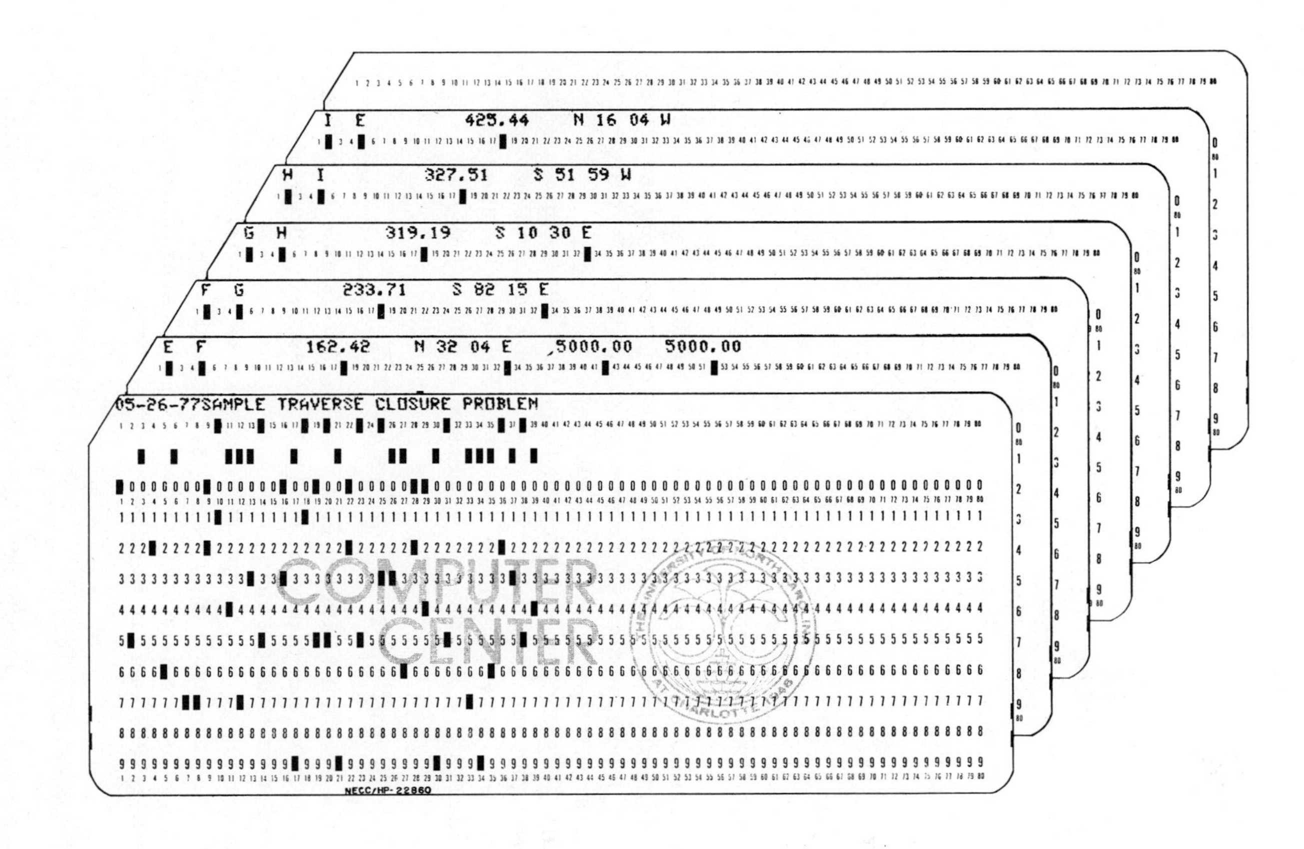

Figure 8-8 Data cards for Example 8-6.

```
SAMPLE TRAVERSE CLOSURE PROBLEM                                                   05-26-77


GIVEN DATA

 LINE       LENGTH       BEARING
 E- F       162.42      N 32   4 E
 F- G       233.71      S 82 15 E
 G- H       319.19      S 10 30 E
 H- I       327.51      S 51 59 W
 I- E       425.44      N 16   4 W


SURVEY ADJUSTED BY COMPASS RULE

POINT LINE        LENGTH   BEARING         LATITUDE   DEPARTURE      X-COORDINATE   Y-COORDINATE
  E                                                                     5000.00        5000.00
     E- F         162.46  N 32   3 E        137.71       86.21
  F                                                                     5086.21        5137.71
     F- G         233.66  S 82 16 E         -31.42      231.54
  G                                                                     5317.75        5106.29
     G- H         319.05  S 10 30 E        -313.71       58.12
  H                                                                     5375.87        4792.58
     H- I         327.46  S 52   0 W       -201.57     -258.07
  I                                                                     5117.80        4591.00
     I- E         425.63  N 16   4 W        409.00     -117.80

ERROR OF CLOSURE =  0.64

PRECISION OF CLOSURE =  1/ 2281

AREA =      134875. SQ FT          OR        3.096 ACRES
```

Figure 8-9 Output for Example 8-6.

In using either the programmed calculator or the digital computer, the first things that should be checked are the computed error of closure and precision of closure. If these are not within allowable limits, all values are suspect, and the entire survey may have to be repeated.

Incidentally, in the computer program (Figure 8-7) the traverse is adjusted by the compass rule, and the area is computed by the coordinate method (discussed in Chapter 9). In the solution of Example 8-5 using the programmed calculator, the traverse was adjusted by the compass rule.

8-8 Errors and Mistakes in Traversing

As far as field work is concerned, errors and mistakes in traversing are those found ordinarily when measuring distances and angles (or bearings). These have been enumerated previously.

With regard to traverse computations, mistakes can be made by confusing north and south latitudes and east and west departures, adding corrections when they should be subtracted (and vice versa), and adding latitudes (instead of departures) to obtain east coordinates.

8-9 Problems

8-1 In the survey of an eight-sided figure, the sum of the interior angles was 1079°58′. Is this acceptable? How about 1079°55′?

8-2 A closed traverse was run and the sum of the latitudes was 1.81 ft while the sum of the departures was 1.15 ft. If the total perimeter traversed is 2080.6 ft, is this acceptable work?

8-3 Repeat Problem 8-2 except that the total perimeter traversed is 8516.2 ft.

In Problems 8-4 through 8-9, compute manually the latitudes, departures, error of closure, and precision of closure. Adjust the latitudes and departures by the compass rule and then compute the adjusted length and bearing of each line. Finally, compute the coordinates of each point, assuming the starting point to have coordinates (N2244.16, E4011.10). Note to instructor: Some parts of these problems may be omitted.

8-4

Line	*Length, m*	*Bearing*
AB	362.55	N 33°10′ W
BC	218.00	N 39°08′ E
CD	163.22	S 10°20′ E
DE	195.95	S 66°50′ E
EA	278.53	S 32°20′ W

8-5

Line	*Length, m*	*Bearing*
FG	98.02	S 42°40′ W
GH	112.16	N 58°00′ W
HI	140.01	N 3°40′ W
IJ	200.90	N 82°55′ E
JK	111.50	S 25°25′ W
KF	54.69	S 20°47′ E

8-6

Line	*Length, ft*	*Bearing*
LM	1000.00	S 4°30′ W
MN	92.18	due west
NO	200.47	N 20°10′ W
OP	772.20	N 10°15′ E
PQ	50.88	S 33°40′ E
QL	118.15	N 39°34′ E

8-7

Line	*Length, m*	*Bearing*
RS	275.55	S 89°00′ E
ST	150.00	S 5°05′ E
TU	198.90	S 22°10′ W
UV	350.80	N 60°50′ W
VR	191.34	N 28°51′ E

8-8

Line	*Length, m*	*Bearing*
AB	6.50	N 85°00′ W
BC	202.02	N 32°30′ W
CD	280.08	N 6°06′ E
DE	281.95	due east
EF	150.10	S 45°10′ W
FG	160.25	S 46°00′ E
GA	310.12	S 41°35′ W

8-9

Line	*Length, ft*	*Bearing*
MN	508.20	S 35°35′ W
NO	716.95	N 1°00′ E
OP	801.55	N 89°00′ E
PQ	77.23	S 40°42′ W
QM	534.42	S 61°06′ W

8-10 Solve Problem 8-4 using a programmed calculator.

8-11 Solve Problem 8-5 using a programmed calculator.

8-12 Solve Problem 8-6 using a programmed calculator.

8-13 Solve Problem 8-7 using a programmed calculator.

8-14 Solve Problem 8-8 using a programmed calculator.

8-15 Solve Problem 8-9 using a programmed calculator.

8-16 Solve Problem 8-4 using the computer program (Fig. 8-7).

8-17 Solve Problem 8-5 using the computer program (Fig. 8-7).

8-18 Solve Problem 8-6 using the computer program (Fig. 8-7).

8-19 Solve Problem 8-7 using the computer program (Fig. 8-7).

8-20 Solve Problem 8-8 using the computer program (Fig. 8-7).

8-21 Solve Problem 8-9 using the computer program (Fig. 8-7).

Note: In Problems 8-10, 8-11, 8-13, 8-14, 8-16, 8-17, 8-19, and 8-20, the computer program and the programmed calculator (maybe) are set up to handle distances in feet (instead of meters). If, however, meters are used, the computed values should be correct except that distances will be in meters instead of feet. The area will be in square meters instead of square feet, but the area in acres will be erroneous.

In Problems 8-22 through 8-25, for the given coordinates (of a closed traverse), calculate the length and bearing of each side.

8-22

Point	x-coordinate	y-coordinate
A	400.33 m	−316.22 m
B	12.23 m	16.77 m
C	543.23 m	301.01 m
D	566.90 m	280.66 m

8-23

Point	N-coordinate, ft	E-coordinate, ft
E	4032.44	3930.77
F	4242.59	4720.80
G	3950.59	4697.99
H	3866.11	4364.88
I	4009.09	3990.44

8-24

Point	x-coordinate, m	y-coordinate, m
J	434.65	−58.77
K	−2.22	3.45
L	−100.47	−123.95
M	−288.64	−10.90
N	509.55	400.22

8-25

Point	N-coordinate, ft	E-coordinate, ft
P	S 209.4	E 2309.6
Q	S 1678.4	E 1675.0
R	S 390.6	E 664.3
S	S 399.8	W 430.8
T	S 1447.7	W 1590.1
U	N 777.4	W 1590.1
V	N 979.6	E 1003.3

8-26 If latitudes and departures for a closed traverse are computed from given coordinates of each point, how accurate do you anticipate the error of closure will be?

In Problems 8-27 and 8-28, for the closed traverse given, compute the missing data.

8-27

Line	Length, ft	Bearing
AB	?	N 46°30′ E
BC	516.4	S 80°10′ E
CD	?	S 25°50′ W
DA	660.8	N 41°40′ W

8-28

Line	Length, m	Bearing
JK	100.00	N 45°50′ W
KL	605.55	N 5°30′ E
LM	95.96	N 88°20′ E
MJ	?	?

8-29 Douglas wanted to run a line (*AE* in Figure 8-10) from one corner of his property to another in order to put a fence on the line. He knew the location of the two corners but neither the bearing nor the length of the line. The line passed through heavily wooded area. Utilizing roads and power line right-of-way, he ran a random line as shown in Figure 8-10. For the given data, compute the bearing to be used in running line *AE*. What is the length of the line? The bearing of *AB* was determined to be S 16°30′ E.

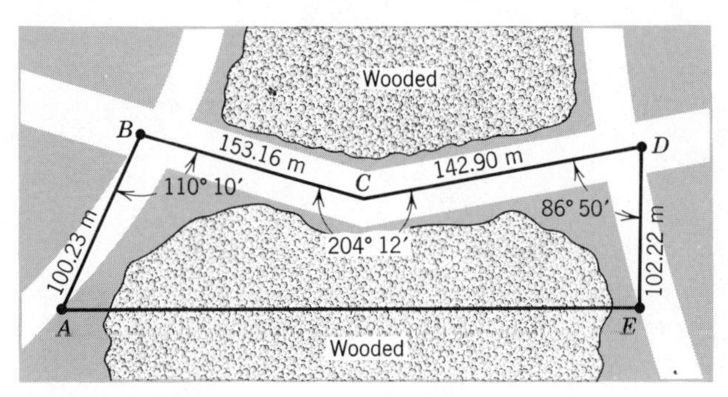

Figure 8-10

8-30 In Problem 8-22 what bearing should be run to survey a cut-off line from point *A* to point *C*?

8-31 The bearings and lengths shown for the closed traverse *ABCDEA* of Figure 8-11 have been balanced. It is desired to divide the parcel into two

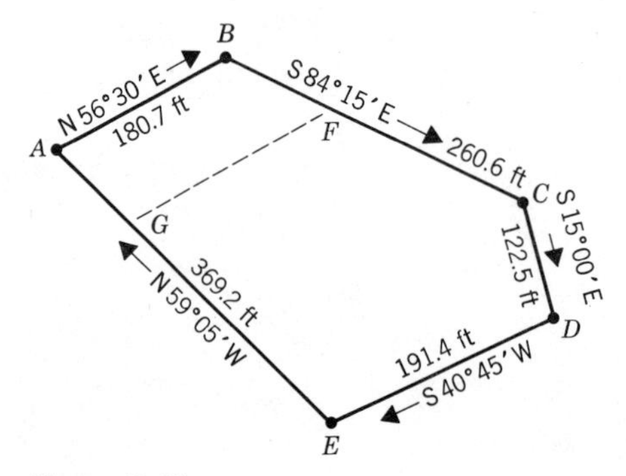

Figure 8-11

tracts by the dashed line shown. The cut-off line is to begin at a point on line *BC*, 100.00 ft from *B* and run parallel to *AB* until it intersects *AE*. What is the length of this cut-off line and how far from *A* will it intersect *AE*?

8-32 A surveyor is running a line in the direction *AB*, the bearing of which is N 10°30′ W. He encountered an area through which it was impossible to continue running the line. He ran a random line around the impossible area as shown in Figure 8-12. What distance *DE* must be measured in order to have point *E* on the line *AB*? What deflection angle must be turned at *E* in order to continue the line *AB*? What is the distance *BE*? The following data were measured.

Point	*Line*	*Deflection Angle*	*Length, m*
B		62°08′ R	
	BC		120.00
C		78°16′ L	
	CD		280.66
D		80°40′ L	

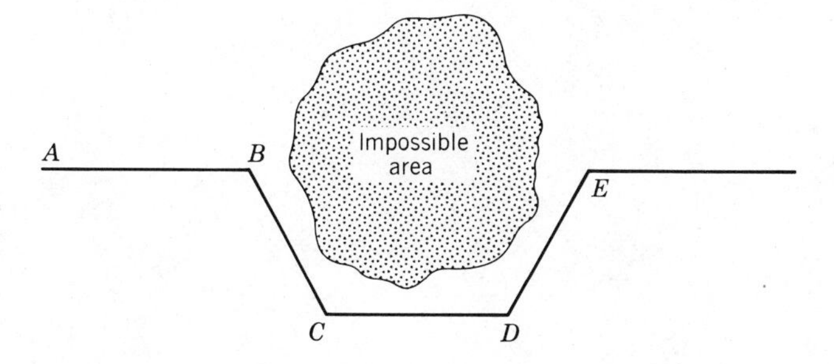

Figure 8-12

AREA

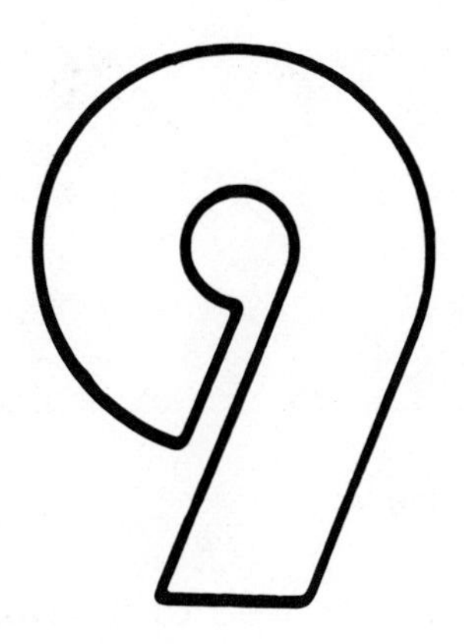

Previous chapters in this book have been devoted to methods used to compile information such as horizontal distances, vertical distances, and directions from field surveys of land. These are basic data required to delineate areas of land, produce general maps and topographic maps (covered in Chapter 10), plan route surveys—to mention a few. Let us now explore some of the ways this information can be utilized to answer specific problems, which were the reasons for making the survey in the first place.

Foremost in the reasons for making land surveys is the determination of area. Land is frequently bought and sold by area (e.g., so much per acre). Building costs are often computed on the basis of so much per square foot. Storm runoff is computed based on the area of the drainage basin (and other parameters). Quantities of water lost to evaporation from a reservoir are functions in part of reservoir area. These are only a few of the examples of the deductions that can be made when the area of land is known.

There are many methods for calculating area. Covered in the remainder of this chapter are the use of triangles, the trapezoid rule, Simpson's rule, the coordinate method, the DMD method, and a machine called a planimeter. In most cases the procedures are presented without derivation. More details regarding these methods may be found in the reference books listed at the end of this book.

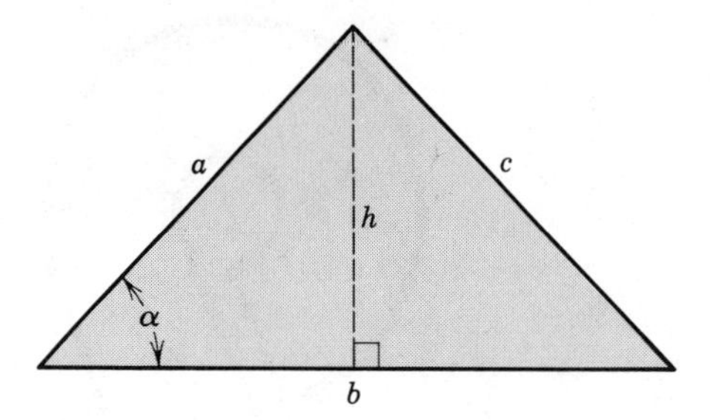

Figure 9-1

9–1 Area of a Triangle

The area of a triangle may be computed using one of several formulas. Referring to the triangle in Figure 9–1, the area may be calculated by using any of the following.

$$A = \tfrac{1}{2}bh \tag{9-1}$$

where A = area

b = base (can be any side)

h = height (perpendicular distance from corner opposite base to the base)

$$A = \tfrac{1}{2}ab \sin \alpha \tag{9-2}$$

where α = angle between sides a and b

$$A = \sqrt{s(s-a)(s-b)(s-c)} \tag{9-3}$$

where $s = \tfrac{1}{2}(a+b+c)$ (i.e., one-half the perimeter)

In these three equations, the unit of a, b, and c may be any unit for measuring linear distance (such as feet or meters) and the area will be in square units of the same type (such as square feet or square meters).

Any simple, closed figure all sides of which are straight lines may be divided into triangles by drawing a sufficient number of diagonals. It follows then that the computation of the area of such a figure can be made by (1) dividing the figure into triangles, (2) determining the required values (lengths and angles), (3) computing the area of each triangle, and (4) adding these areas to determine the total area of the figure. For example, in Figure 9-2 assuming the exterior sides of the five-sided figure have been measured, if the lengths of the diagonals *BE* and *CE* can be determined, the area of each triangle can be calculated by Eq. (9-3) and the results added to determine the area of the pentagon. The lengths of the diagonals may be measured in the field or they may be scaled from a map drawn to scale. Certainly one should realize that the latter method involving scaling distances (and in some cases angles) will sacrifice accuracy in the computed area. Thus the use of this method should be limited to cases where reduced accuracy is acceptable.

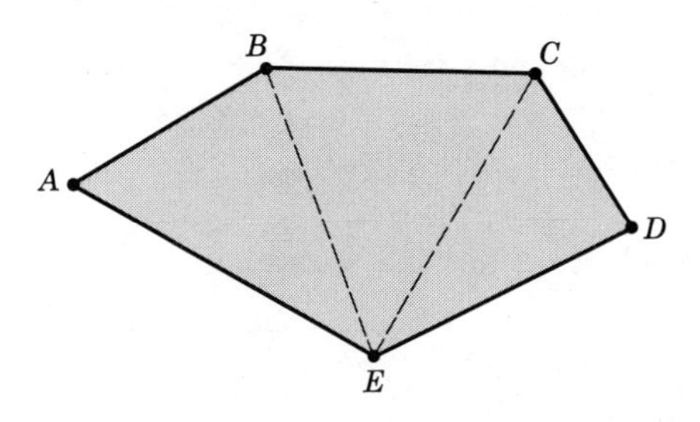

Figure 9-2

9-2 Area by Offsets from a Straight Line

Sometimes it may be necessary to determine the area of a tract of land using offsets from a straight line. For example, in Figure 9-3, *AE* is a straight line separated by *B*, *C*, and *D* into four segments of equal length *b*. *AJ*, *BI*, *CH*, *DG*, and *EF* are perpendicular offsets from *AE*. If length *b* and the offset distances h_1, h_2, h_3, h_4, and h_5 are known, the area within *ABCDEFGHIJA* can be calculated by the formula known as the trapezoid rule.

$$A = b\left(\frac{h_1}{2} + h_2 + h_3 + \cdots + \frac{h_n}{2}\right) \tag{9-4}$$

Equation (9-4) gives the exact area of the tract if lines *JI*, *IH*, *HG*, and *GF* are straight lines. If these lines are curved, the approximate area of the tract

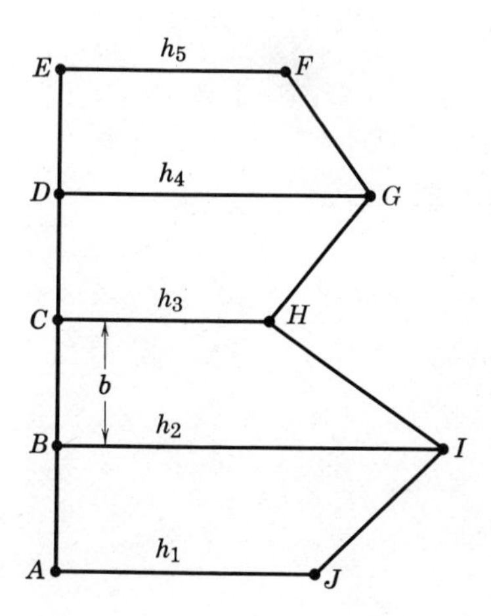

Figure 9-3

may be computed by a formula known as Simpson's rule.

$$A = \frac{b}{3}[h_1 + h_n + 2(h_3 + h_5 + \cdots + h_{n-2}) + 4(h_2 + h_4 + \cdots + h_{n-1})] \quad (9\text{-}5)$$

Note: There must be an odd number of offsets to use this formula.

Example 9-1

A surveyor needed to know the area between the straight line *AB* and a meandering stream as shown in Figure 9-4. At 20 m intervals along *AB* he measured the distances from the line *AB* to the center of the stream with the results as shown on the figure. Calculate the area by both the trapezoid rule and Simpson's rule.

Solution

By the trapezoid rule

$$A = b\left(\frac{h_1}{2} + h_2 + h_3 + \cdots + \frac{h_n}{2}\right) \quad (9\text{–}4)$$

$$A = 20.00\left(\frac{13.44}{2} + 23.15 + 31.00 + 25.95 + 8.08 + 3.66 + 18.77 + 39.18 + \frac{22.24}{2}\right)$$

$$A = 3353 \text{ sq m}$$

By Simpson's rule

$$A = \frac{b}{3}[h_1 + h_n + 2(h_3 + h_5 + \cdots + h_{n-2}) + 4(h_2 + h_4 + \cdots + h_{n-1})] \quad (9\text{–}5)$$

$$A = \frac{20.00}{3}[13.44 + 22.24 + 2(31.00 + 8.08 + 18.77) + 4(23.15 + 25.95 + 3.66 + 39.18)]$$

$$A = 3461 \text{ sq m}$$

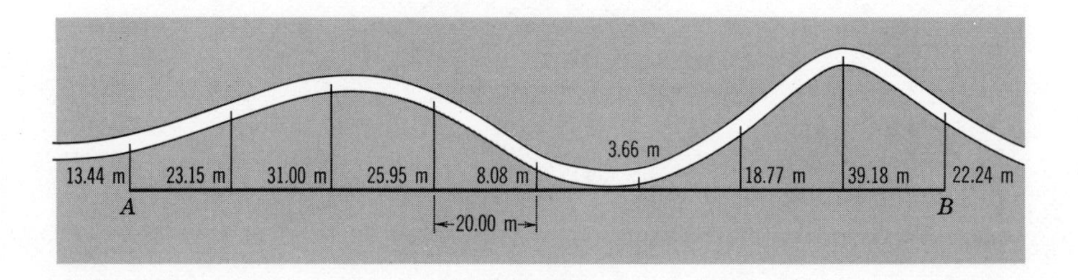

Figure 9-4

9-3 Area by Coordinate Method

If the coordinates of each corner of a closed traverse are known, the area of the traverse may be computed by the coordinate method. Stated in words, the area is determined by the coordinate method by multiplying the x-coordinate of each point by the difference between adjacent y-coordinates, adding the resulting products, and taking half the sum. The y-coordinates must be taken in the same order around the traverse when obtaining the difference between adjacent y-coordinates.

The coordinate method may be stated in equation form as

$$A=\tfrac{1}{2}[x_1(y_2-y_n)+x_2(y_3-y_1)+x_3(y_4-y_2)+\cdots+x_n(y_1-y_{n-1})] \quad (9\text{–}6)$$

A convenient procedure for applying the coordinate method is to list the coordinates in the following form.

$$\frac{x_1}{y_1} \quad \frac{x_2}{y_2} \quad \frac{x_3}{y_3} \quad \cdots \quad \frac{x_n}{y_n} \quad \frac{x_1}{y_1}$$

(Note that the first point is repeated at the end.) The next step is to find the sum of the products of all adjacent diagonal terms taken down to the right (i.e., x_1y_2, x_2y_3, etc.) and, similarly, to find the sum of the products of all adjacent diagonal terms taken up to the right (i.e., y_1x_2, y_2x_3, etc.). One half of the difference between these two sums is the area of the traverse. In applying the coordinate method, should the computed area be negative, the negative sign is ignored.

Example 9–2

Calculate the area of the traverse of Example 8-1 by the coordinate method.

Solution

Arranging the coordinates (determined in Example 8-3) as indicated above gives

$$\frac{5000.00}{5000.00} \quad \frac{5086.21}{5137.71} \quad \frac{5317.75}{5106.29} \quad \frac{5375.87}{4792.57} \quad \frac{5117.80}{4591.00} \quad \frac{5000.00}{5000.00}$$

The sum of the products of adjacent diagonals down to the right is

$$(5000.00)(5137.71)+(5086.21)(5106.29)+(5317.75)(4792.57)$$
$$+(5375.87)(4591.00)+(5117.80)(5000.00)=127{,}415{,}521.5$$

The sum of the products of adjacent diagonals up to the right is

$$(5000.00)(5086.21)+(5137.71)(5317.75)+(5106.29)(5375.87)$$
$$+(4792.57)(5117.80)+(4591.00)(5000.00)=127{,}685{,}273.3$$

The difference between these sums is

$$127{,}685{,}273.3 - 127{,}415{,}521.5 = 269{,}751.8$$

Thus the area is $\dfrac{269{,}751.8}{2} = 134{,}875.9$

This should be rounded off to about 134,900 sq ft. This compares favorably with the areas determined in Examples 8-5 and 8-6.

9-4 Area by the DMD Method

If latitudes and departures of a traverse are known, the area of the traverse may be computed by the DMD method. DMD stands for "double meridian distance." Before proceeding with the DMD method of computing area, the meaning of DMD must be understood. The meridian distance of a line is defined as the distance from the midpoint of the line to the reference meridian (normally the north-south line). Thus in Figure 9-5 the meridian distance of line *AB* is the distance *DE* and that of line *BC* is the distance *HI*.

Observe that the meridian distance of line *BC* (which is distance *HI*) is equal to *DE* plus *EF* plus *FG*. However, *DE* is the meridian distance of line *AB*, *EF* is half the departure of line *AB*, and *FG* is half the departure of line *BC*. Thus, the meridian distance of line *BC* is equal to the meridian distance of line *AB* plus half the departure of line *AB* plus half the departure of line *BC*. In order to avoid having to take half of all the departures encountered in applying the DMD method, it is convenient to multiply the equation stated in the last sentence by 2. The final result (i.e., computation of the area of the traverse) must be divided by 2 to compensate. If that equation is multiplied by

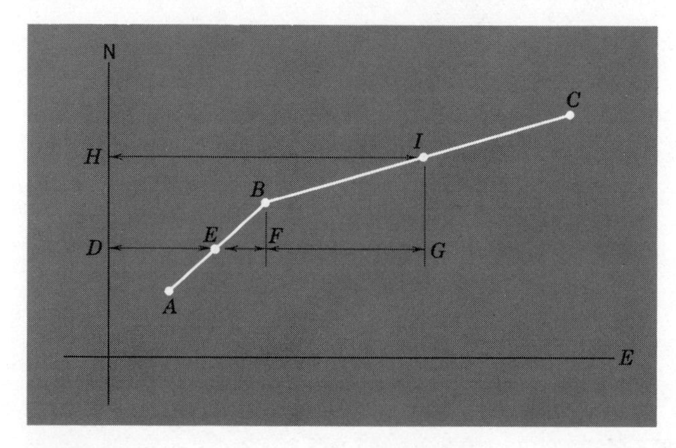

Figure 9-5

2, the double meridian distance (DMD) of line *BC* is equal to the DMD of line *AB* plus the departure of line *AB* plus the departure of line *BC*.

After the DMD of the first line is determined, the DMD of each succeeding line can be determined in the manner described above in terms of the DMD and departure of the preceding line and the departure of the line itself. As a general rule, the DMD of any line (after the first one) of a traverse is equal to the DMD of the previous line plus the departure of the previous line plus the departure of the line itself. With regard to the DMD of the first line, if the reference meridian is placed so that it passes through the starting point (point *A* in Figure 9-5), the meridian distance is equal to half the departure of line *AB*, or the DMD is equal to the departure of line *AB*.

Once the DMD of each line has been computed, the area of the traverse is equal to half the sum of the products of the latitude and the DMD of each line. As in the coordinate method, if the final sum is negative, the negative sign is ignored. Since the DMDs are computed from the departures and the area is computed from the latitudes (and DMDs), it is obvious that this method is useful when the latitudes and departures are known.

Example 9-3

Calculate the area of the traverse of Example 8-1 by the DMD method.

Solution

From Example 8-2, the (adjusted) latitudes and departures of the traverse are given in the second and third columns below.

Line	*Latitude*	*Departure*	*DMD*	*DMD × Latitude*
EF	137.71	86.21	86.21	11,872
FG	− 31.42	231.54	403.96	−12,692
GH	−313.72	58.12	693.62	−217,602
HI	−201.57	−258.07	493.67	−99,509
IE	409.00	−117.80	117.80	48,180
				−269,751

The DMDs are computed and listed in the fourth column. The DMD of the first line (86.21) is equal to the departure of the first line. The DMD of the next line (403.96) is equal to the DMD of the previous line (86.21), plus the departure of the previous line (86.21), plus the departure of the line itself (231.54). DMDs of each succeeding line are computed in the same manner. It will be noted that the DMD of the last line (117.80) is numerically the same but opposite in sign to the departure of the last line (−117.80). This should always be true (assuming the departures have been balanced) and provides a check on the computation of the DMDs. The last column (DMD × Latitude) is obtained by multiplying each DMD (fourth column) by the corresponding latitude (second column). Thus the first entry (11,872) is determined by multiplying 137.71 (latitude) by 86.21 (DMD). Since the sum of the last column is −269,751, the area is half this amount, or 134,875.5. Rounded off, the area is 134,900 sq ft. This compares favorably with the areas determined in Examples 8-5, 8-6, and 9-2.

9-5 Area by Planimeter

With the exception of Simpson's rule, each method presented in this chapter for computing area has assumed straight line boundaries. Often areas are encountered that do not have straight lines for all boundaries. If such areas are drawn to scale as on a map, the value of the area can be determined using a device called a planimeter.

A planimeter (shown in Figure 9-6) is a machine, usually several inches long, that consists essentially of two arms, one of which has an anchor point at one end and the other has a pointer. It has a wheel that is rotated as the operator runs the pointer around the plot of a simple, closed area. One can read from a scale (vernier) a value that, when scaled, gives an area. This device is a graphical integrator of areas.

The procedure for determining an area by planimeter is fairly simple. First the anchor point, which has a sharp point that can be stuck into the paper of the plot to be measured, is anchored at a convenient point outside the area to be planimetered. The movable pointer is positioned over any point of the

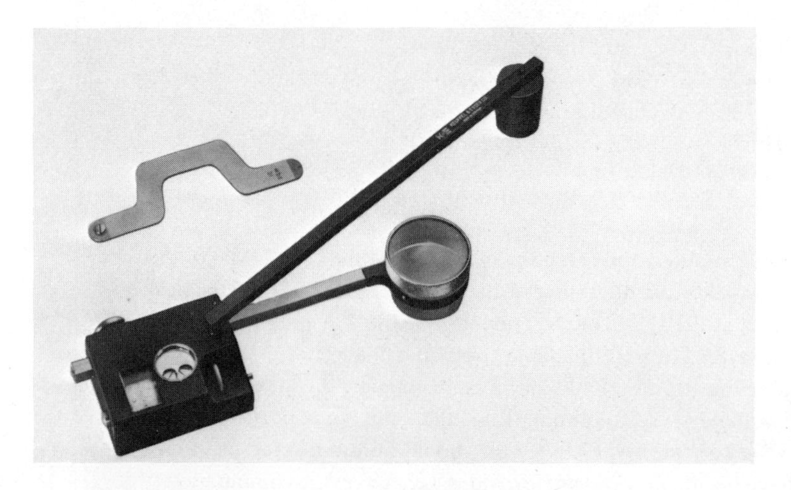

Figure 9-6 Planimeter. (Courtesy of Keuffel & Esser Co.)

traverse and the scale (vernier) reading is made and recorded. The pointer is then carefully traced around the boundary of the area until the pointer returns to the starting point. The tracing is normally done freehand, although a straightedge may be employed to trace along straight lines. When the starting point is reached, the scale reading is made and recorded again. The difference between initial and final scale readings is an indication of the area within the boundary lines.

As described above, the procedure for determining an area by planimeter is fairly simple. A number of additional comments are needed, however. The first concerns the reading of the scale. The reading consists of four digits—the first being read from a circular disk, the next two coming from a drum, and the last coming from a vernier. As shown above, the area is indicated by the difference between starting and ending scale readings. Care must be exercised in this regard. Assuming the numbers are increasing (the numbers increase if proceeding in one direction, such as clockwise, and decrease in the other), if the first reading is 2049 and the second is 7125, the difference is obviously 7125 − 2049, or 5076. However, if the first reading is 9022 and the second reading is 0837, the difference might appear to be 9022 − 0837, or 8185. This is incorrect, however. What has happened is that the initial reading was 9022, and as the pointer is traced around the boundary, the reading increases until it reaches 9999. The next scale reading would be 10000, except that the scale has only four digits. In the example cited, the final reading should have been 10837, and the correct difference would be 10837 − 9022, or 1815. As in all phases of surveying, care must be exercised to avoid mistakes of this type.

In the previous discussion, it has been stated that the difference in scale readings is an indication of the area, but it has been vague as to what units the area is expressed in. Actually, this depends on the planimeter. Some planimeters have an adjustable pointer arm. It can be set to give scale readings that reflect area in units comparable to the scale of the map or drawing being planimetered. Others have a fixed arm and the area is therefore always expressed in the same units. Many planimeters are set so that a difference in scale readings gives area in square inches if a decimal is placed two digits to the left. Thus if the difference in scale readings is 0762, the area would be 7.62 sq in.

In any event when using a given planimeter, it is wise to check the planimeter and, if necessary, apply an adjustment to determine the correct area. This can be done by first drawing a square of specific size on a piece of paper. A 5 inch by 5 inch square (area of 25 square inches) is good. This figure can then be traced using the planimeter to see what the difference between initial and final readings is. If the difference in readings is 2500 the difference between scale readings gives area in square inches if a decimal is placed two digits to the left (i.e., 25.00). For any other difference between readings, a simple ratio can be determined to apply to subsequent differences in readings to determine the correct area. The ratio is

$$\frac{d_t}{A_t} = \frac{d}{A} \tag{9-7}$$

where d_t = difference between planimeter readings during test
A_t = area of test figure
d = difference between planimeter readings for any case
A = area of figure for specific case

Rearranging to solve for area,

$$A = d\left(\frac{A_t}{d_t}\right) \tag{9-8}$$

Example 9-4

To check a planimeter, a 5 inch by 5 inch square was drawn on a piece of paper and planimetered, with a difference in readings of 2482. Later the planimeter was used to determine the area of a drainage basin. The scale on the map is 1 inch = 1000 ft. If the difference in planimeter readings is 6022, what is the area of the drainage basin?

Solution

Equation (9-8) can be used to determine the area of the drainage basin in square inches on the map.

$$A = d\left(\frac{A_t}{d_t}\right) \tag{9-8}$$

$$A = 6022\left(\frac{25.00}{2482}\right)$$

$$A = 60.66 \text{ sq in.}$$

The actual drainage area may be computed by

$$60.66 \text{ sq in.} \left(\frac{1000 \text{ ft}}{1 \text{ in.}}\right)^2 = 60{,}660{,}000 \text{ sq ft}$$

If an area is traced twice with a planimeter, chances are that both differences in readings would not be exactly the same. There are several reasons for this—error of adjustment of planimeter, inability of the operator to trace the boundary perfectly each time, and so on. For this reason, it is good practice to trace the area several times and obtain an average value of the differences in readings. This can be done by making (and recording) an initial reading, tracing the figure, and making the final reading. The next tracing can be done immediately with the final reading for the first tracing being the same as the first reading for the second tracing. This can be repeated as many times as necessary to obtain readings for as many tracings as desired. It is also good practice to trace the area several times in the opposite direction (i.e., counterclockwise if others were done clockwise).

If a map is accurately drawn and the planimeter work done with care, an accuracy of 1 percent is attainable.

9-6 Errors and Mistakes in Area Determination

The accuracy of a computed area depends on the accuracy of the data used to compute the area. The coordinate method of computing area, for example, gives the exact area for the given coordinates. However, if the coordinates of a particular tract of land have been determined inaccurately, the area computed by the coordinate method based on these coordinates will likewise be inaccurate. Errors and mistakes concerning the measurement of distances, angles, and the like, have been discussed in previous chapters.

This reduces the discussion of errors and mistakes more or less to the computation procedure. Errors in computing areas may result from not using a small enough interval when applying the trapezoid rule or Simpson's rule to find the area of an irregular boundary, imperfect tracing of a boundary when using a planimeter, haphazardly setting the planimeter scale bar for a particular scale, and inaccurately determining the planimeter constant. Mistakes may result from incorrect mathematical computation, neglecting the effect of minus signs in the DMD method and the coordinate method, forgetting to divide the final result by two in the DMD method and coordinate method, and failure to repeat the starting point when listing coordinates in using the difference between sums of the products of adjacent diagonal terms taken down to the right and those taken up to the right.

9-7 Problems

9-1 Compute the area of the traverse shown in Figure 4-10.

9-2 Assuming line *ED* in Figure 9-2 is 700.00 m in length, use a scale to determine the length of the other boundary lines and the diagonals and compute the area of the pentagon in hectares.

9-3 If the area of a triangle is 687.5 sq m and the lengths of two sides are 57.18 m and 26.14 m, what is the size of the angle between these two sides?

9-4 The three sides of a triangle are 51.12 m, 64.18 m, and 69.22 m. What is the area of the triangle in hectares?

9-5 Determine the area (in acres) between line YZ and the meandering stream for the offsets indicated in Figure 9-7. Use the trapezoid rule. The offsets are taken every 50 ft along YZ.

9-6 Solve Problem 9-5 by Simpson's rule.

In Problems 9-7 through 9-12, solve for the area of the traverse by the DMD method (use adjusted latitudes and departures).

9-7 Use data given in Problem 8-5.

9-8 Use data given in Problem 8-6.

9-9 Use data given in Problem 8-7.

9-10 Use data given in Problem 8-8.

9-11 Use data given in Problem 8-9.

9-12 Use data given in Problem 8-4.

In Problems 9-13 through 9-22, solve for the area of the traverse by the coordinate method.

9-13 Use data given in Problem 8-5.

9-14 Use data given in Problem 8-6.

9-15 Use data given in Problem 8-7.

9-16 Use data given in Problem 8-8.

9-17 Use data given in Problem 8-9.

9-18 Use data given in Problem 8-22.

9-19 Use data given in Problem 8-23.

9-20 Use data given in Problem 8-24.

9-21 Use data given in Problem 8-25.

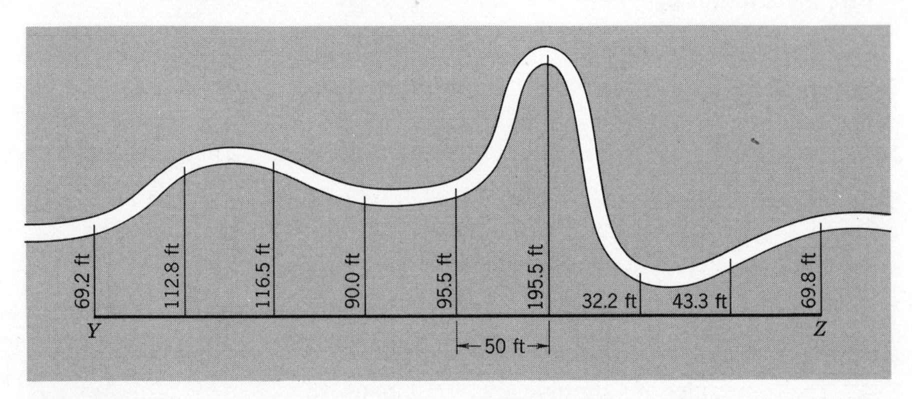

Figure 9-7

9-22 Use data given in Problem 8-4.

9-23 Calculate the area of *ABCDEA* in Figure 8-11 by the DMD method.

9-24 Calculate the area of *ABFGA* and the area of *FCDEGF* in Figure 8-11 by the DMD method. How does the sum of these two areas compare with the area calculated in Problem 9-23? (See Problem 8-31.)

9-25 In Figure 9-8 the coordinates are given in meters with the first number in parentheses being the north coordinate and the second, the east coordinate. Compute the area of the figure in hectares, using the trapezoid rule.

9-26 Solve Problem 9-25 by the coordinate method.

9-27 Solve Problem 9-25 by the DMD method.

9-28 A surveyor drew a 4-inch square on a piece of paper and traced it with a planimeter three times with the following readings: 2074, 3681, 5290, 6900. He then used this planimeter to determine the drainage area marked off on a map. He traced the drainage area four times with the following readings: 2844, 6473, 0107, 3734, 7362. If the map scale is 1 in. = 500 ft, what is the area in acres?

9-29 Using the same planimeter as that given in Problem 9-28, the surveyor traced the cross section of a highway at a certain point. The readings were: 6053, 8053, 0049, 2052, 4057. If the scale is 1 in. = 2 ft, what is the area of the cross section?

9-30 Repeat Problem 9-8 if the distances (given in Problem 8-6) were measured with a tape that is 0.08 ft too short. The nominal length of the tape is 100 ft.

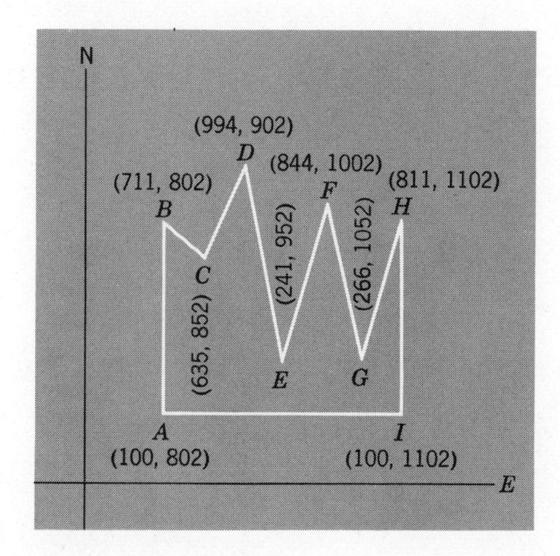

Figure 9-8

TOPOGRAPHIC SURVEYING AND MAPPING

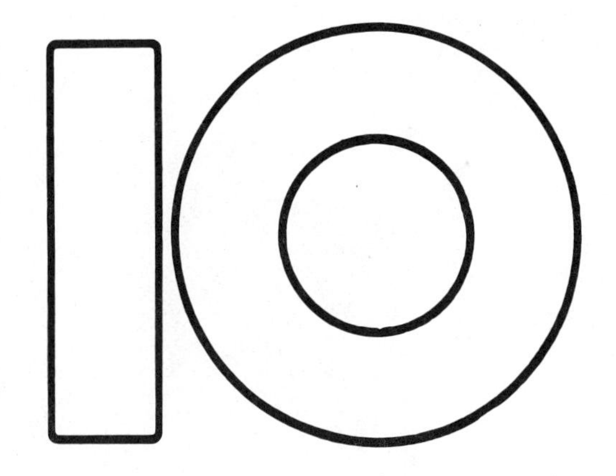

The result of gathering data utilizing the surveying techniques presented in Chapters 4 through 9 is often the production of a topographic (topo, for short) map. A topo map is a scale drawing of a portion of the earth's surface showing natural and artificial features of the land and also showing relief (variation of elevation of the earth's surface).

Topo maps are widely used in engineering work. Maps are used in connection with construction projects, drainage problems, location of sewer lines, location of highways, and many other enterprises. In addition maps have important applications in areas other than engineering. For example, maps are often critically important in military operations. Millions of copies of maps have been required by the United States in fighting recent wars. Maps are used by geologists in exploration for, and evaluation of, mineral and fuel resources. Landscape architects depend on good topo maps. In addition, maps are often used by urban planners, geographers, soil conservationists, lawyers, and many others.

10-1 Availability of Topo Maps

Maps have been prepared for many years by various governmental agencies. Foremost among these, perhaps, is the U.S. Geological Survey (USGS). The USGS has been preparing and publishing topo maps for around 100 years. Presently, most (if not all) of the United States has been mapped, although

Figure 10-1 U.S. Geological topographic quadrangle map. (Courtesy of U.S. Geological Survey.)

some particular maps may be quite old. A part of a USGS topo map is shown in Figure 10-1.

In addition to the USGS, other governmental agencies at federal, state, and local levels prepare maps. Copies of most of these can be easily obtained at nominal cost. For specific information, one should write the Map Information Office of the U.S. Geological Survey.

Many of these government maps cover relatively large areas, and in some applications these are quite useful. For example, these maps are excellent for delineating and quantifying drainage areas for use in determining flood flows for creeks and rivers. On the other hand, the refinement of contour intervals and map scales of these maps will likely not meet standards for smaller, individual projects. Examples are a topo map of an individual lot for locating its septic tank drain field and a topo map for a construction site. In the case where a topo map is needed and a map of sufficient refinement is not available, it becomes the responsibility of the surveyor to collect the necessary data and produce a map for a specific project. The remainder of this chapter deals with the collection of data for, and the preparation of, a topo map.

10-2 Horizontal Control

In collecting data for a topo map and in its preparation, some type of horizontal control is necessary. This means that two or more ground points must be located accurately so that subsequent data collected may be properly oriented. Ordinarily, horizontal control consists of a closed traverse, which in most cases is along the boundary lines of the piece of property. Field methods for running a closed traverse have been discussed in Chapter 8. An illustrative set of field notes is shown in Figure 6-10.

After the preliminary field data have been collected, the traverse is then plotted to begin the map. There are several ways of plotting a traverse: coordinate method, tangent method, chord method, protractor method, and the use of drafting machines. Even the digital computer does an excellent job of plotting traverses.

The coordinate method is probably the most accurate method of plotting a traverse. The computation of coordinates has been explained in Chapter 8. Each corner of the traverse is plotted based on the computed coordinates. Plotting is facilitated by using grid paper, which may be purchased or self-made. After all points have been plotted, they would be connected by lines to form the boundary. The length and bearing of each line can be scaled and compared with the recorded field data to locate any large mistakes in plotting.

The tangent method is used to lay off an angle. In Figure 10-2, suppose it is desired to lay off a deflection angle of 40°38′ R at *B* as shown. (Lines *AB* and *BC* represent successive lines of a traverse.) The first step is to scale off any distance along the extension of line *AB*, locating point *D*. Any distance can be

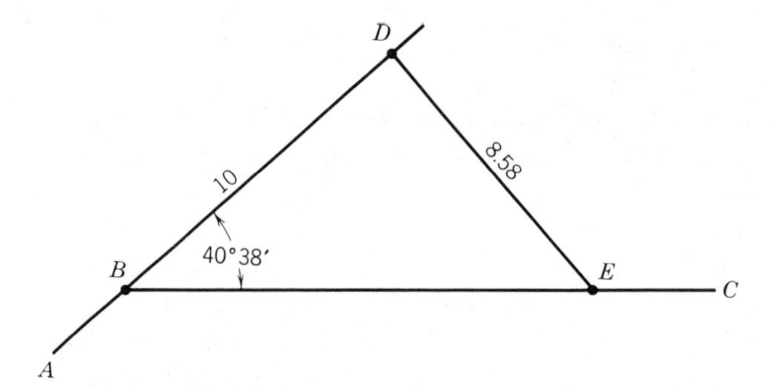

Figure 10-2

laid off, but 10 inches is a good length for computational purposes. At point *D* a line perpendicular to line *ABD* is constructed. Application of Eq. (2-8) indicates that the length *DE* to be laid off to give the required angle *DBE* is *BD* tan *DBE*. The tangent of 40°38′ is 0.858. Thus if *BD* has been laid off as 10 inches, the length of *DE* should be 10(0.858), or 8.58 inches, in order for angle *DBE* to be 40°38′. With point *E* thus established, a line connecting points *B* and *E* (extended beyond *E* if necessary) gives the proper orientation of line *BC* of the traverse. Using an engineer's scale, the length of *BC* can be laid off to establish point *C*.

The chord method of laying off an angle is virtually the same as the procedure described in Section 4-5 and Example 4-4 for laying off an angle with a tape. The only difference is that drafting equipment (scale to measure distance and compass to lay off arcs) is used instead of taping equipment.

Every high school geometry student has probably used a protractor to lay off an angle. A protractor, usually made of plastic or metal, is cut in the shape of a semicircle (or a full circle) with angle graduating marks around the circumference. When used to lay off an angle, the center of the protractor (the center of the circle or semicircle) is placed over the vertex of the angle and the zero mark along the circumference is placed along one line of the angle. A pencil mark is made or a pin stuck at the appropriate graduation marker. If this point is connected to the vertex, the two lines lay off the required angle. The protractor method is relatively quick but is not as accurate as the other methods.

The basic map is plotted by one of the methods described above. Each corner is usually indicated by a small circle, which can be drawn with a compass or circle template. These circles (corners) are connected to form the boundary lines. The length and bearing of each line are placed (written) along the line. These (as well as other notations) should be placed so that they can be read from the bottom or right side of the map. Bearings should be placed on the boundary lines in a clockwise direction. Interior angles and/or coor-

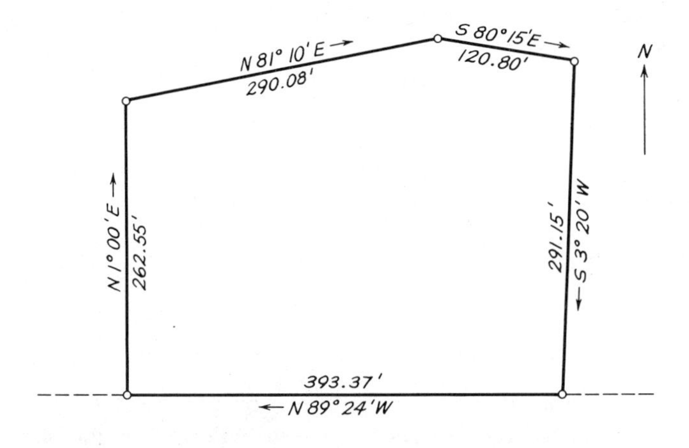

Figure 10-3 Plot of traverse for field data of Figure 6-10.

dinates may also be shown. An example of the plotting of the traverse for which field data are shown in Figure 6-10 is given in Figure 10-3.

10-3 Location and Plotting of Details

One of the important characteristics of a topo map is the field documentation and representation of detail on the map. Detail, as used here, refers to any important points or features on the earth's surface. Whether or not a particular feature is important enough to be included on a topo map depends somewhat on the purpose of the map. A map prepared for landscaping purposes might show individual trees or shrubs; whereas a USGS topo map covers such an immense area that showing individual trees would not be feasible.

The method of locating details in the field is fairly simple and is, in many cases, a matter of common sense. The location of a point (an object such as a tree, a fire hydrant, etc.) can be referenced by several methods. For example, in Figure 10-4 it is assumed that *AB* is known (it may be one line of the boundary traverse) and it is desired to record the location of point *C*. This can be accomplished by several methods: measuring the angles *CBA* and *BAC*; measuring angle *BAC* and distance *AC* (or angle *CBA* and distance *BC*); measuring the distances *AC* and *BC*. In each of these, the bearing or azimuth could be substituted for the angle. There are other methods (which the reader may visualize), but these seem most obvious and most common.

In the case of locating lines (such as a fence) and objects (such as a house), it is a matter of locating two or more points—enough to ensure proper plotting on the map. In Figure 10-5, it is assumed that *AB* and *BC* are known, and it

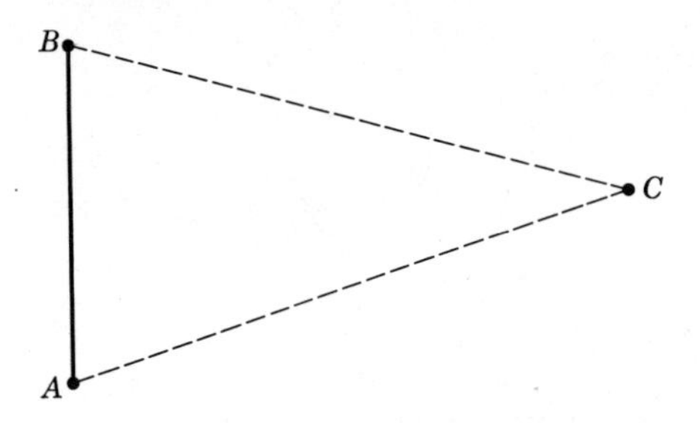

Figure 10-4

is desired to record the location of the house shown. This can be done by locating any two points and measuring all sides of the house. Point *D* could be located by measuring the distance *BD* and the angle *ABD*, and point *E*, by measuring the distance *CE* and the angle *ECB*. If these values plus the dimensions of the house are known, the house can be plotted. This assumes, of course, that all angles of the house are right angles, which is usually the case. If this is not the case, additional measurements would be required. Although theoretically, the location of only two points on the house is adequate to locate (plot) it, it would be good practice and would facilitate plotting to locate one additional point, such as *G*.

The measurement of angles and distances required to locate details by the methods just described can be accomplished by any of the methods presented in earlier chapters. For the purpose of locating and plotting details, the stadia method is usually sufficiently accurate; and since it is relatively fast and efficient, it is frequently employed to locate details. Figure 10-6 shows an illustrative set of field notes for locating details by stadia. These details are for the traverse of Figure 6-10.

With regard to plotting details on the map, the plotting methods are essentially the same as those described in Section 10-2 for plotting a traverse.

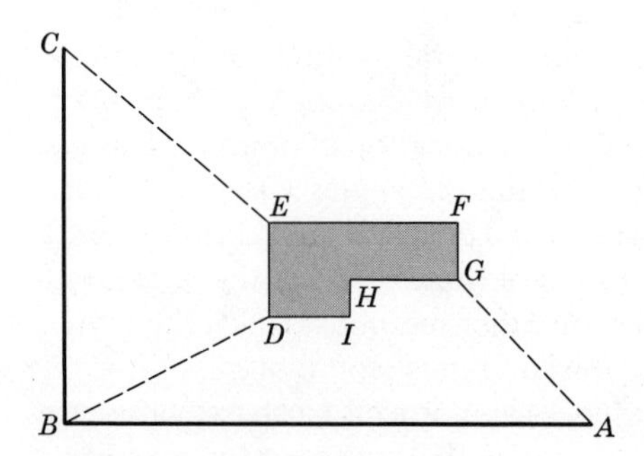

Figure 10-5

LOCATION OF DETAIL BY STADIA

⊼ at	Shot on	Stadia Interval	Vert. Ang.	Dist.	Bearing
A	pine	0.89	+6° 10′	89	N54°30′E
	J	2.62	+11° 50′	252	N38°30′E
	F	2.42	+10° 30′	235	N72°30′E
D	L	2.22	−6° 02′	221	S80°00′W
	M	2.31	−5° 56′	230	S68°00′W
	G	2.33	−6° 00′	231	S52°00′W
	H	1.61	−4° 28′	161	S26°30′W
E	oak	0.76	+0° 16′	77	N61°30′W
	I	0.91	+0° 22′	92	N37°30′W

House	100 ft	by 70	ft
Garage	32 ft	by 46	ft
Drive	20 ft	wide	

K = 100
C = 1.00

Sarah Lane
Charlotte, N.C.
Sept. 28, 1977
Warm, clear 80°F
J. Evett, N
J. Sands, ⊼
R. Jones ϕ
1 hour 30 minutes

N
C
B
D
K
L
Garage
J
M
G
H
Concrete drive
House
I
pine
F
oak
A
Sarah Lane
E

Figure 10-6 Sample field notes for locating details by stadia.

Generally speaking, the objective is to reproduce on the map the measurements made in the field. Thus, if distance *AC* and angle *BAC* were measured in the field to locate point *C* (Figure 10-4), the angle *BAC* would be laid off on the map and the distance *AC* scaled off to locate point *C* on the map. The use of protractor and scale is usually sufficiently accurate for plotting detail. An example of the plotting of details for which field data are shown in Figure 10-6 is given in Figure 10-7.

10-4 Collecting and Plotting Elevation Data

Relief, or variation in elevation of the earth's surface, may be indicated on a map in several ways. Some maps use color; others use contour lines; and there are even maps that actually have the relief scaled on the map itself. However, the most common method is by contour lines. A contour is an imaginary line along the earth's surface, all points of which are at the same elevation. In order to indicate elevations on a map, the imaginary line along the earth's surface is drawn as a solid line on the map. Contour lines will be discussed in more detail later in this section, but first a discussion of the necessary field work involved will be presented.

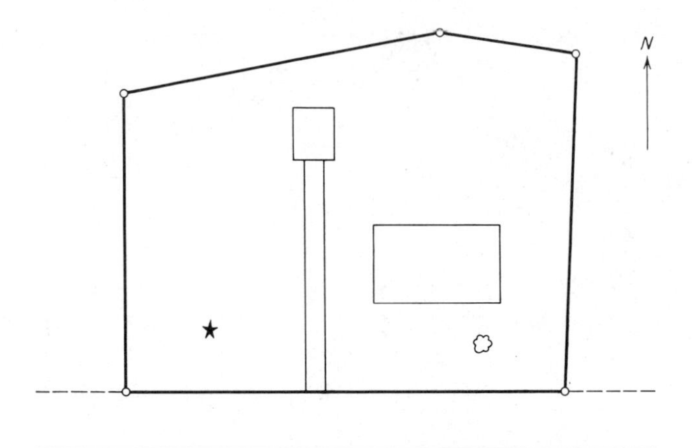

Figure 10-7 Plot of details for field data of Figure 10-6.

Prior to drawing contour lines on a map, it is necessary to determine ground elevations at a number of points. This can be done either by running differential levels or by stadia. For the purpose of drawing contour lines, elevations determined to the nearest 0.1 ft are generally accurate enough. Hence, the stadia method is sufficient. If differential leveling is used, intermediate foresights on ground points may be read only to the nearest 0.1 ft and elevations computed accordingly.

There are two procedures for determining the elevations needed to draw contour lines. One is to have the rodman move around with the rod until he "locates" a specific contour. This is done by having the recorder (note keeper) determine what foresight reading is needed to locate an elevation corresponding to a specific contour. The rodman then locates (by trial and error) points with this foresight reading, which in turn locates points on the contour. When these points are plotted on the map, a line connecting them is the desired contour. The other procedure is to determine the ground elevations at points taken more or less at random, plot these points on the map, and draw the contour lines through and among these points by interpolation.

In determining the elevations of ground points for use in plotting on a map and drawing contours, it is, of course, necessary that the location of each point be known as well as the elevation. Each point can be located by one of the methods related in the last section, or another scheme may be used. One such scheme is to lay out a grid, such as that shown in Figure 10-8. The rodman moves from one point on the grid to another, with a rod reading taken at each point. These data are easily plotted by laying off the grid and writing the respective elevations by each corner of the grid.

Figure 10-9 shows an illustrative set of field notes for determining elevations of ground points to be used in drawing contour lines. These are for the

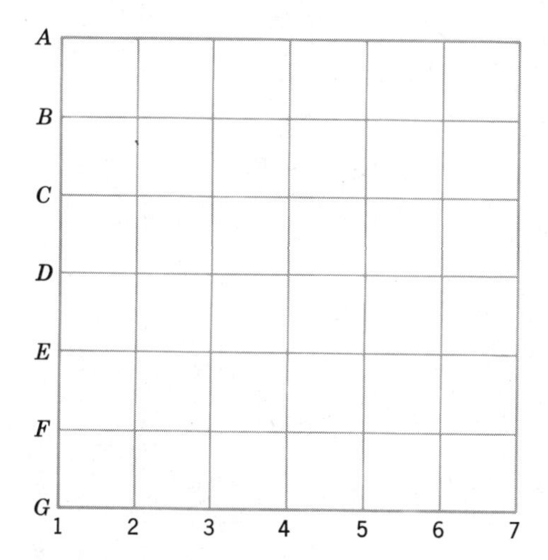

Figure 10-8

traverse of Figure 6-10. These elevations were determined by stadia with the points selected to reflect significant ground points. The second column gives the stadia interval; the third gives the azimuth; and the fourth gives the vertical angle. These are values measured in the field. The fifth column gives the horizontal distance, the sixth gives the vertical distance, and the seventh gives the elevation of each point. These are values computed from the data in columns two, three, and four.

As a demonstration consider the first line of stadia notes in Figure 10-9. With a stadia interval of 1.12 and a vertical angle of $-2°58'$, the horizontal distance is 113 ft and the vertical distance is -5.8 ft. These were determined using the stadia tables, as described in Section 7-3. The notes state that the transit it set up at *A*, the elevation of which is 223.2 ft, and the height of the instrument above the ground is 5.1 ft. Since it is stated that vertical angles were read with the middle cross hair at 5.1 ft on the rod, the elevation of point 1 is simply $223.2+(-5.8)$, or 217.4 ft. The distance 113 ft is, of course, the distance from point *A*, the location of the transit, to point 1. Elevations of, and distances to, other points are determined in the same manner.

With regard to drawing the contour lines on the map, the first step is to plot all the points for which ground elevations were obtained and to write the elevation of each point beside its plotted location. As an example, the points and elevations for which data are shown in Figure 10-9 are plotted and shown in Figure 10-10. The next step is to select a particular contour and sketch the contour line. For example, in Figure 10-10 two adjacent points in the lower left of the figure have elevations of 216.9 and 217.5. In moving from one point at elevation 216.9 to another at elevation 217.5, it is imperative that the elevation at some point in between be 217.0. Hence the 217-ft contour line

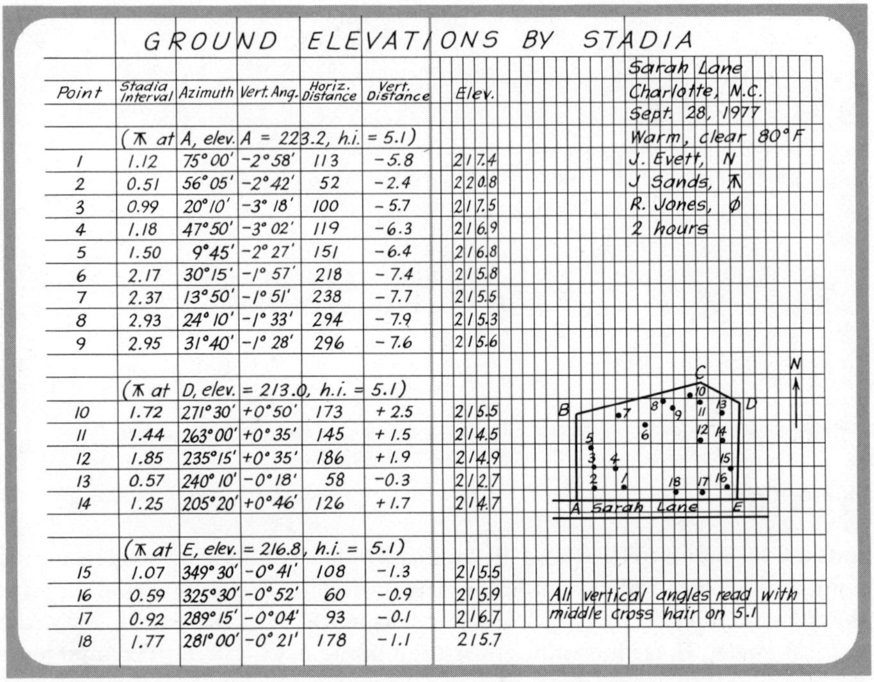

GROUND ELEVATIONS BY STADIA

Point	Stadia Interval	Azimuth	Vert. Ang.	Horiz. Distance	Vert. Distance	Elev.
	(⊼ at A, elev. A = 223.2, h.i. = 5.1)					
1	1.12	75° 00′	−2° 58′	113	−5.8	217.4
2	0.51	56° 05′	−2° 42′	52	−2.4	220.8
3	0.99	20° 10′	−3° 18′	100	−5.7	217.5
4	1.18	47° 50′	−3° 02′	119	−6.3	216.9
5	1.50	9° 45′	−2° 27′	151	−6.4	216.8
6	2.17	30° 15′	−1° 57′	218	−7.4	215.8
7	2.37	13° 50′	−1° 51′	238	−7.7	215.5
8	2.93	24° 10′	−1° 33′	294	−7.9	215.3
9	2.95	31° 40′	−1° 28′	296	−7.6	215.6
	(⊼ at D, elev. = 213.0, h.i. = 5.1)					
10	1.72	271° 30′	+0° 50′	173	+2.5	215.5
11	1.44	263° 00′	+0° 35′	145	+1.5	214.5
12	1.85	235° 15′	+0° 35′	186	+1.9	214.9
13	0.57	240° 10′	−0° 18′	58	−0.3	212.7
14	1.25	205° 20′	+0° 46′	126	+1.7	214.7
	(⊼ at E, elev. = 216.8, h.i. = 5.1)					
15	1.07	349° 30′	−0° 41′	108	−1.3	215.5
16	0.59	325° 30′	−0° 52′	60	−0.9	215.9
17	0.92	289° 15′	−0° 04′	93	−0.1	216.7
18	1.77	281° 00′	−0° 21′	178	−1.1	215.7

Sarah Lane
Charlotte, N.C.
Sept. 28, 1977
Warm, clear 80° F
J. Evett, N
J Sands, ⊼
R. Jones, φ
2 hours

All vertical angles read with middle cross hair on 5.1

Figure 10-9 Sample field notes for determining elevations of ground points.

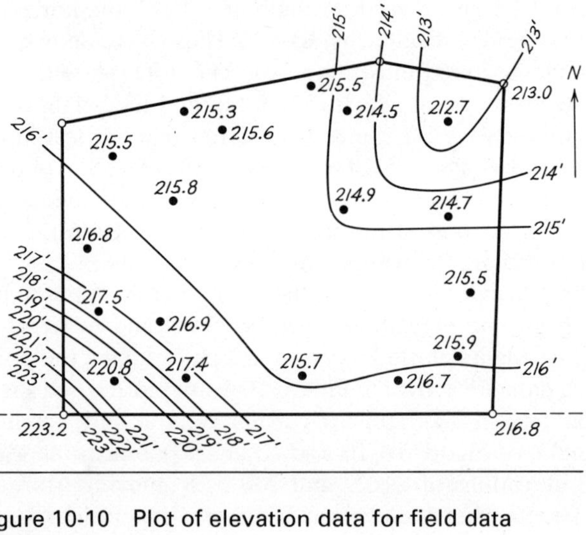

Figure 10-10 Plot of elevation data for field data of Figure 10-9.

should be drawn between these two points. The point with elevation 216.9 is also adjacent to a point of elevation 217.4, and by the same argument, the 217-ft contour must pass between these two points. Thus the 217-ft contour line is drawn in as shown. Two adjacent points in the lower left have elevations of 217.5 and 220.8. Thus, if contour lines are being drawn at 1-ft intervals, it should be obvious that three contour lines (218, 219, and 220) must be drawn between these two points. These three contour lines are shown passing between the points indicated and continuing on in accord with plotted elevations for other points. Remaining contour lines are plotted in the same manner.

In the foregoing discussion contour lines were drawn at 1-ft intervals. For maps representing areas with large differences in elevation, it is not usually feasible to use 1-ft contour intervals—the contour lines would be too close together. In such cases another contour interval, such as 2 ft, 5 ft, 10 ft, 20 ft, or 50 ft, might be used. The contour interval must be chosen based on the maximum difference in elevation on the map and to some extent on the purpose of the map. The interval selected should give a reasonable number of contour lines without crowding the map.

Some additonal remarks about contour lines are itemized below.

1. Contour lines must never meet or cross (except in rare cases, such as a vertical or overhanging cliff).
2. Contour lines must eventually close on themselves. In some cases this will occur on the drawing, but it often occurs outside the area of interest and thus does not show on the drawing.
3. Contour lines must be one continuous line and should never fork or divide.
4. Closely spaced contour lines indicate steep slopes; widely spaced ones indicate flatter slopes.
5. Contour lines form more or less U (or V) shapes when crossing streams, with the U's (or V's) pointing upstream.
6. Contour lines form more or less U (or V) shapes when crossing ridge lines, with the U's (or V's) pointing downhill.
7. Contour lines are normally drawn freehand, and every fifth one may be made a little darker for easier reading.

10-5 Mapping

Figures 10-3, 10-7, and 10-10 have illustrated the plotting of the traverse, plotting of details, and drawing of contour lines, respectively, for a particular problem. In practice these would probably be done all on the same drawing in the form of a rough draft. The final drawing is then traced from the rough draft onto tracing paper. Normally, only the contour lines (not the points and elevations used to draw the contour lines) are shown on the final drawing.

It should go without saying that the appearance of the final drawing is of considerable importance. A well-planned, well-arranged, neat drawing tends to inspire the user's confidence in the map; a poorly planned, poorly arranged, sloppy one does not. To begin with, an appropriate scale must be chosen to produce a map that is not too large but is large enough to indicate the necessary details. The map should be oriented so that the traverse is balanced on the map. If possible the map should be oriented so that north is straight upward (i.e., toward the top of the map). A meridian arrow should be placed on every map, preferably near the top of the map. The meridian may show magnetic north or true north, or both.

One of the most important features of a map is the lettering on it. Good lettering is necessary for a good overall appearance. Obviously, some people are better than others at lettering, but everyone can use guide lines and exercise care to do the best lettering he is capable of doing. The use of lettering sets or press-on letters gives a nice appearance, although it may take longer to use them.

Every map should have a title. Often the title is placed in a block in the lower right corner. Information in the title box should include at least an overall title, name of the project (if applicable), name of owner or client, the date, the scale, and the name of the person preparing the map. Additional material may be included as needed.

There are three ways of indicating the scale of a drawing. One is by an indication such as 1 in. = 100 ft. Another is by an indication such as $\frac{1}{50,000}$. This means that 1 unit measured on the map is equal to 50,000 units on the ground. For example, 1 in. on the map is equal to 50,000 in. on the ground, or 1 cm on the map is equal to 50,000 cm on the ground. A third method is to use a graphical scale, such as that shown in Figure 10-11. A graphical scale has the advantage that distances on the map may be determined (by scaling) even if the map is reproduced photographically at a size different from that at which it was drawn. Many maps use one of the first two indications for scale together with the graphical scale.

The plottings in Figures 10-3, 10-7, and 10-10 have been combined and title information, meridian arrow, etc., added to form the final product shown in Figure 10-12. For illustrative purposes, this problem has been quite simple with a simple traverse and few details and contour lines shown.

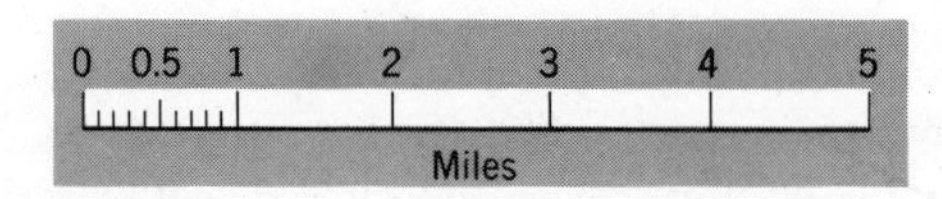

Figure 10-11 Graphical scale.

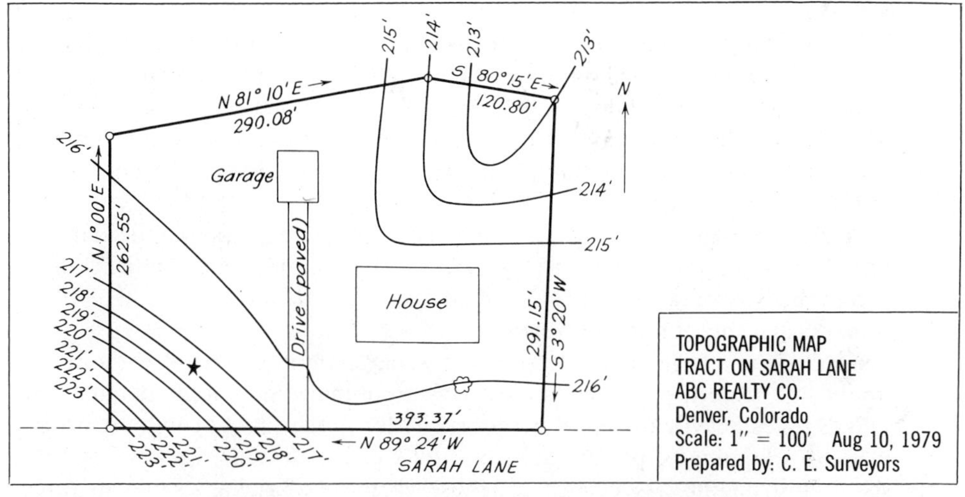

Figure 10-12 Combined plotting of Figures 10-3, 10-7, and 10-10.

10-6 Topographic Details by Plane Table and Aerial Photography

The methods presented in this chapter for collection of data for, and preparation of, a topo map are commonly used. There are, however, other methods for accomplishing these tasks. One method is the use of a plane table and alidade, and another is the use of aerial photography.

A plane table is, in effect, a drawing board mounted on a tripod. An alidade is similar in function to a transit. The telescope is mounted on a flat base plate and rests directly on the plane table. The base table has a straight edge and scale for plotting locations parallel to the line of sight.

In using the plane table, drawing paper is mounted on the drawing board. As measurements are made in the field, instead of recording them in the field book, they are drawn directly on the paper mounted on the drawing board. For example, if the plane table is set up over one point of known elevation for the purpose of determining ground point elevations, one would set the alidade over the plotted position of the instrument and sight on the rod held at a specific ground point. Based on the stadia interval read, the vertical and horizontal distances can be computed. The location of the point can be plotted by drawing a line adjacent to the base plate in the direction of the telescope and scaling the distance as determined by stadia. The elevation of the point determined by stadia can be written on the map next to the plotted point. When all points of interest have been mapped, the surveyor may draw the contour lines based on the plotted points while standing there looking at the lay of the land.

Use of the plane table has the obvious advantage that the map (draft) can be made in the field while observing the area. Because of this and because few if any notes need be taken in a field book, a good, representative map should result. It has the disadvantages that more field time is necessary, it is not well suited to wooded and/or rugged areas, and a larger number of equipment items must be carried.

Aerial photography refers to a means of obtaining topographic data by taking photographs from an aircraft. This is a very sophisticated subject, and a comprehensive treatment is beyond the scope of this book. A few general remarks are appropriate, however.

For large and rugged, inaccessible areas, aerial photography is, without any doubt, the fastest, most economical way to obtain topographic details. The USGS relies heavily on aerial photography for preparation of topo maps. By using overlapping of photographs and by viewing through a special stereoscopic instrument, a specialist can observe the underlying terrain in three dimensions. Special stereoscopic plotters are available for making topo maps.

For additional information on both the plane table and aerial photography, the reader is referred to the references listed at the end of this book.

10-7 Problems

10-1 Obtain the USGS topo map for the area where you live. Some of these maps show individual houses. Can you locate your house? Which way does runoff from rainfall drain in the area where you live? If possible, determine the slope of the street in front of your house. At what elevation is your house above mean sea level?

10-2 Plot the (adjusted) traverse of Problem 8-4 by each of the methods listed below. Keep account of the time required by each method. Is there any correlation between time required and attainable accuracy by each method?

- **(a)** coordinate method
- **(b)** tangent method
- **(c)** chord method
- **(d)** protractor

10-3 Repeat Problem 10-2 for the traverse of Problem 8-5.

10-4 Repeat Problem 10-2 for the traverse of Problem 8-6.

In each of Problems 10-5 through 10-9, trace the figure and draw contour lines as indicated. Be sure to label each contour line.

10-5 Figure 10-13. Use 2-ft contour interval.

10-6 Figure 10-14. Use 1-m contour interval.

10-7 Figure 10-15. Use 10-ft contour interval.

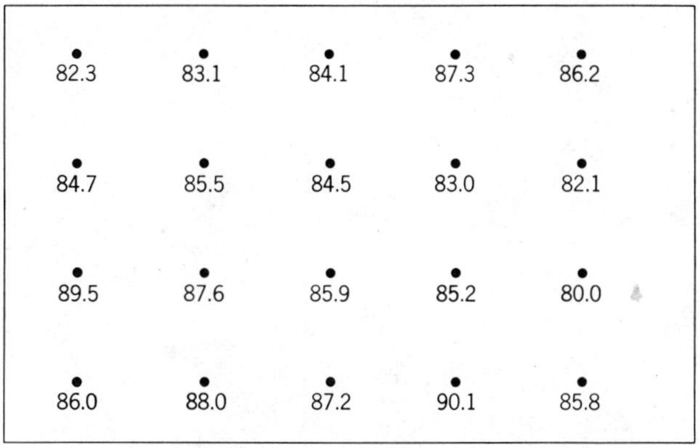

82.3 83.1 84.1 87.3 86.2

84.7 85.5 84.5 83.0 82.1

89.5 87.6 85.9 85.2 80.0

86.0 88.0 87.2 90.1 85.8

Figure 10-13

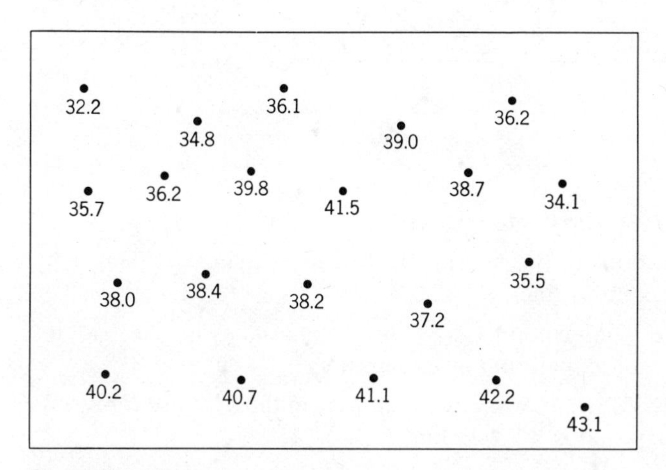

Figure 10-14

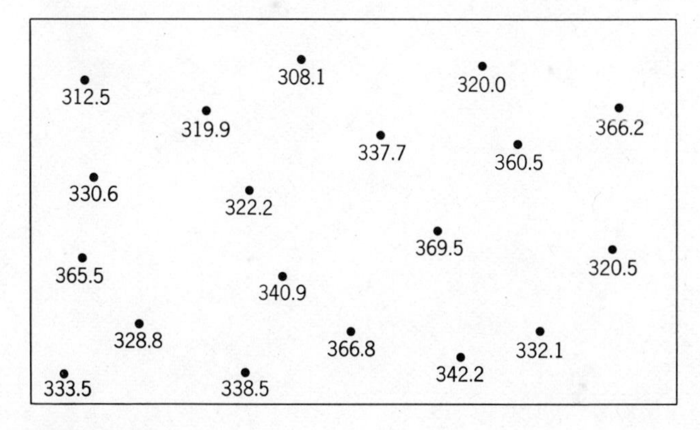

Figure 10-15

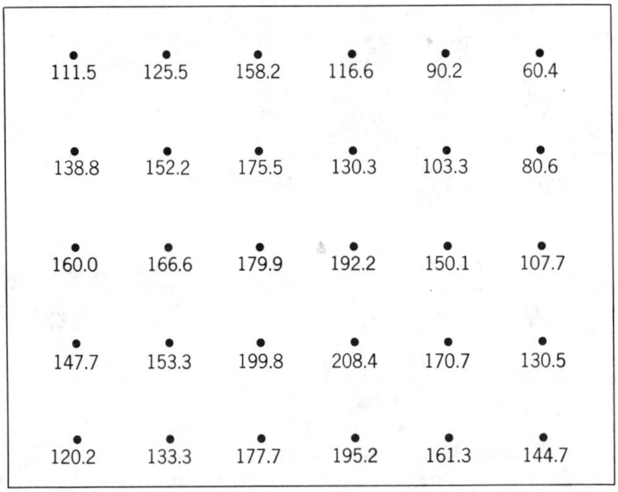

111.5
125.5
158.2
116.6
90.2
60.4
138.8
152.2
175.5
130.3
103.3
80.6
160.0
166.6
179.9
192.2
150.1
107.7
147.7
153.3
199.8
208.4
170.7
130.5
120.2
133.3
177.7
195.2
161.3
144.7

Figure 10-16

10-8 Figure 10-16. Use 10-m contour interval.

10-9 Figure 10-17. Use 5-m contour interval.

10-10 Why are elevations to the nearest 0.1 ft (rather than 0.01 ft) usually adequate for drawing contour lines?

10-11 If a map scale is indicated as $\frac{1}{50,000}$, how many miles in the field are represented by a line on the map that is 4.32 in. long?

10-12 If a map scale is $\frac{1}{50,000}$, how many kilometers in the field are represented by a line on the map that is 7.52 cm long?

35.5
48.8
66.6
73.3
42.2
50.6
72.2
65.0
38.8
62.1
69.9
78.8
89.2
27.5
71.2
82.2
29.2

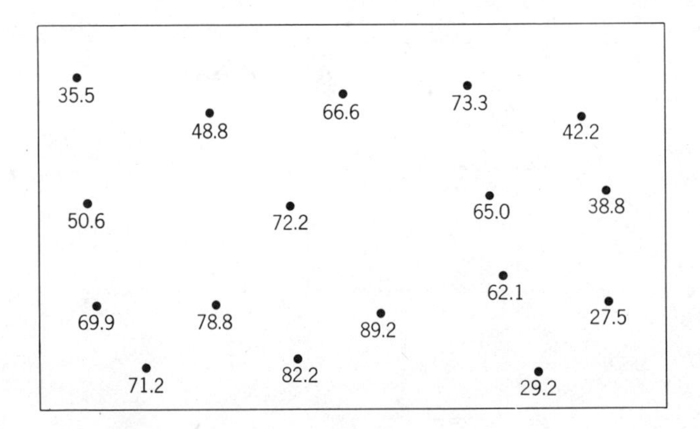

Figure 10-17

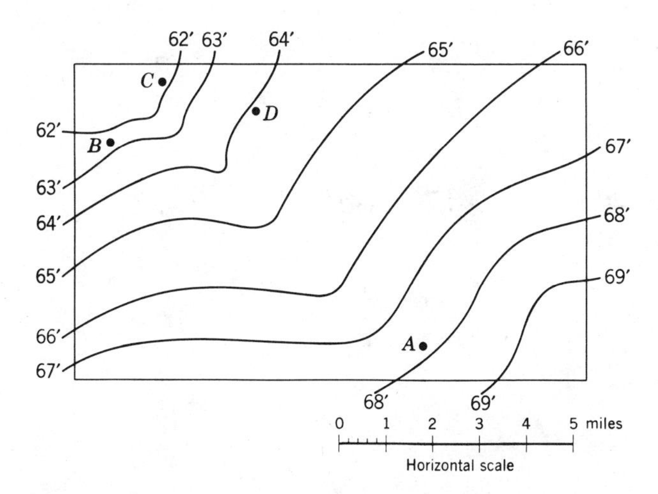

Figure 10-18

10-13 Repeat Problem 10-11 if the scale is $\frac{1}{62,500}$.

10-14 Repeat Problem 10-12 if the scale is $\frac{1}{62,500}$.

10-15 For the contour lines shown in Figure 10-18, find the average slope of the land between points *A* and *B*.

10-16 Repeat Problem 10-15 for points *C* and *D*.

10-17 Make a perspective drawing (sketch) to show why contour lines point upstream when crossing a creek or ditch.

10-18 On a map of scale 1 in. = 200 ft, how far apart would 1 ft contour lines be drawn to represent a uniform slope of 2.6 percent?

10-19 For an area selected by your instructor, collect data for a closed traverse, location of details, and ground elevations for drawing contour lines. Record these data in good form in your field book. Using these data prepare a good topo map for the area.

CONSTRUCTION SURVEYING

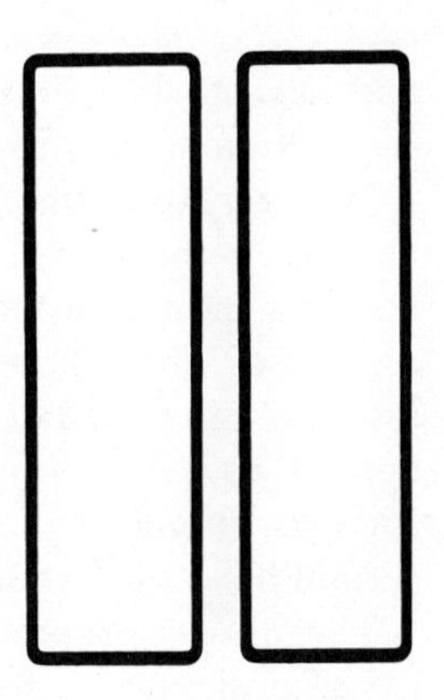

A great deal of surveying is done in conjunction with construction projects of various types. Generally speaking, it is the responsibility of the surveyor to see that all structures are located in the correct positions—both horizontally and vertically—in accordance with the plans prepared by the engineer and/or architect.

Construction surveying involves a number of varied surveying operations. It would not be possible to cover in this book every conceivable surveying problem which might arise. Instead a few of the more common and important ones are presented briefly in this chapter.

The reader will note that there are few if any new surveying techniques presented in this chapter. It is mainly application of general surveying techniques (measurement of horizontal and vertical distances, angles, etc.) to specific problems—problems that happen to arise in conjunction with construction.

11-1 Laying Out a Building

One important facet of construction surveying is that of laying out a building. There are several phases involved in this. First, the building must be located at the correct location on the property. Second, the building must be laid out with the correct dimensions. (Often building components are prefabricated and have to fit.) Finally, grade elevations must be determined so that the foundation of the proposed structure will be level.

With regard to the first requirement mentioned above (the building must be located at the correct location), perhaps the first consideration is that the building should be located on the correct piece of property. The author has seen cases where houses were built on property lines and even on the wrong lot. This can raise headaches for the owner as well as legal problems. (If you build a house on someone else's property, who owns the house?) Thus the surveyor should make certain he is on the correct property and should make certain he knows where the property corners and lines are.

The next consideration is to position the building properly on the land. Local ordinances often specify minimum distances between building and property lines. Such regulations must be known and adhered to. Usually the exact location of the building on the property is indicated in the building plans, and the surveyor's job is to make the proper measurements to lay out the building according to the plans.

In orienting the building, a wooden stake with tack would likely be driven into the ground to mark each corner of the structure. Such stakes would be temporary as they would of necessity be removed during the digging and construction of the foundation. In order to provide more permanent marking of corners during construction of the foundation, "batter boards" are placed near each corner.

Batter boards consist of three or four stakes driven into the ground and two boards placed horizontally at some known elevation and nailed to the driven stakes. The horizontal boards are set level and the vertical distance to grade level specified. Batter boards are generally set a couple feet off the line in each direction at each corner. Nails are driven (or notches cut) at proper

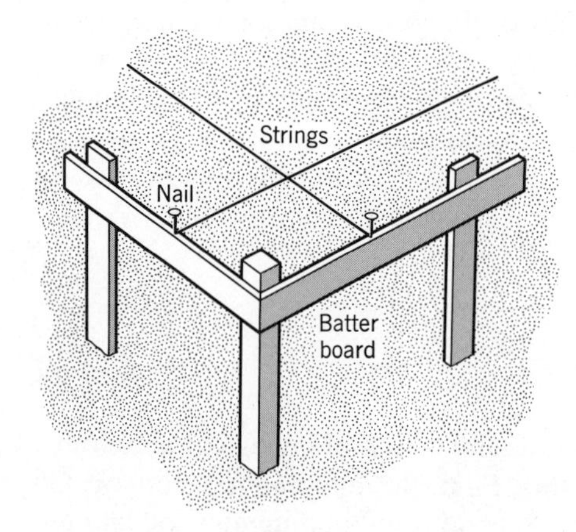

Figure 11-1

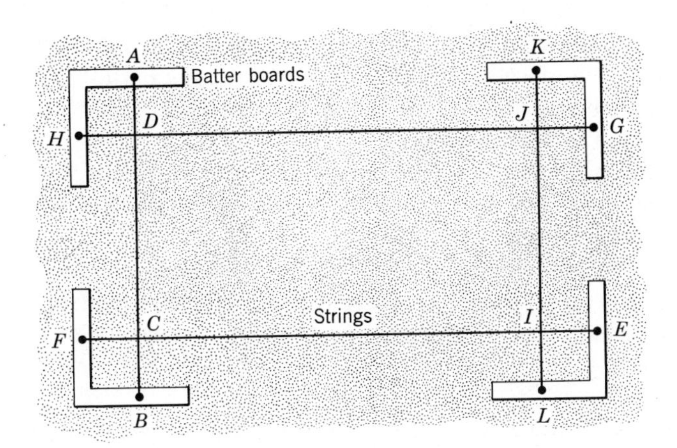

Figure 11-2

locations on the horizontal boards so that strings can be attached at the nails (or notches) and stretched from one corner to another. A corner is defined where two strings cross. The concept of batter boards may be better understood by referring to Figure 11-1 and Figure 11-2. Figure 11-1 shows the batter boards (in perspective) at one corner. Figure 11-2 shows batter boards (in plan view) used to define the four corners of a simple building.

Batter boards should be established firmly in the ground and braced, if necessary, to prevent any movement. The strings may be taken down and replaced as necessary during construction. For example, the strings might be needed to determine (by measuring down from the string) how much cut is needed for the foundation trench. The strings would be removed to allow a person or machine to dig in the trench. Later they could be replaced for use by the brick mason.

The procedure for setting batter boards would be as follows:

1. Locate each corner of the building on the ground by temporary stake and tack point.
2. Drive batter board stakes at each corner and nail horizontal boards to batter board stakes at the desired elevation. They can be set at the desired elevation by using a level and level rod.
3. Place the nails for one string (e.g., points A and B in Figure 11-2) defining one side of the building, so that the string passes directly over the desired corners (points C and D in Figure 11-2). Stretch the string tightly between A and B.
4. Set the transit up over the corner at C, sight on A, turn a right angle, and place a nail at E so that ACE is a right angle.
5. Stretch a string tightly from E maintaining it directly under the transit (i.e., over corner at C) and locate point F on the batter board. Drive a

nail at *F* and attach the string at *F*. The string can be stretched directly under the transit (over corner at *C*) by moving the string until it just touches the plumb bob string attached to the transit. The intersection of strings *EF* and *AB* defines corner *C*.

6. Lay off the building dimension *CD* to establish corner *D* exactly.
7. Move the transit to *D*, sight on *B*, turn a right angle, and place a nail at *G* so that *GDB* is a right angle.
8. Locate point *H* by stretching a string from *G* through *D* (same procedure as step 5 above), and place a nail at *H*. Attach string from *G* to *H*. The intersection of strings *AB* and *GH* defines corner *D*.
9. Lay off the building dimensions *CI* and *DJ* to establish corners *I* and *J*.
10. Locate two points *K* and *L* so that a string stretched between them crosses the corners *I* and *J* exactly. Place nails at *K* and *L* and stretch string. The intersection of strings *KL* and *FE* defines corner *I*, and that of *KL* and *HG* defines corner *J*.

If these steps have been done perfectly, the building should be properly laid out. One should certainly check the measurement of *IJ* to make sure it is correct, and it is also good practice to measure the diagonals to see if they are correct.

The preceding discussion has assumed a very simple building with right angle corners and with the same elevation throughout. Certainly many buildings will be more complicated than this—some of them very much so. Nevertheless, the general procedure described above gives the fundamental steps required.

11-2 Setting Grade Stakes

In some types of construction requiring excavation and filling of earth, grade stakes are used to indicate to the contractor how much cut or fill is required to obtain the elevation desired at a specific location. Sometimes grade stakes may be driven into the ground until the top of the stake is at the elevation required. If this is done, the top of the stake is commonly colored blue with a lumber crayon (keel). If the required elevation is far below or above the existing ground elevation, it is not feasible to drive the stake so that the top of the stake is at the elevation required. (If the required elevation is far below the existing ground elevation, the stake would have to be driven into the ground and out of sight. If it is far above, the stake would have to extend upward very high and would be unwieldy.) In such cases the stakes may be driven to a convenient height above the ground and the required cut or fill needed to obtain the required elevation may be written directly on the side of the stake. It will make matters simpler for the contractor if the stakes are placed at a height such that the cut or fill is a whole number.

The manner in which grade stakes are placed is not too difficult. A transit may be used for horizontal alignment. A level would be used to obtain elevation. The surveyor would know or would compute the desired elevation and subtract it from the height of instrument to determine what rod reading is required. The stake would then be driven (more or less by trial and error) until the required rod reading is obtained with the rod resting on the stake. In the case of the grade stake set with an indicated cut or fill written on it, the rod reading would have to be adjusted by the amount of cut or fill.

Consider the case of determining grade for laying a sewer line. The liquid normally flows by gravity, and the proper grade (slope) must be maintained carefully. The person digging (or operating a piece of digging machinery) must know how deep a trench to dig at points along the trench. The person actually laying the pipe sections in the dug trench must know the exact elevations at which the ends of the pipe section should be laid.

This information can be furnished by the surveyor by setting grade stakes at equal intervals (commonly 50 ft) along the line of the sewer. These would be set off line by several feet so they would not be dug up during excavation. The station (distance from starting point), offset distance, and amount of cut are written on the stakes. The person digging the trench can measure down from the grade stake to determine when he has dug the trench deep enough.

The persons laying the pipe sections must be able to measure down from the grade stake very accurately. They must also be able to maintain accurate horizontal alignment. This can be accomplished by setting a batter board across the trench at each stake. Each batter board should be set at the same height above the final elevation of the pipeline. These batter board heights can be determined at each point by noting the cut on the grade stake and measuring up (or down) as necessary. Once the batter boards have been set up at each station and at the proper elevations, a transit can be used to delineate horizontal alignment by driving a nail into the batter board at the proper location. By stretching a string tightly between the nails in the batter boards, the exact location of the sewer line is determined. The persons laying the pipe simply measure the proper amount vertically downward from the stringline to determine the point to which the pipe should be laid.

An alternative method is to set the transit up on a platform across the trench. By setting the telescope at the desired grade and sighting directly along the line of the sewer, the exact location and elevation of a pipe section can be determined by maneuvering the level rod until it is on line at the required rod reading. This method has the advantage that no batter boards are required. It has the disadvantage that the presence of the surveyor is required throughout the excavation and pipe laying process. If batter boards are present, the contractor can work from them without having the surveyor present.

A relatively new method of maintaining alignment and grade in laying pipelines is by means of a laser. The laser device may eventually replace the other methods completely.

11-3 Setting Slope Stakes

In staking out a highway, stakes are set on the center line at 50-ft or 100-ft intervals and at other critical points (beginning and end of curves, for example). In addition the surveyor must set stakes to each side of the roadway to indicate to the contractor how the roadway is to be shaped at that location. These stakes are called slope stakes.

The shape of the cross section of the final roadway generally takes one of the forms shown in Figure 11-3. In Figure 11-3*a* the roadway must be completely filled in; in Figure 11-3*b* the roadway must be completely cut out; in Figure 11-3*c* the roadway is partly filled in and partly cut. Each profile consists of a more or less flat roadway with sides shaped according to a given slope until they intersect the original ground surface. A side slope of $1\frac{1}{2}$ horizontally to 1 vertically is common, although it may be flatter or steeper (depending on the type of soil).

The job of the surveyor is to set slope stakes to each side of the center line at the points where the side slopes intersect the original ground surface. The slope stakes are placed with the flat side facing the center line and with the stake leaning away from the center line to indicate fill and toward the center line to indicate cut. The station number may be written on the outside of the stake, and the amount of cut or fill on the inside. Knowing the width and (final) elevation of the roadway, the side slopes, and the location of the slope stakes, the contractor can excavate or fill in as needed in order to shape the desired roadway. It should be noted that there is no actual cut or fill at the slope stake. The cut or fill indicated on the slope stake is the difference between ground elevation at the slope stake and the (final) elevation of the roadway.

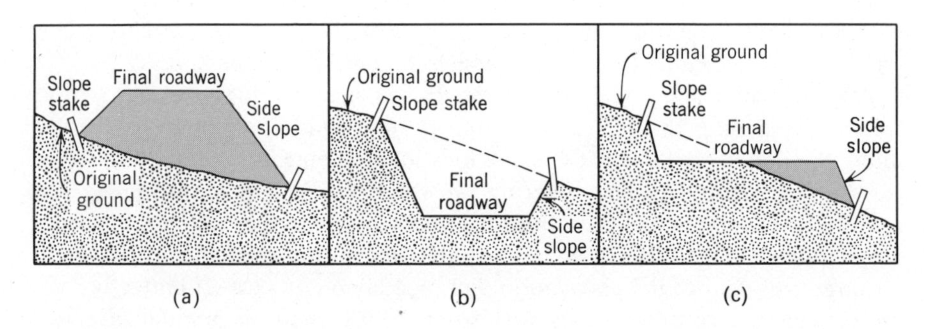

Figure 11-3

In setting slope stakes a level, level rod, and tape may be used to measure the required vertical and horizontal distances. In dealing with cuts and fills of earth, measurement of distances to the nearest 0.1 ft is usually adequate. Therefore, in order to speed up the field work, a 50-ft cloth tape and a hand level may be used.

If the original ground surface is a horizontal line, the location of the slope stake is rather easily determined by geometry. If the original ground surface is not horizontal, a trial and error solution is required. For example, assume in Figure 11-4 that it is desired to set the slope stake to the right side. The instrumentman might estimate the fill to be 7.0 ft. With a fill of 7.0 ft, the distance out from the center is computed as $15+1\frac{1}{2}(7.0)$, or 25.5 ft (i.e., half the roadway width plus the slope times the fill). The rodman is sent to a point 25.5 ft out, a rod reading made, and the fill computed. If this computed fill is equal to the fill estimated initially, then this is the correct location for the slope stake. Suppose, however, this computed fill is 9.2 ft. Obviously this is not the correct point, so the procedure is repeated for another estimated fill. A second estimate of fill might be 10.0 ft. The distance out would be $15+1\frac{1}{2}(10.0)$, or 30.0 ft. The rodman is sent this distance out, and a rod reading indicates a fill of 9.9. ft. This is close enough, and the slope stake could be driven here. Had the computed fill not been equal to (or close enough to) the estimated one, another trial would be necessary. Usually the correct point can be located in two or three attempts.

11-4 Computation of Volume

Sometimes the surveyor will be called on to compute volumes. In highway construction, for example, large amounts of earth must be moved, and the contractor is often paid on the basis of the amount of earth moved. Also, if earth is taken from private property, the land owner would likely be paid on

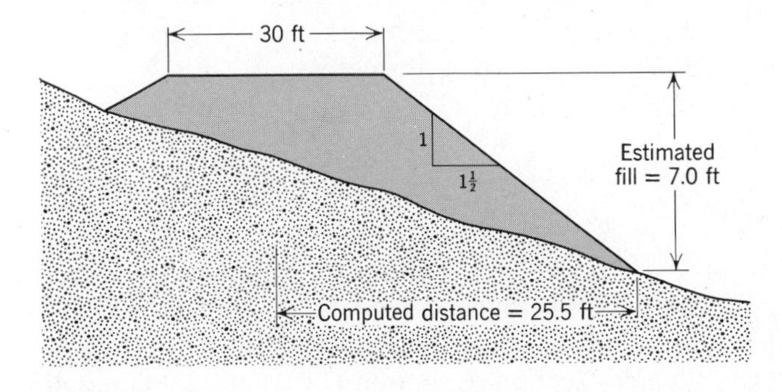

Figure 11-4

the basis of the amount of earth taken. Two examples of volume computation are given in this section—volume of cut or fill in a roadway between adjacent cross sections and volume taken from borrow pits.

The volume of cut or fill in a roadway between adjacent sections is most easily computed by multiplying the average of the end areas by the distance between cross sections. In equation form

$$V=\left(\frac{A_1+A_2}{2}\right)L \qquad (11\text{-}1)$$

where V = volume
A_1 and A_2 = area at either end cross section
L = distance between cross sections

In Eq. (11-1), if L is in feet and A_1 and A_2 are in square feet, the computed volume will be in cubic feet. As indicated in Chapter 1, volumes of earthwork are usually expressed in cubic yards. Since there are 27 cu ft in 1 cu yd, Eq. (11-1) may be modified to give the volume in cubic yards as follows:

$$V=\frac{L(A_1+A_2)}{54} \qquad (11\text{-}2)$$

Equation (11-2) gives volume in cubic yards if A_1 and A_2 are in square feet and L is in feet.

In applying Eqs. (11-1) or (11-2) it is, of course, necessary to determine the cross section area at each end. These areas can be determined mathematically based on the geometry of the cross section, or they may be determined by planimeter (Chapter 9).

Equations (11-1) and (11-2) are approximate because they are based on the invalid assumption that the area varies linearly from one end to the other. They will generally give a volume slightly larger than the volume of a true prismoid. Nevertheless, these formulas are often used because of their simplicity, and because, when dealing with earthwork, it is not likely that volumes between cross sections will be perfect prismoids.

If greater accuracy is desired, the prismoidal formula may be used.

$$V=\frac{L(A_1+4A_m+A_2)}{162} \qquad (11\text{-}3)$$

A_m is the cross section area of a section midway between A_1 and A_2. It is not the average of A_1 and A_2, but it may be computed based on average values of width and depth at each end cross section. Equation (11-3) gives volume in cubic yards if A_1, A_2, and A_m are in square feet and L is in feet.

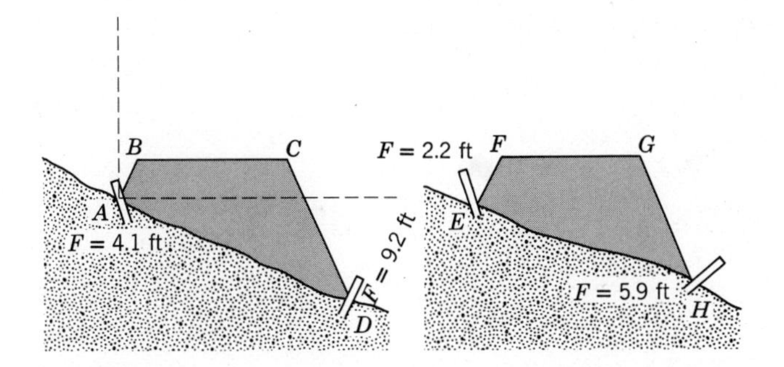

Figure 11-5

Example 11-1

The cross sections shown in Figure 11-5 are for the ends of a 50 ft segment of proposed roadway. Calculate the volume of fill required for this segment. The road surface is 30 ft wide and slopes are 2 to 1.

Solution

The area of each cross section must be computed first. This may be done by several methods, but the coordinate method (Chapter 9) will be used here. If the origin of a coordinate system is assumed at point *A* (shown by dashed line), the coordinates of point *A* will be (0, 0). The indicated fill of 4.1 at the slope stake at point *A* is the vertical distance from point *A* to the roadway surface *BC*. In moving from point *A* to point *B* at a 2 to 1 slope, one would move 4.1 ft vertically and 8.2 ft horizontally. Thus the coordinates of point *B* are (8.2, 4.1). Assuming a level road surface, the coordinates of point *C* are (38.2, 4.1), since the width of the road surface is 30 ft. In moving from point *C* to point *D*, at a 2 to 1 slope, one would move 9.2 ft vertically and 18.4 ft horizontally. Thus the coordinates of *D* are (56.6, −5.1). The area of *ABCDA* can now be computed by the coordinate method (Section 9-3).

$$\frac{0}{0} \quad \frac{8.2}{4.1} \quad \frac{38.2}{4.1} \quad \frac{56.6}{-5.1} \quad \frac{0}{0}$$

The sum of the products of adjacent diagonals down to the right is

$$0(4.1)+8.2(4.1)+38.2(-5.1)+56.6(0)=-161.2$$

The sum of the products of adjacent diagonals up to the right is

$$0(8.2)+4.1(38.2)+4.1(56.6)+(-5.1)(0)=388.7$$

The area is half the difference between these two sums.

$$A = \frac{388.7 - (-161.2)}{2}$$

$$A = 275 \text{ sq ft}$$

The area of the other cross section is computed in the same manner.

$$\frac{0}{0} \quad \frac{4.4}{2.2} \quad \frac{34.4}{2.2} \quad \frac{46.2}{-3.7} \quad \frac{0}{0}$$

$$0(2.2) + 4.4(2.2) + 34.4(-3.7) + 46.2(0) = -117.6$$

$$0(4.4) + 2.2(34.4) + 2.2(46.2) + (-3.7)(0) = 177.3$$

$$A = \frac{177.3 - (-117.6)}{2}$$

$$A = 147 \text{ sq ft}$$

The (approximate) volume is computed from Eq. (11-2).

$$V = \frac{L(A_1 + A_2)}{54} \qquad (11\text{-}2)$$

$$V = \frac{50(275 + 147)}{54}$$

$$V = 391 \text{ cu yd}$$

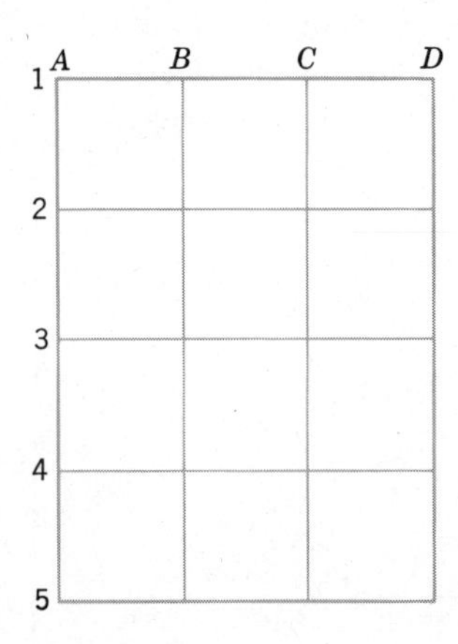

Figure 11-6

Sometimes in construction work it is necessary to obtain earth. Often it is less expensive to obtain earth from nearby property than to haul it from distant locations. The area from which earth is removed is known as a "borrow pit," and the surveyor's job is to determine the volume of earth removed.

The usual method for determining the volume of earth removed from a borrow pit is to lay off a grid, such as that shown in Figure 11-6. The grid can be lettered and numbered as shown in order to refer to each grid point. Levels can be run before any earth is removed and then again after earth has been removed. By subtracting the final elevation at a point from the initial elevation at the same point, the depth of earth removed at the point can be computed. If the depth of earth removed at each of the four corners of a square is known, the approximate volume of earth removed from that square can be computed by multiplying the average of the four depths by the area of the square. Thus in Figure 11-6, if each square is 50 ft by 50 ft and the depths of cut at $A1$, $B1$, $A2$, and $B2$ are 4.0 ft, 3.5 ft, 6.2 ft, and 5.8 ft, respectively, the volume of earth removed would be

$$\left(\frac{4.0+3.5+6.2+5.8}{4}\right)(50\times 50)=12{,}200 \text{ cu ft, or } 451 \text{ cu yd}$$

The total volume can be determined by computing the volume for each square and adding them together. The procedure can be simplified somewhat by multiplying the cut at each corner by the number of squares containing that particular corner, adding these products, dividing the sum by 4, and multiplying the result by the area of a square.

Example 11-2

Calculate the volume of earth removed from the borrow pit shown in Figure 11-6 if each square is 50 ft by 50 ft and the cuts at each corner are as shown in the second column below.

Solution

Point	*Cut*	n^*	$n \times$ cut
$A1$	4.0	1	4.0
$A2$	6.2	2	12.4
$A3$	5.7	2	11.4
$A4$	5.9	2	11.8
$A5$	8.3	1	8.3
$B1$	3.5	2	7.0
$B2$	5.8	4	23.2
$B3$	6.1	4	24.4
$B4$	6.9	4	27.6

$B5$	5.0	2	10.0
$C1$	3.9	2	7.8
$C2$	4.8	4	19.2
$C3$	5.5	4	22.0
$C4$	6.1	4	24.4
$C5$	6.2	2	12.4
$D1$	6.9	1	6.9
$D2$	7.2	2	14.4
$D3$	7.4	2	14.8
$D4$	6.8	2	13.6
$D5$	6.2	1	6.2
			281.8

*n is the number of squares containing that particular corner.

$$V = \frac{281.8}{4}(50 \times 50)$$

$V = 176{,}000$ cu ft, or 6520 cu yd

In Example 11-2 the cuts at the four corners of the overall figure ($A1$, $A5$, $D1$, and $D5$) are multiplied by 1, since these points appear in only one individual square. The cuts at the remaining points on the perimeter of the overall figure ($A2$, $A3$, $A4$, $B1$, $B5$, $C1$, $C5$, $D2$, $D3$, and $D4$) are multiplied by 2, since these points appear in two individual squares. The cuts at the remaining points are multiplied by 4, since these points appear in four individual squares.

The grid shown in Figure 11-6 consists of 12 squares. If the borrow pit has an irregular shape, it can be marked off in squares as much as possible with the remaining area marked off in triangles or other shapes. (See Figure 11-7.)

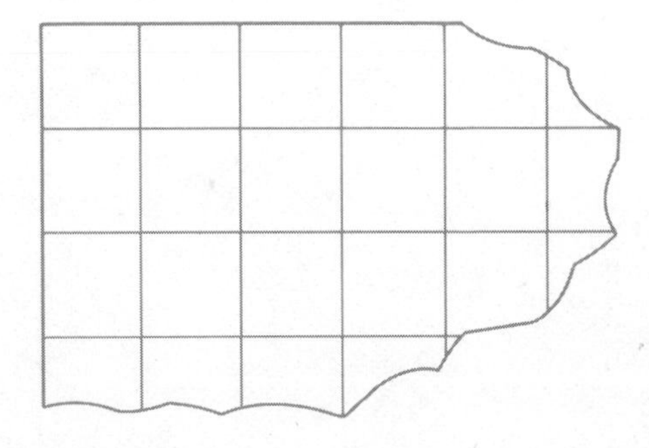

Figure 11-7

The volume of the borrow pit in the area covered by squares can be calculated in the manner described previously. The remaining volume can be calculated as special cases involving the volumes of the nonsquare portions.

Careful vertical and horizontal control must be observed when determining the volume of a borrow pit by the method described. A nearby bench mark can provide the necessary vertical control. Stakes placed to mark grid points will be knocked out during excavation. Hence a means of reproducing grid point locations must be devised by the surveyor. This could be done by referencing two or three of the corner points on the grid to nearby permanent points.

11-5 Horizontal and Vertical Curves

When two straight sections of a transportation route (highway, railroad, etc.) intersect, it is generally desirable to connect them by smooth transition curves in both a horizontal and a vertical plane. Curves in a horizontal plane (horizontal curves) are normally either circular curves or spiral curves. Circular curves are covered briefly in this section. For treatment of spiral curves, the reader is referred to reference 4 at the end of this book. Curves in a vertical plane (vertical curves) are usually parabolic, and the equal tangent parabolic (vertical) curve is covered in this section.

Before discussing the (horizontal) circular curve in detail, it is necessary to define the term "degree of curve." Actually, there are two definitions of the degree of curve—the chord definition and the arc definition. If the chord definition is used, the degree of curve is the angle at the center of a circular arc that is subtended by *a chord* of 100 ft. This is shown in Figure 11-8*a*. It is evident that the degree of curve D (chord definition) is related to the radius of the curve R by the formula

$$R = \frac{50}{\sin D/2} \tag{11-4}$$

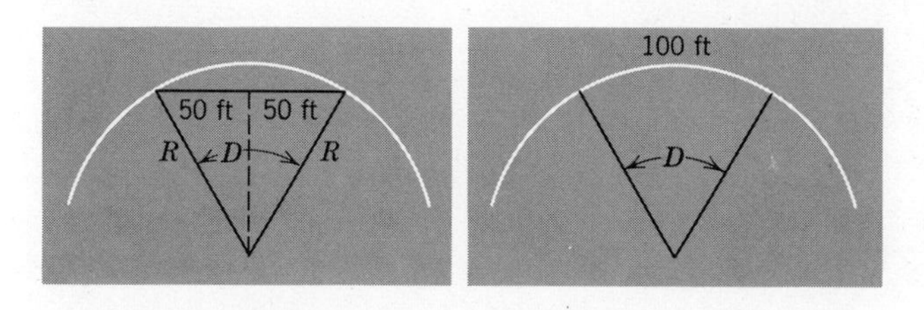

Figure 11-8 Degree of curve. (*a*) Chord definition. (*b*) Arc definition.

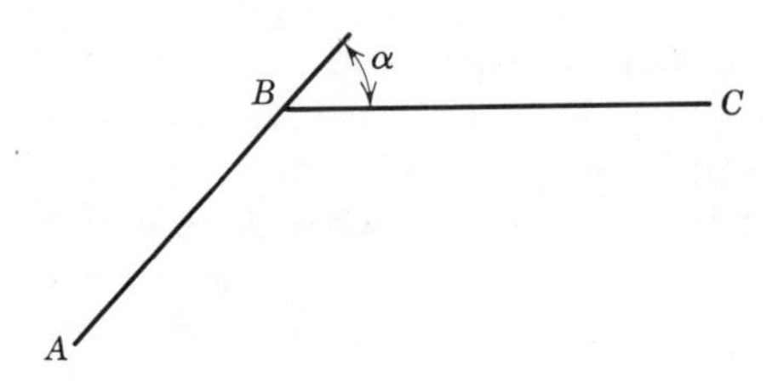

Figure 11-9

If the arc definition is used, the degree of curve is the angle at the center of a circular arc that is subtended by *an arc* of 100 ft. This is shown in Figure 11-8*b*. It is evident that the degree of curve (arc definition) is related to the radius of the curve by the formula

$$R = \frac{18{,}000}{\pi D} \tag{11-5}$$

[In both Eqs. (11-4) and (11-5), D should be expressed in degrees.] If the degree of curve is specified for a particular application, the required radius can be computed using either Eq. (11-4) or Eq. (11-5).

To introduce the (horizontal) circular curve, assume that a transportation route is progressing from A to B (Figure 11-9) and at B the route changes direction, makes a deflection angle to the right, and then proceeds in the direction B to C. It is desired to use a circular curve as a transition from AB to BC. The circular curve is shown in Figure 11-10. The deflection angle α

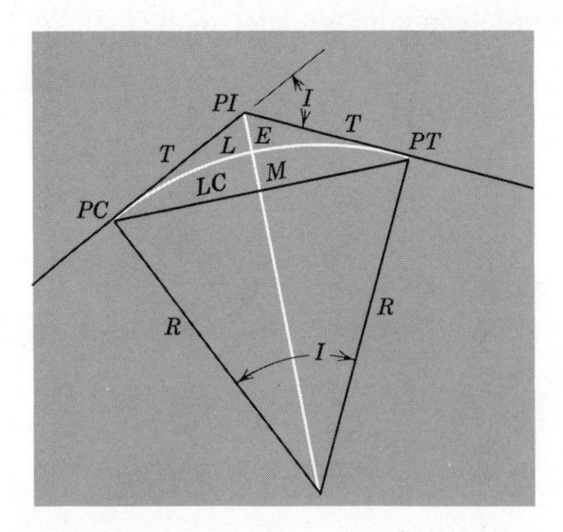

Figure 11-10 Circular curve components

(Figure 11-9) is traditionally designated by I, as shown in Figure 11-10. It should be noted that this deflection angle (I) is equal to the central angle of the circular curve. The point on line AB (Figure 11-9) where the circular curve begins is called the "point of curvature" (PC in Figure 11-10), and the point on line BC (Figure 11-9) where the circular curve ends is called the "point of tangency" (PT in Figure 11-10). Point B (Figure 11-9) is called the "point of intersection" (PI in Figure 11-10).

The distance from the PC to the PI and from the PI to the PT is called the "tangent distance" and is designated by T. The tangent distance T can be determined using the formula

$$T = R \tan \frac{I}{2} \tag{11-6}$$

The length of the curve (arc distance from PC to PT) (L in Figure 11-10) can be determined using the formula

$$L = \frac{\pi R I}{180} \tag{11-7}$$

The length of the straight line from PC to PT, called the "long chord" (LC in Figure 11-10), can be determined using the formula

$$LC = 2R \sin \frac{I}{2} \tag{11-8}$$

Also of interest are the "external distance" (E in Figure 11-10), which is the distance from the PI to the curve measured along the (radial) line from the PI to the center of the circle, and the "middle ordinate" (M in Figure 11-10), which is the distance from the curve to the long chord measured along the same (radial) line from the PI to the center of the circle. These parameters can be determined using the formulas

$$E = R\left(\sec \frac{I}{2} - 1\right) \tag{11-9}$$

$$M = R\left(1 - \cos \frac{I}{2}\right) \tag{11-10}$$

The use of the preceding equations to design a circular curve will be illustrated by an example problem.

Example 11-3

Two segments of a transportation route (AB and BC in Figure 11-9) intersect at station 33+88.75, with a deflection angle of 9°30′ R(angle α in Figure 11-9).

Design a circular curve, assuming a maximum degree of curve of 2°00′ (arc definition).

Solution
Given:

$$PI = 33 + 88.75$$
$$I = 9°30'$$
$$D = 2°00'$$

Use Eq. (11-5) to compute the radius.

$$R = \frac{18{,}000}{\pi D} \tag{11-5}$$

$$R = \frac{18{,}000}{\pi(2.00)}$$

$$R = 2864.79 \text{ ft}$$

Use Eq. (11-6) to compute the tangent distance.

$$T = R \tan\frac{I}{2} \tag{11-6}$$

$$T = 2864.79 \tan\frac{9.50°}{2}$$

$$T = 238.05 \text{ ft}$$

The *PC* can now be determined by establishing a point 238.05 ft from the *PI* along the initial line (i.e., line *AB* in Figure 11-9). The station of the *PC* is (33+88.75)−(2+38.05), or 31+50.70. The *PT* can be determined by establishing a point 238.05 ft from the *PI* along the final line (i.e., line *BC* in Figure 11-9). Its station cannot be determined until the length of the curve is known. The latter value can be determined using Eq. (11-7).

$$L = \frac{\pi R I}{180} \tag{11-7}$$

$$L = \frac{\pi(2864.79)(9.50)}{180}$$

$$L = 475.00 \text{ ft}$$

The station of the *PT* is (31+50.70)+(4+75.00), or 36+25.70. The long chord, external distance, and middle ordinate can be computed using Eqs. (11-8), (11-9), and (11-10).

$$LC = 2R \sin\frac{I}{2} \tag{11-8}$$

$$LC = 2(2864.79)\left(\sin\frac{9.50°}{2}\right)$$

$$LC = 474.46 \text{ ft}$$

$$E = R\left(\sec\frac{I}{2} - 1\right) \tag{11-9}$$

$$E = 2864.79\left(\sec\frac{9.50°}{2} - 1\right)$$

$$E = 9.87 \text{ ft}$$

$$M = R\left(1 - \cos\frac{I}{2}\right) \tag{11-10}$$

$$M = 2864.79\left(1 - \cos\frac{9.50°}{2}\right)$$

$$M = 9.84 \text{ ft}$$

For specific information on laying out points along a circular curve in the field, the reader is referred to reference 1 at the end of this book.

The *vertical curve* is used to effect a smooth transition when two segments of a transportation route at different grades intersect. The procedure for designing an equal tangent parabolic (vertical) curve is as follows, with references made to Figure 11-11.

Normally, the initial and final grades (G_1 and G_2), the station and elevation of the point of intersection (*PI*), and the maximum change in grade per station would be known initially. The total length of the curve can be determined based on the grades and the maximum change in grade per station. For an equal tangent curve, this total length is divided in half; half is subtracted from the station of the *PI* to locate the beginning of the curve (*PC*) back along the initial grade, and half is added to the station of the *PI* to locate the end of the curve (*PT*) ahead along the final grade.

Since the grades are known, the elevations of the *PC* and *PT* can be computed. The average of the elevations of the *PC* and *PT* gives the elevation of a point below (or above) the *PI*, this point being designated as "*A*"

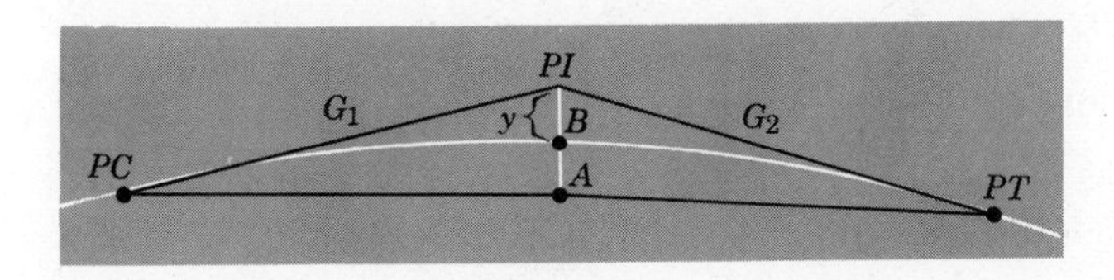

Figure 11-11 Vertical curve components.

in Figure 11-11. The average of the elevations of the *PI* and the point *A* gives the elevation of the midpoint of the vertical curve. This point is designated as "*B*" in Figure 11-11. The "offset" from the *PI* to the vertical curve is designated by "*y*" in Figure 11-11. The offset distance from the tangent at any other point is computed by multiplying the offset at the *PI* by the square of the ratio of (1) the distance from the *PC* (or *PT*) to the point in question to (2) the distance from the *PC* (or *PT*) to the *PI*. Elevations along the curve for any point can be computed by (1) computing the tangent elevation at the point, (2) computing the offset from the tangent as described above, and (3) subtracting (or adding) the offset from the tangent elevation. The location of the summit (or sag) point can be determined by locating the distance along the curve from the *PC* at which the grade is zero.

Example 11-4

On a railroad, a grade of +0.7 percent meets a grade of −0.9 percent at station 69+00 and elevation 303.66 ft. The maximum change in grade per station is 0.2 percent. Design a vertical curve and determine the location and elevation of the summit point.

Solution

The total length of the curve is determined by dividing the total change in grade by the maximum change in grade per station. Since the grade changes from +0.7 to −0.9 percent, the change in grade is 1.6 percent. Thus the length of the curve is

$$\frac{1.6\%}{\dfrac{0.2\%}{\text{sta}}} = 8 \text{ stations, or } 800 \text{ ft}$$

The *PC* is determined by subtracting 400 ft from the *PI*, and the *PT* is determined by adding 400 ft to the *PI*. Thus the *PC* is at station 65+00, and the *PT* is at station 73+00. With a slope of +0.7 percent on the initial grade, the elevation of the *PC* is

$$303.66 - 0.007(400) = 300.86 \text{ ft}$$

With a slope of −0.9 percent on the final grade, the elevation of the *PT* is

$$303.66 - 0.009(400) = 300.06 \text{ ft}$$

In order to determine the elevations of several points along the vertical curve, the elevations of points on the tangents at 100 ft intervals will be determined next in the same manner that the elevations of the *PC* and the *PT* were determined. These elevations are listed in column 2 below.

Station	*Tangent Elevation, ft*	*Offset, ft*	*Curve Elevation, ft*
65+00(*PC*)	300.86	0	300.86
66+00	301.56	−0.10	301.46
67+00	302.26	−0.40	301.86
68+00	302.96	−0.90	302.06
69+00(*PI*)	303.66	−1.60	302.06
70+00	302.76	−0.90	301.86
71+00	301.86	−0.40	301.46
72+00	300.96	−0.10	300.86
73+00(*PT*)	300.06	0	300.06

The next step is to compute the offsets from the tangent elevations for each station listed in the table. The offset at the *PI* (station 69+00) is determined by first averaging the elevations of the *PC* and the *PT* to determine the elevation of point *A* in Figure 11-11 and then averaging the elevations of point *A* and the *PI* to determine the elevation of point *B* in Figure 11-11. The offset at the *PI* is then the difference between the elevations of *B* and the *PI*.

$$\frac{300.86+300.06}{2}=300.46 \text{ ft (elevation of } A)$$

$$\frac{300.46+303.66}{2}=302.06 \text{ ft (elevation of } B)$$

$$303.66-302.06=1.60 \text{ ft (offset at } PI)$$

The offset at station 66+00 is determined by

$$(1.60)\left(\frac{100}{400}\right)^2=0.10 \text{ ft}$$

This is also the offset at station 72+00. (Why?) The offset at stations 67+00 and 71+00 is

$$(1.60)\left(\frac{200}{400}\right)^2=0.40 \text{ ft}$$

The offset at stations 68+00 and 70+00 is

$$(1.60)\left(\frac{300}{400}\right)^2=0.90 \text{ ft}$$

These offsets are listed in column 3 of the table, and each curve elevation is determined by adding algebraicly the offset distances to the tangent elevations. The curve elevations are listed in column 4. In order to locate the summit point, it is necessary to determine where the slope of the curve is zero. Since the initial slope is +0.7 percent and the slope is changing at the rate of 0.2 percent per station, it would take

$$\frac{0.7\%}{\frac{0.2\%}{\text{sta}}} = 3.50 \text{ stations, or } 350 \text{ ft}$$

to get to the point of zero slope (the summit). Thus the summit occurs at station (65+00)+(3+50), or 68+50. The tangent elevation at station 68+50 is

$$300.86 + 0.007(350) = 303.31 \text{ ft}$$

The offset at station 68+50 is

$$(1.60)\left(\frac{350}{400}\right)^2 = 1.22 \text{ ft}$$

Thus the summit elevation is

$$303.31 - 1.22 = 302.09 \text{ ft}$$

This summit elevation of 302.09 ft occurs at station 68+50. Pertinent information pertaining to this curve is illustrated in Figure 11-12.

11-6 Problems

11-1 A surveyor is trying to set a grade stake at elevation 112.23 ft. He set up his level and read a backsight of 9.16 ft on a bench mark of 106.55 ft. What foresight reading should be read on the grade stake?

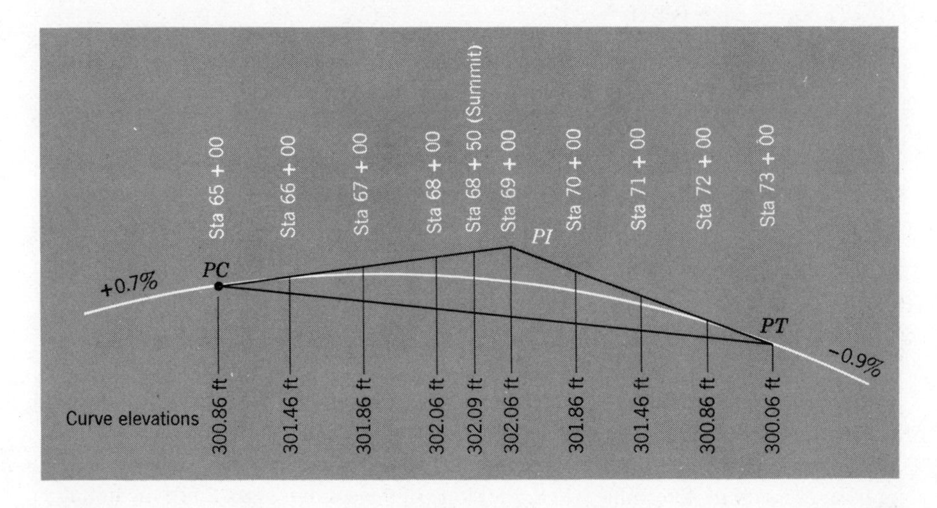

Figure 11-12

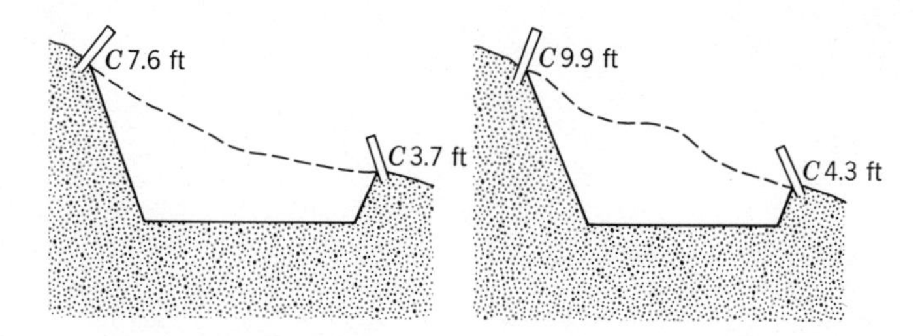

Figure 11-13

11-2 The cross sections shown in Figure 11-13 are for the ends of a 50-ft segment of proposed roadway. Calculate the volume of earth to be removed. The road surface is 32 ft wide and slopes are $1\frac{1}{2}$ to 1.

11-3 The cross sections shown in Figure 11-14 are for the ends of a 20-m segment of proposed roadway. Calculate the volume of fill required. The road surface is 10 m wide and slopes are 1 to 1.

11-4 Calculate the volume of fill required between cross sections 50 ft apart if one cross section is a rectangle 40 ft wide and 8 ft high and the other cross section is a rectangle 30 ft wide and 6 ft high. Use Eq. (11-2). Now find the average cross section at the midpoint of the segment by averaging lengths and widths at the given cross sections. Use Eq. (11-3) to calculate the volume. By what percent do the two computed volumes differ?

11-5 The grid shown in Figure 11-15 was laid off to compute the volume of earth removed from a borrow pit. Each square is 30 ft by 30 ft and the numbers given at each corner are the depths of cut, in feet. Calculate the volume of earth removed.

11-6 Repeat Problem 11-5 for the data shown in Figure 11-16.

11-7 Repeat Problem 11-5 for the data shown in Figure 11-17, except that each square is 10 m by 10 m and cuts are given in meters.

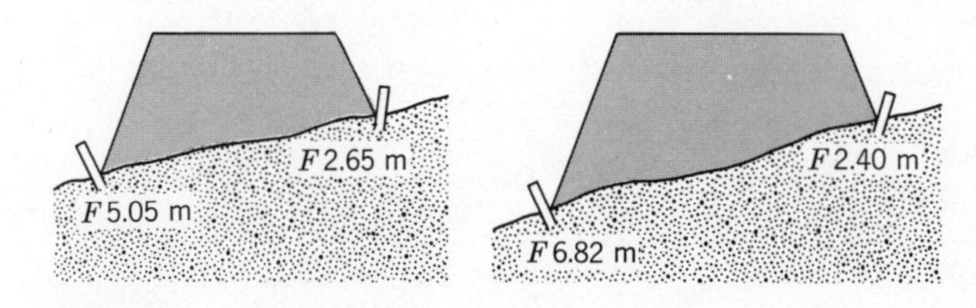

Figure 11-14

2.2	4.6	5.1	3.8	3.0	2.5
3.3	5.9	8.2	2.9	4.0	4.6
2.7	5.5	5.9	5.9	4.1	4.3
3.9	3.8	4.8	4.2	5.1	5.3
1.5	4.5	5.1	5.0	7.1	7.7

Figure 11-15

11.2	10.1	9.5	7.7	6.2	6.0	5.1	5.5
12.0	8.2	8.5	6.2	6.1	5.1	5.2	7.7
10.0	9.5	8.1	7.0	7.7	7.7	4.1	4.0
9.5	8.8	7.9	6.3	6.3	5.7	5.0	5.2
8.0	8.2	7.6	7.0	5.2	5.1	4.2	4.3

Figure 11-16

6.2	6.6	9.9	10.2	10.9	11.1
8.8	9.2	9.6	8.2	5.1	11.6
6.6	8.0	8.2	9.1	9.6	10.2
5.8	5.6	7.7	7.9	6.9	8.2

Figure 11-17

98.5	98.3	99.3	111.4	104.3	100.0
99.2	99.5	100.4	101.3	101.5	102.2
100.7	99.8	101.9	102.9	102.7	102.7
103.0	104.0	103.7	103.9	106.2	107.5

Figure 11-18

11-8 The grid shown in Figure 11-18 was laid out in 50 ft by 50 ft squares. The original ground elevations are shown. What volume of earth will be removed to grade the area to an elevation of 96.2 ft?

11-9 Repeat Problem 11-8 for the data shown in Figure 11-19, except that the final elevation is to be 250.0 ft.

11-10 The grid shown in Figure 11-20 was laid out in 40 ft by 40 ft squares. The original ground elevations are shown. To what elevation must excavation be made in order to remove 7000 cu yd? (The final elevation is to be level throughout.)

11-11 The grade stake for a sewer pipe is at elevation 56.52 ft. It is desired to set grade stakes 52.0 ft, 86.5 ft, and 103.3 ft along the sewer pipe from the given grade stake (in the uphill direction). The slope of the sewer pipe is to be 1.20 percent. If a backsight of 5.25 ft is taken on a bench mark of 55.09 ft, what foresights should be made to set the desired grade stakes?

255.0	258.3	259.6	260.0	258.0	256.6	257.7
253.3	259.9	260.2	261.1	261.1	262.1	259.8
255.1	257.7	261.1	262.2	259.7	258.1	259.0
257.7	259.8	260.5	261.3	262.2	260.0	258.3
255.2	259.9	260.9	263.3	264.4	265.1	261.3

Figure 11-19

650.0	651.1	649.8	649.9	650.2	650.4	650.6
651.1	652.2	649.4	649.8	650.2	650.1	651.0
651.0	651.5	651.6	651.4	650.2	650.0	651.2
651.8	650.8	650.2	649.7	649.8	650.2	650.4
651.0	651.2	650.8	651.5	651.0	651.2	651.1

Figure 11-20

11-12 What vertical angle should be set on a transit if the line of sight is to be used to set grade for a pipeline to be laid at a slope of 2.1 percent?

11-13 It is desired to set a slope stake to the right side of the cross section shown in Figure 11-21. If the estimated fill was 8.0 ft, how far out from the center line should this slope stake be if the 8.0 ft fill is correct?

In each of Problems 11-14 through 11-18, design a circular curve, giving all pertinent data.

	Station of PI	*Deflection Angle*	*Degree of Curve (Arc Definition)*
11-14	36+00	8°15′	2°00′
11-15	125+88.8	9°20′	1°30′
11-16	18+25.57	5°05′	1°30′
11-17	9+05.55	10°00′	2°00′
11-18	26+92.2	7°33′	2°00′

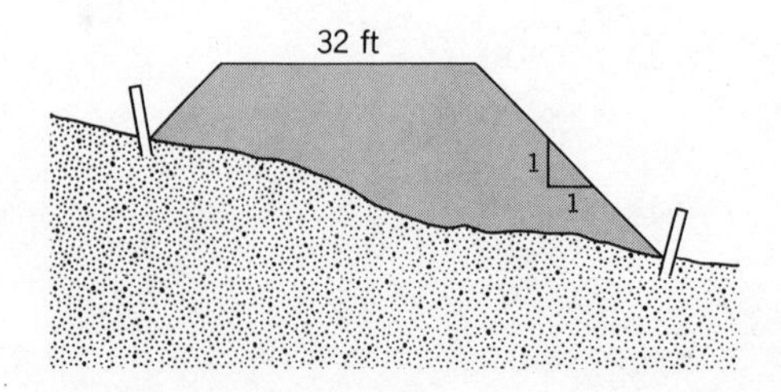

Figure 11-21

In each of Problems 11-19 through 11-23, design a parabolic vertical curve, giving all pertinent data including location and elevation of sag point or summit point. Show all data on a neat sketch.

	Initial Grade (%)	*Final Grade (%)*	*Station of PI*	*Elevation of PI (ft)*	*Maximum Change of Grade per Station*
11-19	−2.5	+2.5	36+75.0	316.00	1.0
11-20	+2.0	−1.0	88+00	400.25	0.5
11-21	+1.5	−2.5	19+25.5	99.25	0.5
11-22	−2.8	+1.2	55+18.2	101.77	0.5
11-23	−4.0	−1.0	66+60.3	209.99	0.5

LAND SURVEYING

One phase of surveying that requires certain knowledge and abilities beyond the elementary principles of surveying is *land surveying*. Land surveying refers generally to the delineation of boundary corners and lines of parcels of land. As will be evident in this chapter, land surveying may on occasion require some knowledge of legal principles and local customs. A good land surveyor must have experience, patience, and good judgment.

In-depth treatment of land surveying is beyond the scope of this book; this chapter gives an overall view of land surveying in general. For more details refer to one of the books on land surveying. The best method (perhaps the only one) to learn how to do land surveying is to work under an experienced land surveyor for a period of time.

12-1 Property Corners

Property corners established in the field should be marked in a more or less permanent manner. Concrete monuments are excellent, but because of their bulk and cost, they are not usually used for routine surveys of property. A common property corner is the metal pipe (or metal rod). These can be driven near, or flush to, the ground; and, barring being knocked out by construction equipment, they are fairly permanent.

In some cases property corners occur at locations at which it would not be feasible to place a metal pipe or rod. Examples would be the center of a road or stream. In such cases the location of the corner may be referenced by setting a witness corner at some distance away from the true corner. Probably the witness corner should be set on one of the boundary lines, as indicated in Figure 12-1. The location of the witness corner with respect to the actual corner should, of course, be determined, recorded in the field book, and noted on the map (plat).

Corner markers other than concrete monuments and metal pipes have been used. Centuries ago a corner might have been marked by a pile of sand or a pile of rocks. Fifty to 100 years ago corners were often marked by rounded wooden stakes, cut from the center of trees. Although wooden stakes will rot in time, these stakes have lasted many years, and sometimes surveyors will locate corners exactly by finding the hole containing the remains of the rotted stake. Other possible corner markers include trees, stones, springs, and the like.

The general locations of corners are sometimes referenced by placing three horizontal slash marks (with ax or machete) on a tree or trees within a 5 or 6 ft radius of the corner. This is done particularly on rural surveys of large areas where corners may be located in the middle of a forest and are therefore difficult to locate. If a tree is itself a corner, it may be marked with an "X" in addition to the three slashes. (Note: These and other descriptions of marks on trees may vary from one location to another, according to local customs.)

12-2 Property Lines

Knowing where one's property corners are is certainly important, but landowners also want to know where their property lines are. There are many reasons for this. As stated in Chapter 11, there is often a minimum distance at which a building may be set from the property line. A person fencing his property does not want the fence to stray off his property.

In relatively small urban lots, the location of property lines is often not too difficult if the location of the corners is known. The lines may be so short that

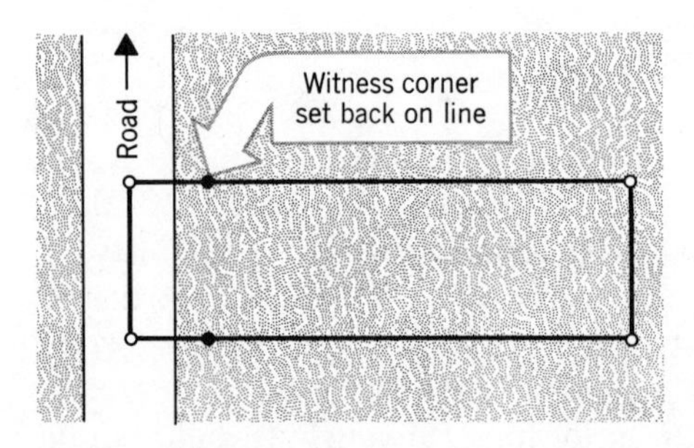

Figure 12-1

one can simply sight a line by eye from one corner to another, or a string may be stretched between corners. If the area is wooded or has heavy underbrush, a transit and range poles may be used to run the line and perhaps mark it at intervals with wooden stakes.

On rural surveys of large areas where corners may be hundreds or thousands of meters apart, the problem is more difficult. A common method of indicating a line is to mark trees. Trees to one side or the other of the line may be marked with two slash marks placed on the side of the tree adjacent to the line. If a tree is on the line, it may be marked with a blaze on each side of the tree, the blazes being in line with the property line. A blaze is made by using the ax or machete to shave off the bark to form an approximate circle of 5 or 6 in. diameter.

Sometimes the marks on trees are painted a bright color, such as yellow. When trees are properly marked (especially if the marks are painted), an experienced surveyor can often "walk" a line in the field years after the survey was made. One must be careful, however, not to mistake extraneous marks for surveyors' marks.

12-3 Resurveys

Often the land surveyor is called on to resurvey property. There are many possible reasons for this. If the property is being sold, an updated legal description of the property may be needed. Resurveys may be needed to settle disputes between adjacent land owners.

On first consideration, one would probably feel that resurveying a piece of property would be quite easy—just locate the corners and verify the measurements of the original survey. Sometimes—occasionally—it is that easy, but more often than not it is considerably more difficult. There are two major potential problems—(1) the property corners may not be readily located and (2) measurements made in the resurvey may not agree with those of the original survey.

Before going to the field to begin a resurvey job, the surveyor should "do his homework." Obviously, all available data on the property to be resurveyed must be gathered. Often, however, the surveyor may have to survey parts of adjacent property in order to locate a corner or a line on the property he is being paid to survey. Thus it is helpful to have data on adjacent property. Such information may be obtained from the proper government official. Usually this is a local (county) governmental office called the Clerk of Court, Register of Plats and Deeds, or the like.

Usually the first step on reaching the resurvey site is to try to find the corners. Sometimes they may be spotted quickly, if, for example, the corner is marked by a metal pipe sticking up two or three inches above ground. Sometimes they can be found with a little digging with a shovel. Some surveyors use metal detectors to search for metal pipe corners. Often local residents can be helpful in pointing out the locations of corners.

If the corners at each end of a line can be found, the surveyor may measure the distance and bearing of the line to see how these compare with the corresponding values on the plat. If the measurements agree closely, lines can be run from each corner according to the data on the plat to try to locate other corners. If the measurements do not agree, possibly corrections may be computed and applied to data on the map to run (approximately) other lines in an attempt to locate other corners.

As indicated previously, sometimes it may be necessary to survey parts of adjacent property. For example, in Figure 12-2 suppose the surveyor is resurveying tract *ABCDA*, that he has located point *D*, but nothing else checks out. Suppose a local resident is able to point out corner *F* on the adjacent tract and the surveyor has accurate data on the length and direction of *FE*. By running line *FE*, point *E* can be established (the corner marker may be located or point *E* may just be marked temporarily). A straight line can then be run from point *D* through point *E* and on to the vicinity of point *C*. If the distance *DC* is measured off along this line, the marker for point *C* may be located.

Once the corners are located, the surveyor measures accurately the lengths and directions of each line. One would expect these measured values to compare favorably with corresponding values on the plat; quite often, however, they will not. There are many possible reasons why. One is that registration of surveyors is relatively new in some areas. Thus not too many years ago, some surveying was done by persons not properly qualified. Another reason is that the equipment available years ago was inferior to modern equipment. The well-equipped survey party of today with a registered land surveyor as party chief and trained crew members as rodmen and tapemen is relatively new. Fifty years ago the land surveyor was a "lone

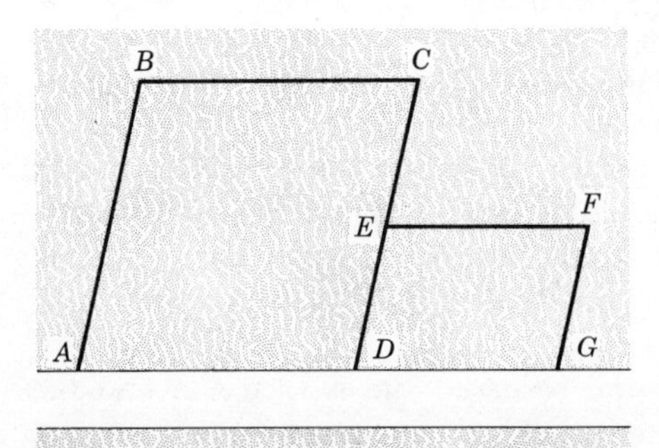

Figure 12-2

wolf." He appeared at the survey locale with his surveying instrument (usually a Jacob's staff) and surveyor's chain. The landowner furnished the help to cut lines, measure the distances, and act as rodman. For many of these helpers it was the first time they had seen a land survey in progress. One final reason is that years ago land was not nearly as valuable as it is now. Accordingly, it was just not considered necessary or worth the time and effort required to try to do accurate surveying.

When the surveyor encounters such a situation where measured data do not compare favorably with corresponding data on an old plat, he must take some kind of action. Also, when the surveyor is working toward resolving a property dispute between adjacent landowners and the data do not check out properly, the surveyor must take some kind of action. The exact action to take is not necessarily cut and dried. Generally speaking, however, the surveyor should always try to restake in the field the same corners and lines that were staked out in the original survey, regardless of whether or not measured values agree with those recorded on the old plat or deed. The objective should be to walk in the footsteps of the original surveyor and to carry out the *intent* of the original survey.

In the case of trying to resolve a property dispute, the surveyor should do his best to restake in the field the same corners and lines that were staked out in the original survey and then attempt to get all parties involved to agree to a settlement. If this is not possible, the dispute will likely have to be settled in a court of law. In this case the surveyor's function is that of expert witness. He must present his findings to the court and let the court settle the dispute. The surveyor may then be called upon to stake out the property line according to the court decree.

It should be emphasized that the discussion in this section is intended to be very general. Almost every resurvey job is going to be unique in some ways, and in effect each must be treated as a special case. The surveyor must approach each job not on the basis of following a prescribed set of steps but rather based on what he finds after he gets there—using his personal knowledge of local procedures, local customs, and local law.

12-4 Legal Descriptions of Property

When property is bought and sold, inherited, or otherwise changes hands, the title to the land is transferred by means of a written legal document called a *deed.* A deed contains a written description of the boundaries of the property. There are several methods used in preparing the written description—metes and bounds, indicating corners by coordinates, indicating lot and block number of a subdivision, by townships and sections, and so on. The first three of these are covered briefly in this section. The last one is covered briefly in the next section.

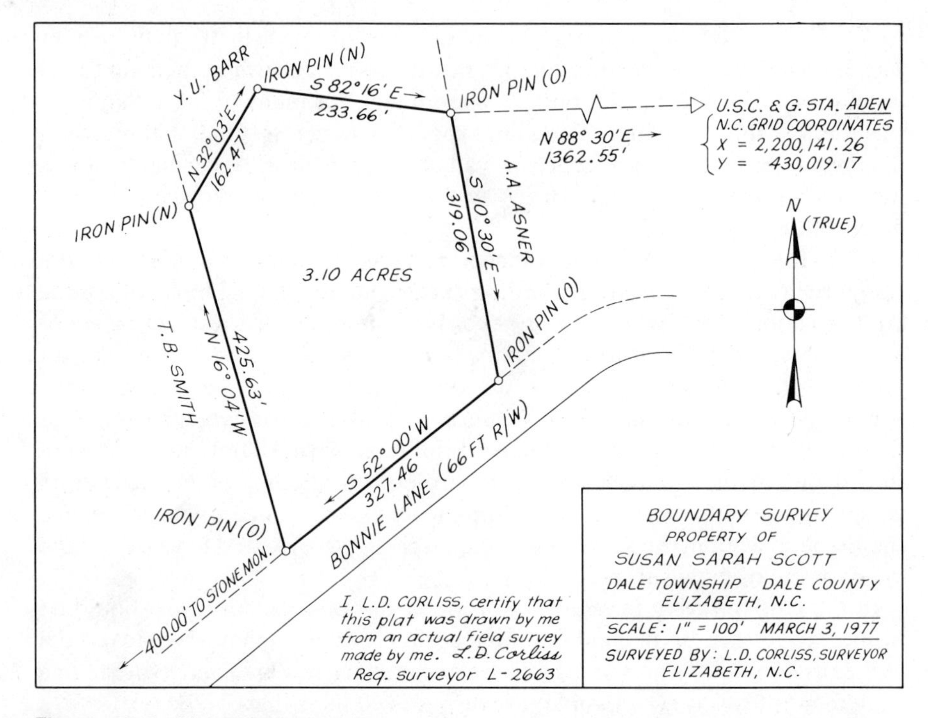

Figure 12-3 Hypothetical plat.

The metes and bounds method is observed largely in the 13 original states plus a few other states. It consists essentially of starting at a corner of the property and giving the length and direction of the line from that corner to the next corner. At the next corner the length and direction of the next line are given. This continues until all lines have been so described.

When a piece of property is to be transferred, the surveyor may be called upon to survey (resurvey) the property and prepare a plat. From the plat the metes and bounds description may be prepared for the lawyer's use in drawing up the deed. A hypothetical plat has been prepared using the (adjusted) traverse of Example 8-2 (see Figure 12-3), and a metes and bounds description of this property follows.

"Beginning at an iron pin at the southerly corner of the property located 33 ft from the center line of Bonnie Lane and 400.00 ft, N 52°00′ E, from a stone monument and running thence along the line of T. B. Smith property N 16°04′ W, 425.63 ft to an iron pin; thence along the line of Y. U. Barr property N 32°03′ E, 162.47 ft to an iron pin; thence along the line of Y. U. Barr property S 82°16′ E, 233.66 ft to an iron pin; thence along the line of A. A. Asner property S 10°30′ E, 319.06 ft to an iron pin; thence along Bonnie Lane S 52°00′ W, 327.46 ft to the point of beginning. The tract contains 3.10 acres and is shown on a plat dated

March 3, 1977, by L. D. Corliss, registered land surveyor. All bearings are referred to the true meridian."

An alternative to the metes and bounds description is to give the coordinates of each corner of the property in lieu of the length and direction of each line. The coordinates should be referred to the state plane coordinate system (see Section 13-5).

If a lot is one in a large subdivision, and a subdivision plat giving official dimensions for each lot has been recorded, it is possible in some places to describe property by referring to the lot number, block number (if necessary), and subdivision name or number.

12-5 U.S. Public Land Surveys

Because of problems and undesirable conditions observed to exist in the 13 original states and a few other eastern states, the Continental Congress established in 1784 the U.S. System of Public Land Surveys. A very brief description of this system is presented here. The reader interested in additional details is referred to the references listed at the end of this book.

Basically, the U.S. System of Public Land Surveys divides land into *quadrangles*, approximately 24 mi on a side. Each quadrangle is divided into 16 *townships*, each approximately 6 mi on a side. Each township is divided into 36 *sections*, each approximately 1 mi on a side. Each section may be further divided into halves, quarters, eighths, or whatever is desired. These distances are indicated as being approximate because of the effect of the earth's curvature.

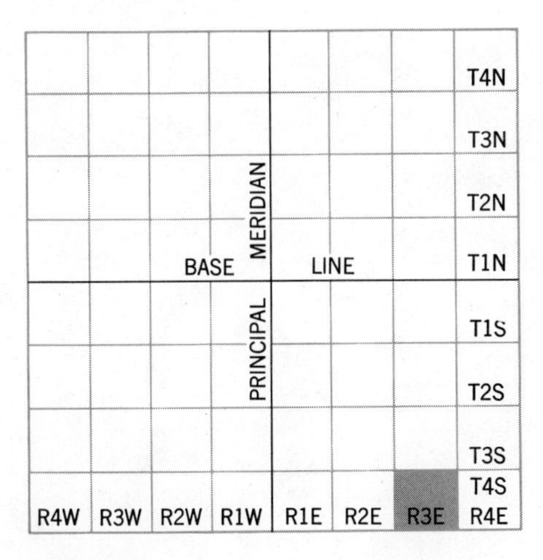

Figure 12-4

Figure 12-4 shows four quadrangles, each of which is divided into 16 townships. Note that a principal meridian running north-south and a base line running east-west are used to mark what might be thought of as reference axes. A particular township is referred to by indicating which column and row it falls in. A column of townships (north-south) is called a range, and a row (east-west) is called a tier. These are numbered as shown in Figure 12-4, and any township can be identified by stating its range and tier. For example, the shaded township in Figure 12-4 would be identified as T4S, R3E.

Figure 12-5 shows the division of a township into 36 sections. The sections are identified by number according to the numbering scheme shown. Each section is 1 sq mi, or 640 acres (not accounting for convergence).

Figure 12-6 shows how a section may be further divided into halves, quarters, eighths, and sixteenths (quarter-quarters). Each sixteenth of a section is 40 acres. Fractions of a section are identified as indicated in Figure 12-6.

This system allows for easy description of property being sold by designating the quarter-quarter section, the quarter section, the section number, the township, and the principal meridian. As a specific example, a tract of land might be identified as the NE$\frac{1}{4}$SE$\frac{1}{4}$, Section 26, T4S, R3E, of the second principal meridian.

12-6 Problems

12-1 Prepare a plat and metes and bounds description for the traverse of Problem 8-4.

6	5	4	3	2	1
7	8	9	10	11	12
18	17	16	15	14	13
19	20	21	22	23	24
30	29	28	27	26	25
31	32	33	34	35	36

Figure 12-5

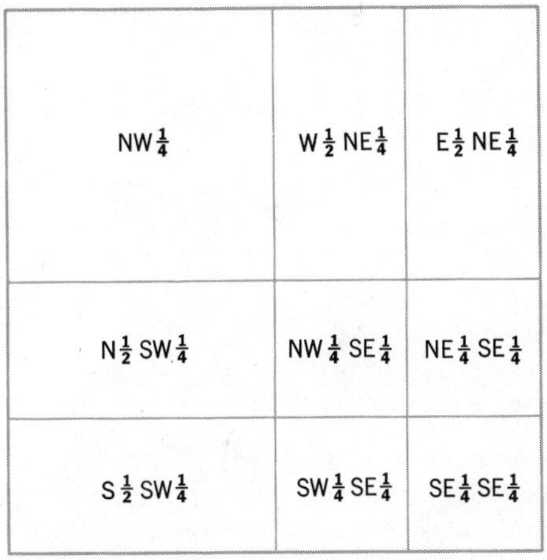

NW$\frac{1}{4}$	W$\frac{1}{2}$ NE$\frac{1}{4}$	E$\frac{1}{2}$ NE$\frac{1}{4}$
N$\frac{1}{2}$ SW$\frac{1}{4}$	NW$\frac{1}{4}$ SE$\frac{1}{4}$	NE$\frac{1}{4}$ SE$\frac{1}{4}$
S$\frac{1}{2}$ SW$\frac{1}{4}$	SW$\frac{1}{4}$ SE$\frac{1}{4}$	SE$\frac{1}{4}$ SE$\frac{1}{4}$

Figure 12-6

12-2 Repeat Problem 12-1 for the traverse of Problem 8-5.

12-3 Repeat Problem 12-1 for the traverse of Problem 8-6.

12-4 Go to the courthouse and find the plat and deed for the property where you live. Try to find the property corners.

12-5 What kind(s) of property descriptions is (are) used in the area where you live?

12-6 Go to the courthouse and locate three plats of surveys that are at least 50 years old. Using methods of Section 8-3, compute the error of closure and the precision of closure for the traverses shown on the plats.

12-7 Repeat Problem 12-6 except that the plats should be no more than 2 years old. Compare the results of Problems 12-6 and 12-7.

TERMINATION

The preceding 12 chapters have covered some of the more technical aspects of surveying. A knowledge of these is absolutely essential for anyone planning to do general surveying. Also important are certain less technical, more practical aspects of surveying, including ethics, clearing brush, hazards, record keeping, the State Plane Coordinate Systems, registration as a land surveyor, and the future. These topics are covered in this terminating chapter so that the reader will be exposed to these more practical aspects and not just the technical.

13-1 Ethics

The words "ethics" and "ethical" are not easily defined precisely. Ethics is a noun while ethical is an adjective. Ethics might be thought of as a set of moral principles that, when followed, result in ideal ends. Ethical might be thought of as describing one whose conduct is characterized by excellence and results in right (correct) and good ends. The phrase "right and good" is somewhat subjective, but most people know "right and good" when they see it.

It should go without saying that a surveyor should act in an ethical manner at all times. He owes this to the public, to his surveying colleagues, and to himself. He owes it to himself as one means of maintaining an unquestionable reputation—with respect to both the public and his colleagues.

Many organized groups have "codes of ethics," which generally consist of things one should and should not do. The author is not going to reproduce one of these but will instead give several general thoughts on the subject.

First, the surveyor should be honest. He should record all data as measured and never "fake" the data in any way. He should not bias results in favor of his client. For example, in setting a corner between two pieces of property, the corner should be set where the evidence indicates it should be and not arbitrarily with a motive of favoring the client. Also on the subject of honesty, the surveyor should never sign and place his seal on a plat unless the survey and plat were done by him or under his immediate supervision.

Another consideration is that the surveyor should maintain confidentiality with his clients. Sometimes in the course of his work, a surveyor learns certain information about his client which the client does not want to be made public—at least not until a later time. For example, someone might hire a surveyor to survey some property which he is considering purchasing for the purpose of placing a business on it. The nature of the business might be such that the client does not want his competitors to learn about his proposed new location. A surveyor should honor such confidential knowledge.

Finally, the surveyor should be responsible. He should accept responsibility for his work, including his mistakes. Also, he should recognize his own limitations and should not hesitate to obtain outside expertise when the occasion arises.

These three qualities—honesty, confidentiality, and responsibility—although far from complete, do go a long way toward setting forth some general standards of ethics. The reader is referred to one of the various codes of ethics for additional and more specific information.

13-2 Clearing Brush

For this and the next section, I am going to relate some personal experiences.

One thing that has always bothered me in the teaching of surveying is the fact that field work is usually learned and practiced by students during a specified laboratory period, with the field work usually done around campus in wide open spaces. This is out of convenience and consideration of restraints on available time for teaching surveying in a lecture/laboratory format. In all my years of experiencing field work in a laboratory situation (both as a student and a teacher), there is one important piece of equipment that I have never seen—a bush ax. I once had a student return to campus for the fall semester after having taken my surveying course during the previous spring and having worked for a surveying party during the summer. When he first saw me, he said, "Dr. Evett, I worked for a survey party this summer, but, you know, all I did was cut brush. We didn't learn how to do that in our surveying class!" How true! (I am aware that some schools have mandatory summer surveying camps where bush axes may be used in field work. However, I have not been personally involved with any of these.)

The fact of the matter is that, in running a line from point *A* to point *B*, it is generally necessary to be able to sight along the line through the transit in order to determine angles, bearings, or azimuths, and/or to set intermediate points on line. It is also necessary to be able to stretch the tape tightly along the line so that an accurate measurement can be obtained in accord with established procedure. These may not be problems if the surveying is being done in an open area. However, if it is being done in a wooded area underlain by heavy underbrush, it will be necessary to clear trees and brush along the line. The degree to which the line should be cleared out would depend to some extent on the purpose of the survey.

Several words of caution should be offered with regard to clearing a line. First a surveyor should realize that he has no inherent right to trespass on someone else's property. Thus extreme caution should be exercised when it is necessary to enter upon such property. I have found that few people will object to a surveyor's entering on their property when necessary, provided they know who the surveyor is and why he is there and provided the surveyor (including his assistants) is careful not to damage property. Accordingly, I recommend that a surveyor advise, where possible, anyone whose property he must enter to carry out a survey as to why he is there. It takes only a minute to knock on someone's door and tell him what you are doing, and it may prevent you from finding yourself looking down the barrel of a gun. With regard to damaging property, the surveyor must be careful to see that he and his assistants do not damage fences, flowers, shrubbery, and so on. Common sense can be helpful here. For example, cutting down a small dogwood tree in the middle of a forest would likely cause no harm; but cutting down one that someone has planted and nurtured in his front yard would quite likely cause harm (the harm being to the surveyor).

Getting back to the clearing of a line, as stated previously, the purpose is generally to open a line sufficiently for sighting and measuring distance. This is generally done by clearing out underbrush and smaller trees and "delimbing" larger trees. What does one do when the line of sight passes directly through a tree that is too large to cut down? Well, it depends. If a line is being run by bearing, the transit can simply be picked up and set up again on the other side of the tree. The transit can then be set on the proper bearing and the range poles placed on line to continue the running of the line. Using this method the line is extended "through" the tree without having to sight back on the line. If a line is being run by something other than magnetic bearing, it is generally necessary to actually sight along the line from one end to the other (usually by means of intermediate setups of the transit), and a large tree on line presents a difficulty. One solution is to run some kind of an offset line, such as moving over a foot or more to run a parallel line long enough to get by the tree.

Another question arises: "When cutting brush, how does one know where the line is?" The instrumentman, after orienting the transit in the proper direction, "lines in" two or more range poles along the line. Then by simply sighting back so that the range poles line up, the location of the line can be quickly determined. That sounds simple, but sometimes it is difficult to get inexperienced help to understand.

13-3 Hazards

Surveying would probably not be found on the list of dangerous occupations; nevertheless, there are some potential hazards of which the person going out to survey should be aware. These can be broadly classified and discussed under the following headings: plants, animals, people, and weather.

When considering plants, the first things that immediately come to my mind are poison oak, poison ivy, and poison sumac. Anyone doing much surveying should be able to recognize these plants and avoid them. If you are conscientious in your work, however, you cannot always avoid them. As a matter of fact, during one survey, I recognized that poison ivy was present; but in order to do what was required to run a line, I had to walk through it; and so I did. I have found that, if I take a shower immediately after returning from surveying when these plants are present and also wash the clothes I was wearing, I usually do not have an adverse reaction to them. This did not work in the experience cited above, as I developed a rash (in spite of a shower) and then compounded the problem by scratching it (mostly in my sleep), resulting in numerous infected sores. These rashes can be treated by a physician. For more specific information on these plants, see reference 2 in the list at the end of this book.

Another plant that causes problems is briars. Frequently, very thick briar patches are encountered, and the results can be scratches on the skin and tears on the clothing. Even the innocent tree limb can be hazardous if a person walking in front of another inadvertently pushes a tree limb forward and then releases it in passing, resulting in a whiplash action. This can be painful to the person following if struck by the limb, and serious head and eye injuries are possible if struck in this area.

When considering animals, the first that comes to my mind is the snake. Frankly, I do not like snakes. Although I have actually seen only a very few snakes while surveying, I have "imagined," or anticipated, hundreds. More dangerous in terms of fatalities are stings from bees, wasps, yellowjackets, and the like. It is not too uncommon to cut into a nest of these, and the consequences can be serious. Some people suffer serious (even fatal) reactions to such stings. Needless to say, care should be exercised. More in the nuisance category would be bites from mosquitoes and redbugs (chiggers). The latter cause severe itching. Ticks, some of which transmit fever, are another problem. (My father has just recently recovered from a case of Rocky Mountain spotted fever, apparently contracted by contact with ticks while surveying in the woods. He diagnosed his own case while watching a television program in which the symptoms of the disease were being discussed.) Check your body carefully for ticks after being out surveying in the woods. With regard to larger animals, dogs are often a problem, and occasionally when walking through a barnyard, horses or a curious bull can be a potential hazard.

With regard to the hazards from both plants and animals described above, I would advise: (1) that extreme care be exercised at all times, (2) that a first aid kit (including snake bite kit) be carried, (3) that some kind of bug spray be applied to the body prior to entering the woods, and (4) that a shower (or bath) be taken and the clothes washed after completing field work.

Another potential hazard, oddly, is people. Often it is necessary to survey along a road or even down the center line of a road. Do not assume that people will slow down and exercise caution. Most will, but sooner or later some clown will come along and see how close he can come without hitting you or your equipment. Warning signs (SURVEY PARTY AHEAD) placed along the highway and orange vests worn by the workers provide some protection. Another potential problem results from the fact that property is valuable and people can get emotional about it. As related in Chapter 12, the surveyor is sometimes called upon to settle boundary disputes and, as such, one can get caught in the middle. I have, on one or two occasions, seen guns or knives drawn. Usually, however, these are drawn to impress the other disputer, rather than against the surveyor. Nevertheless, I do not feel comfortable in the presence of knives or guns and would just as soon leave when they are present.

Another hazard to consider is extreme weather. I have personally surveyed in weather ranging from so hot that dizziness was experienced to so cold that the feet became numb. It is a good idea to carry water in a thermos or canteen if surveying is to be done where water or other drink cannot be purchased. In hot weather, salt tablets are beneficial, and the wearing of a hat affords protection from the sun. Other hazardous weather conditions include hail, thunderstorms, and tornadoes, all of which can happen suddenly with little prior warning. Never survey when there is lightning present. Usually the safest procedure is to sit in your car or truck until the lightning storm is over.

13-4 Record Keeping

One important aspect of the surveying process that is often taken for granted is record keeping. It is essential that complete, accurate records be maintained. A good rule of thumb might be "Save everything!"

The individual going to work for an established surveying company or for an agency such as a highway department will find an established procedure for preparing, filing, and maintaining records. Others, however, may not. For these, the following is offered.

Most surveying jobs result in some kind of drawing—a map, plat, or profile sheet, for example—being prepared for the client. Generally, these drawings are prepared on tracing paper with copies made therefrom (such as blueprints) for the client. The originals should be kept on file by the surveyor for future reference. In order to be able to locate a given drawing quickly, some type of filing system should be utilized. One method would be to simply number each drawing consecutively by placing a number at the bottom right corner and then to file the drawings in consecutive order (or by townships) in a drawer or file cabinet.

It is also important that any field notes plus other supporting documents be kept on file. Supporting documents would include plats of adjacent property, previous plats of the property being surveyed at present, copies of deed descriptions, and the like. Often, former clients will request that property be surveyed that is adjacent to property previously surveyed by the surveyor. It is certainly helpful to be able to go to the file and retrieve readily all notes and other supporting documents pertaining to the particular property. They may also be needed for subsequent testimony in court by the surveyor. Such records can be conveniently filed by placing them in regular file folders and filing by number or client name.

Another item that can be quite useful if saved is the draft of the plat. Most surveyors will make a rough draft of a plat on plain paper and then prepare the final plat by tracing the draft on tracing paper. Again considering the situation of going back to do more surveying later where a surveyor has previously done work in the area, it is helpful to continue the drafting of the new work on the same draft. Such drafts can be saved easily by numbering them and placing them in a drawer.

In view of the three different filing entities alluded to above [i.e., filing of (1) plats, (2) folders containing field notes and supporting documents, and (3) rough drafts of plats], it is essential to have some kind of system of cross referencing the three filing entities. One solution is to keep an index card file, with one (or more, if needed) card for each client. On each card could be listed (after the name) the plat number, the file folder number (file containing field notes and supporting documents), and the rough draft number. It would also be helpful to give the date and a short description of the job. Different jobs for the same client could, of course, be entered as separate listings on the same file card. Under a system such as that just described, when a surveyor is interested in all information about a particular job, he can look up the name in the card file, note the three file numbers, and then pull the available information from the three respective files.

The system described above would be suitable for a small firm with a more elaborate system (probably utilizing computers) being needed by large firms. Variations could be made also from this system. For example, some might prefer to number plats for a period of one year and then begin numbering with number one again at the beginning of each new year, using a prefix number to indicate the year (e.g., 78-1, 78-2, 78-3, . . . , 78-n; 79-1, 79-2, 79-3, . . . , 79-m; etc.). Others might simply prefer to file alphabetically in each case. One drawback to doing that, however, is that, in the case of plats and drafts of plats, it is probably easier to file consecutively (on top of each other in a drawer) than alphabetically. In other words, when filing consecutively, or numerically, each new plat is just placed on top of the previous one; whereas when filing alphabetically, one has to shuffle through the plats each time when filing a new plat in order to find the proper location. In any event, the important point to be made here is that it is of the utmost importance to maintain a good, up-to-date record-keeping system—one in which virtually everything is kept, and one from which information concerning any given job can be readily retrieved.

13-5 State Plane Coordinate Systems

There has been established over a period of time the state plane coordinate systems. This overall system establishes a number of stations across the United States whose coordinates are known. Since the earth's surface is not flat, the systems are developed using two types of projections—the transverse Mercator projection and the Lambert conformal projection. Each state is covered by one or more such projection.

This coordinate system is extremely useful to the surveyor. A land surveyor can tie in a property survey with this coordinate system by running a traverse from one of the coordinate system stations to one of the corners of his property survey. He can then calculate the coordinates of each point in his survey and in effect ties his survey in with the system. The significance of this is that the corners of the survey should be easily reproducible on the ground at any time in the future, even if all corners of the survey are destroyed.

The system is also useful to anyone doing surveying covering a wide area, such as highway surveying. In running miles of highway, one can start on one coordinate station and end on another, thereby in effect having a closed traverse, which can be checked for closure.

This information on the state plane coordinate systems is presented here so that you will be aware of it. For further details refer to the references listed at the end of this book.

13-6 Registration as a Land Surveyor

Today the various states require a person to be a "registered (licensed) land surveyor" in order to perform (or be in responsible charge of) surveying. (Some states may allow a registered professional engineer to perform surveying.) Registration, or licensing, is done by the individual states whose requirements vary somewhat. Generally the requirements are satisfactory education, experience, and character.

Any person interested or involved in surveying (such as the surveyor's helper, the newly graduated civil engineer, etc.) should be encouraged to find out the local requirements and begin to work toward registration.

13-7 The Future

The future of surveying is bright. More and more high precision, time saving pieces of equipment come on the market every day. Electronic distance measuring devices, lasers to run levels and perform additional tasks, computers to do what was previously time consuming (and mistake prone) hand calculations, and computers to do drafting may be used by everyone at some time in the future.

Nevertheless, it is felt that the person calling himself a surveyor should be thoroughly familiar with the fundamentals of surveying and that for the near future at least, much of the work will still be done with the old standbys—the tape, transit, and level. Thus the emphasis in this book has been on these instruments and the fundamentals.

The author hopes this book has been an effective introduction to surveying and that the reader will be stimulated to go on to further study and practice.

BIBLIOGRAPHY (Abridged)

1. Brinker, Russell C. and Paul R. Wolf, *Elementary Surveying*, IEP—A Dun-Donnelley Publisher, New York, 1977.

2. Crooks, Donald M. and Dayton L. Klingman, *Poison Ivy, Poison Oak and Poison Sumac*, Farmers' Bulletin No. 1972, U.S. Department of Agriculture, Washington, D.C., 1976.

3. Davis, Raymond E. and Joe W. Kelly, *Elementary Plane Surveying*, McGraw-Hill, New York, 1967.

4. Hickerson, Thomas F., *Route Location and Design*, McGraw-Hill, New York, 1967.

5. Moffitt, Francis H. and The Late Harry Bouchard, *Surveying*, Intext Educational Publishers, New York, 1975.

6. Pafford, F. William, *Handbook of Survey Notekeeping*, John Wiley, New York, 1962.

7. Wood, James F., *Autocurve*, James F. Wood, El Paso, 1973.

ANSWERS TO EVEN-NUMBERED PROBLEMS

1

(1-2) 931.9 ft; 284.0 m
(1-4) 2634.5 acres; 4.1164 sq mi; 10.661 sq km; 1066.1 ha
(1-6) 1000 acres
(1-8) 117.8 ft
(1-10) 8900 cu yd
(1-12) 217,000 sq m; 21.7 ha
(1-14) 21,000,000 cu ft

2

(2-2) $y_1 = -1.61$; $y_2 = -6.39$
(2-4) $a = 9.50$; $b = 0.500$
(2-6) $a = 0$; $b = 1.00$
(2-8) $(x-3)^2 + (y+1)^2 = 144$
(2-10) Since there are two equations in answer to Problem 2-9, there are two pairs of answers to Problem 2-10:

$$\left.\begin{matrix} x_1 = 6.50 & x_2 = 6.50 \\ y_1 = 4.78 & y_2 = -18.74 \end{matrix}\right\} \text{for one equation}$$

$$\left.\begin{matrix} x_1 = 6.50 & x_2 = 6.50 \\ y_1 = 27.5 & y_2 = 4.45 \end{matrix}\right\} \text{for the other equation}$$

(2-12) 4.23 ft
(2-14) 77°50′
(2-16) $C = 40°46'$; $A = 80°24'$; $B = 58°50'$
(2-18) 121.75 ± 0.02
(2-20) 69°10′; 81°59′; 131°19′; 77°32′
(2-22) 318.35 ± 0.14 ft
(2-24) 95.13 ± 0.11 ft
(2-26) 629 ft
(2-28) $x_1 = 378.5$ ft; $y_1 = 392.5$ ft; $x_2 = 352.6$ ft; $y_2 = 263.0$ ft
(2-30) 169 yd
(2-32) 420.8 ± 0.1 m
(2-34) ± 0.15 ft

4

(4-2) 155
(4-4) It cannot be paced with this precision. It should be rounded off to 299 ft.
(4-6) 48.78 m (by both formulas)
(4-8) 413.54
(4-10) 11.63 m
(4-12) 51,910 ft
(4-14) 81.92 ft; yes; 81.93 ft
(4-16) 477.11 ft
(4-18) 477.22 ft
(4-20) 478.05 ft
(4-22) 221.94 ft
(4-24) 188.60 ft
(4-26) 643.16 ft
(4-28) 110.04 ft by 88.03 ft; 109.98 ft by 87.98 ft
(4-30) 141.23 m by 70.62 m
(4-32) 103.56 m

5

(5-2) elevation of $BM_3 = 78.325$ m
(5-4) elevation of $BM_2 = 224.95$ ft
(5-6) elevation of $BM_2 = 58.257$ m
(5-8) elevation of $BM_2 = 108.87$ ft
(5-10) yes; no
(5-12) Probably misread the rod one time by one foot. Send the crew back out to check their work.
(5-14) 220 ft
(5-16) yes; S, 32.06 ft; T, 36.40 ft; U, 20.11 ft
(5-18) 193.57 ft

(5-20) 132.017 m
(5-22) 34,500 ft

6

(6-2) S 20°45′ W
(6-4) S 2°30′ W
(6-6) N 86°00′ W
(6-8) S 87°30′ E
(6-10) N 15°00′ W
(6-12) 309°45′
(6-14) *AB*, S 55°56′ E; *BC*, S 45°50′ E
(6-16) 13°30′; 147°45′; 3°00′; 195°45′
(6-18) *JK*, 79°57′; *KL*, 322°04′
(6-20) *FG*, 62°15′; *GH*, 94°30′; *HI*, 271°30′; *IF*, 255°45′
(6-22) 28°54′
(6-24) *A*, 111°30′; *B*, 56°54′; *C*, 88°06′; *D*, 69°19′; *E*, 214°11′; *BC*, N 67°24′ W; *CD*, N 24°30′ E; *DE*, S 44°49′ E; *EA*, S 79°00′ E

7

(7-2) 0.16 percent
(7-4) $H = 327$ ft; $V = +68.2$ ft
(7-6) $H - 483$ ft; $V =$ 187.4 ft
(7-8) $H = 627$ ft; $V = -87.9$ ft
(7-10) 152.0 ft
(7-12) 221.0 ft
(7-14) 87.6 ft
(7-16) $RS = 852$ ft; $ST = 292$ ft; $TU = 893$ ft; $UV = 533$ ft; Elev. $S = 775.3$ ft; Elev. $T = 863.7$ ft; Elev. $U = 1075.3$ ft; Elev. $V = 1188.6$ ft

8

(8-2) no
(8-4) $e_c = 0.51$ m; $P_c = \frac{1}{2400}$; *DE* adjusted: 195.96 m, S 66°51′ E; Coordinates of *D*: 2556.42 m (N), 3979.81 m (E)
(8-6) $e_c = 1.14$ ft; $P_c = \frac{1}{1960}$; *PQ* adjusted: 50.87 ft, S 33°39′ E; Coordinates of *P*: 2195.42 ft (N), 3907.72 ft (E)
(8-8) $e_c = 0.46$ m; $P_c = \frac{1}{3000}$; *EF* adjusted: 150.11 m, S 45°09′ W; Coordinates of *E*: 2693.42 m (N), 4207.97 m (E)
(8-22) *CD*: 31.22 m, S 49°19′ E
(8-24) *MN*: 897.85 m, N 62°45′ E
(8-26) It should close
(8-28) 680.21 m; S 6°57′ W

(8-30) N 13°03′ E
(8-32) $DE = 28.29$ m; $EB = 322.33$ m; deflection angle at $E = 96°48' R$

9

(9-4) 0.157 ha
(9-6) 0.916 acres
(9-8) 2.98 acres
(9-10) 9.38 ha
(9-12) 6.67 ha
(9-14) 2.98 acres
(9-16) 9.38 ha
(9-18) 15.2 ha
(9-20) 17.3 ha
(9-22) 6.67 ha
(9-24) 0.297 acres; 1.43 acres
(9-26) 15.7 ha
(9-28) 207 acres
(9-30) 2.97 acres

10

(10-12) 3.76
(10-14) 4.70
(10-18) 0.192 in.

11

(11-2) 476 cu yd
(11-4) 463 cu yd; 457 cu yd; 1.30%
(11-6) 6530 cu yd
(11-8) 8070 cu yd
(11-10) 645.7 ft
(11-12) 1°12′
(11-14) $R = 2864.79$ ft; $T = 206.61$ ft; $PC = 33 + 93.39$; $L = 412.50$ ft; $PT = 38 + 05.89$; $LC = 412.14$ ft; $E = 7.44$ ft; $M = 7.42$ ft
(11-16) $R = 3819.72$ ft; $T = 169.56$ ft; $PC = 16 + 56.01$; $L = 338.89$ ft; $PT = 19 + 94.90$; $LC = 338.78$ ft; $E = 3.76$ ft; $M = 3.76$ ft
(11-18) $R = 2864.8$ ft; $T = 189.0$ ft; $PC = 25 + 03.2$; $L = 377.5$ ft; $PT = 28 + 80.7$; $LC = 377.2$ ft; $E = 6.2$ ft; $M = 6.2$ ft
(11-20) Curve is 600 ft long. $PC = 85 + 00$; Elevation = 394.25 ft; Summit at station 89 + 00; Elevation = 398.25 ft
(11-22) Curve is 800 ft long; $PC = 51 + 18.2$; Elevation = 112.97 ft; Sag at station 56 + 78.2; Elevation = 105.13 ft

TRIGONOMETRIC TABLES

A

Natural Trigonometric Functions

0°

M	Sine	Cosine	Tan.	Cotan.	M
0	0.00000	1.0000	0.00000	Infinite	60
1	.00029	1.0000	.00029	3437.7	59
2	.00058	1.0000	.00058	1718.9	58
3	.00087	1.0000	.00087	1145.9	57
4	.00116	1.0000	.00116	859.44	56
5	0.00145	1.0000	0.00145	687.55	55
6	.00175	1.0000	.00175	572.96	54
7	.00204	1.0000	.00204	491.11	53
8	.00233	1.0000	.00233	429.72	52
9	.00262	1.0000	.00262	381.97	51
10	0.00291	1.0000	0.00291	343.77	50
11	.00320	.99999	.00320	312.52	49
12	.00349	.99999	.00349	286.48	48
13	.00378	.99999	.00378	264.44	47
14	.00407	.99999	.00407	245.55	46
15	0.00436	0.99999	0.00436	229.18	45
16	.00465	.99999	.00465	214.86	44
17	.00495	.99999	.00495	202.22	43
18	.00524	.99999	.00524	190.98	42
19	.00553	.99998	.00553	180.93	41
20	0.00582	0.99998	0.00582	171.89	40
21	.00611	.99998	.00611	163.70	39
22	.00640	.99998	.00640	156.26	38
23	.00669	.99998	.00669	149.47	37
24	.00698	.99998	.00698	143.24	36
25	0.00727	0.99997	0.00727	137.51	35
26	.00756	.99997	.00756	132.22	34
27	.00785	.99997	.00785	127.32	33
28	.00814	.99997	.00815	122.77	32
29	.00844	.99996	.00844	118.54	31
30	0.00873	0.99996	0.00873	114.59	30
31	.00902	.99996	.00902	110.89	29
32	.00931	.99996	.00931	107.43	28
33	.00960	.99995	.00960	104.17	27
34	.00989	.99995	.00989	101.11	26
35	0.01018	0.99995	0.01018	98.218	25
36	.01047	.99995	.01047	95.489	24
37	.01076	.99994	.01076	92.908	23
38	.01105	.99994	.01105	90.463	22
39	.01134	.99994	.01135	88.144	21
40	0.01164	0.99993	0.01164	85.940	20
41	.01193	.99993	.01193	83.844	19
42	.01222	.99993	.01222	81.847	18
43	.01251	.99992	.01251	79.943	17
44	.01280	.99992	.01280	78.126	16
45	0.01309	0.99991	0.01309	76.390	15
46	.01338	.99991	.01338	74.729	14
47	.01367	.99991	.01367	73.139	13
48	.01396	.99990	.01396	71.615	12
49	.01425	.99990	.01425	70.153	11
50	0.01454	0.99989	0.01455	68.750	10
51	.01483	.99989	.01484	67.402	9
52	.01513	.99989	.01513	66.105	8
53	.01542	.99988	.01542	64.858	7
54	.01571	.99988	.01571	63.657	6
55	0.01600	0.99987	0.01600	62.499	5
56	.01629	.99987	.01629	61.383	4
57	.01658	.99986	.01658	60.306	3
58	.01687	.99986	.01687	59.266	2
59	.01716	.99985	.01716	58.261	1
60	0.01745	0.99985	0.01746	57.290	0
M	Cosine	Sine	Cotan.	Tan.	M

89°

Natural Trigonometric Functions

1°

M	Sine	Cosine	Tan.	Cotan.	M
0	0.01745	0.99985	0.01746	57.290	60
1	.01774	.99984	.01775	56.351	59
2	.01803	.99984	.01804	55.442	58
3	.01832	.99983	.01833	54.561	57
4	.01862	.99983	.01862	53.709	56
5	0.01891	0.99982	0.01891	52.882	55
6	.01920	.99982	.01920	52.081	54
7	.01949	.99981	.01949	51.303	53
8	.01978	.99980	.01978	50.549	52
9	.02007	.99980	.02007	49.816	51
10	0.02036	0.99979	0.02036	49.104	50
11	.02065	.99979	.02066	48.412	49
12	.02094	.99978	.02095	47.740	48
13	.02123	.99977	.02124	47.085	47
14	.02152	.99977	.02153	46.449	46
15	0.02181	0.99976	0.02182	45.829	45
16	.02211	.99976	.02211	45.226	44
17	.02240	.99975	.02240	44.639	43
18	.02269	.99974	.02269	44.066	42
19	.02298	.99974	.02298	43.508	41
20	0.02327	0.99973	0.02328	42.964	40
21	.02356	.99972	.02357	42.433	39
22	.02385	.99972	.02386	41.916	38
23	.02414	.99971	.02415	41.411	37
24	.02443	.99970	.02444	40.917	36
25	0.02472	0.99969	0.02473	40.436	35
26	.02501	.99969	.02502	39.965	34
27	.02530	.99968	.02531	39.506	33
28	.02560	.99967	.02560	39.057	32
29	.02589	.99966	.02589	38.618	31
30	0.02618	0.99966	0.02619	38.188	30
31	.02647	.99965	.02648	37.769	29
32	.02676	.99964	.02677	37.358	28
33	.02705	.99963	.02706	36.956	27
34	.02734	.99963	.02735	36.563	26
35	0.02763	0.99962	0.02764	36.178	25
36	.02792	.99961	.02793	35.801	24
37	.02821	.99960	.02822	35.431	23
38	.02850	.99959	.02851	35.070	22
39	.02879	.99959	.02881	34.715	21
40	0.02908	0.99958	0.02910	34.368	20
41	.02938	.99957	.02939	34.027	19
42	.02967	.99956	.02968	33.694	18
43	.02996	.99955	.02997	33.366	17
44	.03025	.99954	.03026	33.045	16
45	0.03054	0.99953	0.03055	32.730	15
46	.03083	.99952	.03084	32.421	14
47	.03112	.99952	.03114	32.118	13
48	.03141	.99951	.03143	31.821	12
49	.03170	.99950	.03172	31.528	11
50	0.03199	0.99949	0.03201	31.242	10
51	.03228	.99948	.03230	30.960	9
52	.03257	.99947	.03259	30.683	8
53	.03286	.99946	.03288	30.412	7
54	.03316	.99945	.03317	30.145	6
55	0.03345	0.99944	0.03346	29.882	5
56	.03374	.99943	.03376	29.624	4
57	.03403	.99942	.03405	29.371	3
58	.03432	.99941	.03434	29.122	2
59	.03461	.99940	.03463	28.877	1
60	0.03490	0.99939	0.03492	28.636	0
M	Cosine	Sine	Cotan.	Tan.	M

88°

Natural Trigonometric Functions

2°

M	Sine	Cosine	Tan.	Cotan.	M
0	0.03490	0.99939	0.03492	28.636	60
1	.03519	.99938	.03521	28.399	59
2	.03548	.99937	.03550	28.166	58
3	.03577	.99936	.03579	27.937	57
4	.03606	.99935	.03609	27.712	56
5	0.03635	0.99934	0.03638	27.490	55
6	.03664	.99933	.03667	27.271	54
7	.03693	.99932	.03696	27.057	53
8	.03723	.99931	.03725	26.845	52
9	.03752	.99930	.03754	26.637	51
10	0.03781	0.99929	0.03783	26.432	50
11	.03810	.99927	.03812	26.230	49
12	.03839	.99926	.03842	26.031	48
13	.03868	.99925	.03871	25.835	47
14	.03897	.99924	.03900	25.642	46
15	0.03926	0.99923	0.03929	25.452	45
16	.03955	.99922	.03958	25.264	44
17	.03984	.99921	.03987	25.080	43
18	.04013	.99919	.04016	24.898	42
19	.04042	.99918	.04046	24.719	41
20	0.04071	0.99917	0.04075	24.542	40
21	.04100	.99916	.04104	24.368	39
22	.04129	.99915	.04133	24.196	38
23	.04159	.99913	.04162	24.026	37
24	.04188	.99912	.04191	23.859	36
25	0.04217	0.99911	0.04220	23.695	35
26	.04246	.99910	.04250	23.532	34
27	.04275	.99909	.04279	23.372	33
28	.04304	.99907	.04308	23.214	32
29	.04333	.99906	.04337	23.058	31
30	0.04362	0.99905	0.04366	22.904	30
31	.04391	.99904	.04395	22.752	29
32	.04420	.99902	.04424	22.602	28
33	.04449	.99901	.04454	22.454	27
34	.04478	.99900	.04483	22.308	26
35	0.04507	0.99898	0.04512	22.164	25
36	.04536	.99897	.04541	22.022	24
37	.04565	.99896	.04570	21.881	23
38	.04594	.99894	.04599	21.743	22
39	.04623	.99893	.04628	21.606	21
40	0.04653	0.99892	0.04658	21.470	20
41	.04682	.99890	.04687	21.337	19
42	.04711	.99889	.04716	21.205	18
43	.04740	.99888	.04745	21.075	17
44	.04769	.99886	.04774	20.946	16
45	0.04798	0.99885	0.04803	20.819	15
46	.04827	.99883	.04833	20.693	14
47	.04856	.99882	.04862	20.569	13
48	.04885	.99881	.04891	20.446	12
49	.04914	.99879	.04920	20.325	11
50	0.04943	0.99878	0.04949	20.206	10
51	.04972	.99876	.04978	20.087	9
52	.05001	.99875	.05007	19.970	8
53	.05030	.99873	.05037	19.855	7
54	.05059	.99872	.05066	19.740	6
55	0.05088	0.99870	0.05095	19.627	5
56	.05117	.99869	.05124	19.516	4
57	.05146	.99867	.05153	19.405	3
58	.05175	.99866	.05182	19.296	2
59	.05205	.99864	.05212	19.188	1
60	0.05234	0.99863	0.05241	19.081	0
M	Cosine	Sine	Cotan.	Tan.	M

87°

Natural Trigonometric Functions

3°

M	Sine	Cosine	Tan.	Cotan.	M
0	0.05234	0.99863	0.05241	19.081	60
1	.05263	.99861	.05270	18.976	59
2	.05292	.99860	.05299	18.871	58
3	.05321	.99858	.05328	18.768	57
4	.05350	.99857	.05357	18.666	56
5	0.05379	0.99855	0.05387	18.564	55
6	.05408	.99854	.05416	18.464	54
7	.05437	.99852	.05445	18.366	53
8	.05466	.99851	.05474	18.268	52
9	.05495	.99849	.05503	18.171	51
10	0.05524	0.99847	0.05533	18.075	50
11	.05553	.99846	.05562	17.980	49
12	.05582	.99844	.05591	17.886	48
13	.05611	.99842	.05620	17.793	47
14	.05640	.99841	.05649	17.702	46
15	0.05669	0.99839	0.05678	17.611	45
16	.05698	.99838	.05708	17.521	44
17	.05727	.99836	.05737	17.431	43
18	.05756	.99834	.05766	17.343	42
19	.05785	.99833	.05795	17.256	41
20	0.05814	0.99831	0.05824	17.169	40
21	0.5844	.99829	.05854	17.084	39
22	.05873	.99827	.05883	16.999	38
23	.05902	.99826	.05912	16.915	37
24	.05931	.99824	.05941	16.832	36
25	0.05960	0.99822	0.05970	16.750	35
26	.05989	.99821	.05999	16.668	34
27	.06018	.99819	.06029	16.587	33
28	.06047	.99817	.06058	16.507	32
29	.06076	.99815	.06087	16.428	31
30	0.06105	0.99813	0.06116	16.350	30
31	.06134	.99812	.06145	16.272	29
32	.06163	.99810	.06175	16.195	28
33	.06192	.99808	.06204	16.119	27
34	.06221	.99806	.06233	16.043	26
35	0.06250	0.99804	0.06262	15.969	25
36	.06279	.99803	.06291	15.895	24
37	.06308	.99801	.06321	15.821	23
38	.06337	.99799	.06350	15.748	22
39	.06366	.99797	.06379	15.676	21
40	0.06395	0.99795	0.06408	15.605	20
41	.06424	.99793	.06437	15.534	19
42	.06453	.99792	.06467	15.464	18
43	.06482	.99790	.06496	15.394	17
44	.06511	.99788	.06525	15.325	16
45	0.06540	0.99786	0.06554	15.257	15
46	.06569	.99784	.06584	15.189	14
47	.06598	.99782	.06613	15.122	13
48	.06627	.99780	.06642	15.056	12
49	.06656	.99778	.06671	14.990	11
50	0.06685	0.99776	0.06700	14.924	10
51	.06714	.99774	.06730	14.860	9
52	.06743	.99772	.06759	14.795	8
53	.06773	.99770	.06788	14.732	7
54	.06802	.99768	.06817	14.669	6
55	0.06831	0.99766	0.06847	14.606	5
56	.06860	.99764	.06876	14.544	4
57	.06889	.99762	.06905	14.482	3
58	.06918	.99760	.06934	14.421	2
59	.06947	.99758	.06963	14.361	1
60	0.06976	0.99756	0.06993	14.301	0
M	Cosine	Sine	Cotan.	Tan.	M

86°

Natural Trigonometric Functions

4°

M	Sine	Cosine	Tan.	Cotan.	M
0	0.06976	0.99756	0.06993	14.301	60
1	.07005	.99754	.07022	14.241	59
2	.07034	.99752	.07051	14.182	58
3	.07063	.99750	.07080	14.124	57
4	.07092	.99748	.07110	14.065	56
5	0.07121	0.99746	0.07139	14.008	55
6	.07150	.99744	.07168	13.951	54
7	.07179	.99742	.07197	13.894	53
8	.07208	.99740	.07227	13.838	52
9	.07237	.99738	.07256	13.782	51
10	0.07266	0.99736	0.07285	13.727	50
11	.07295	.99734	.07314	13.672	49
12	.07324	.99731	.07344	13.617	48
13	.07353	.99729	.07373	13.563	47
14	.07382	.99727	.07402	13.510	46
15	0.07411	0.99725	0.07431	13.457	45
16	.07440	.99723	.07461	13.404	44
17	.07469	.99721	.07490	13.352	43
18	.07498	.99719	.07519	13.300	42
19	.07527	.99716	.07548	13.248	41
20	0.07556	0.99714	0.07578	13.197	40
21	.07585	.99712	.07607	13.146	39
22	.07614	.99710	.07636	13.096	38
23	.07643	.99708	.07665	13.046	37
24	.07672	.99705	.07695	12.996	36
25	0.07701	0.99703	0.07724	12.947	35
26	.07730	.99701	.07753	12.898	34
27	.07759	.99699	.07782	12.850	33
28	.07788	.99696	.07812	12.801	32
29	.07817	.99694	.07841	12.754	31
30	0.07846	0.99692	0.07870	12.706	30
31	.07875	.99689	.07899	12.659	29
32	.07904	.99687	.07929	12.612	28
33	.07933	.99685	.07958	12.566	27
34	.07962	.99683	.07987	12.520	26
35	0.07991	0.99680	0.08017	12.474	25
36	.08020	.99678	.08046	12.429	24
37	.08049	.99676	.08075	12.384	23
38	.08078	.99673	.08104	12.339	22
39	.08107	.99671	.08134	12.295	21
40	0.08136	0.99668	0.08163	12.251	20
41	.08165	.99666	.08192	12.207	19
42	.08194	.99664	.08221	12.163	18
43	.08223	.99661	.08251	12.120	17
44	.08252	.99659	.08280	12.077	16
45	0.08281	0.99657	0.08309	12.035	15
46	.08310	.99654	.08339	11.992	14
47	.08339	.99652	.08368	11.950	13
48	.08368	.99649	.08397	11.909	12
49	.08397	.99647	.08427	11.867	11
50	0.08426	0.99644	0.08456	11.826	10
51	.08455	.99642	.08485	11.785	9
52	.08484	.99639	.08514	11.745	8
53	.08513	.99637	.08544	11.705	7
54	.08542	.99635	.08573	11.664	6
55	0.08571	0.99632	0.08602	11.625	5
56	.08600	.99630	.08632	11.585	4
57	.08629	.99627	.08661	11.546	3
58	.08658	.99625	.08690	11.507	2
59	.08687	.99622	.08720	11.468	1
60	0.08716	0.99619	0.08749	11.430	0
M	Cosine	Sine	Cotan.	Tan.	M

85°

Natural Trigonometric Functions

5°

M	Sine	Cosine	Tan.	Cotan.	M
0	0.08716	0.99619	0.08749	11.430	60
1	.08745	.99617	.08778	11.392	59
2	.08774	.99614	.08807	11.354	58
3	.08803	.99612	.08837	11.316	57
4	.08831	.99609	.08866	11.279	56
5	0.08860	0.99607	0.08895	11.242	55
6	.08889	.99604	.08925	11.205	54
7	.08918	.99602	.08954	11.168	53
8	.08947	.99599	.08983	11.132	52
9	.08976	.99596	.09013	11.095	51
10	0.09005	0.99594	0.09042	11.059	50
11	.09034	.99591	.09071	11.024	49
12	.09063	.99588	.09101	10.988	48
13	.09092	.99586	.09130	10.953	47
14	.09121	.99583	.09159	19.918	46
15	0.09150	0.99580	0.09189	10.883	45
16	.09179	.99578	.09218	10.848	44
17	.09208	.99575	.09247	10.814	43
18	.09237	.99572	.09277	10.780	42
19	.09266	.99570	.09306	10.746	41
20	0.09295	0.99567	0.09335	10.712	40
21	.09324	.99564	.09365	10.678	39
22	.09353	.99562	.09394	10.645	38
23	.09382	.99559	.09423	10.612	37
24	.09411	.99556	.09453	10.579	36
25	0.09440	0.99553	0.09482	10.546	35
26	.09469	.99551	.09511	10.514	34
27	.09498	.99548	.09541	10.481	33
28	.09527	.99545	.09570	10.449	32
29	.09556	.99542	.09600	10.417	31
30	0.09585	0.99540	0.09629	10.385	30
31	.09614	.99537	.09658	10.354	29
32	.09642	.99534	.09688	10.322	28
33	.09671	.99531	.09717	10.291	27
34	.09700	.99528	.09746	10.260	26
35	0.09729	0.99526	0.09776	10.229	25
36	.09758	.99523	.09805	10.199	24
37	.09787	.99520	.09834	10.168	23
38	.09816	.99517	.09864	10.138	22
39	.09845	.99514	.09893	10.108	21
40	0.09874	0.99511	0.09923	10.078	20
41	.09903	.99508	.09952	10.048	19
42	.09932	.99506	.09981	10.019	18
43	.09961	.99503	.10011	9.9893	17
44	.09990	.99500	.10040	9.9601	16
45	0.10019	0.99497	0.10069	9.9310	15
46	.10048	.99494	.10099	9.9021	14
47	.10077	.99491	.10128	9.8734	13
48	.10106	.99488	.10158	9.8448	12
49	.10135	.99485	.10187	9.8164	11
50	0.10164	0.99482	0.10216	9.7882	10
51	.10192	.99479	.10246	9.7601	9
52	.10221	.99476	.10275	9.7322	8
53	.10250	.99473	.10305	9.7044	7
54	.10279	.99470	.10334	9.6768	6
55	0.10308	0.99467	0.10363	9.6493	5
56	.10337	.99464	.10393	9.6220	4
57	.10366	.99461	.10422	9.5949	3
58	.10395	.99458	.10452	9.5679	2
59	.10424	.99455	.10481	9.5411	1
60	0.10453	0.99452	0.10510	9.5144	0
M	Cosine	Sine	Cotan.	Tan.	M

84°

Natural Trigonometric Functions

6°

M	Sine	Cosine	Tan.	Cotan.	M
0	0.10453	0.99452	0.19510	9.5144	60
1	.10482	.99449	.10540	9.4878	59
2	.10511	.99446	.10569	9.4614	58
3	.10540	.99443	.10599	9.4352	57
4	.10569	.99440	.10628	9.4090	56
5	0.10597	0.99437	0.10657	9.3831	55
6	.10626	.99434	.10687	9.3572	54
7	.10655	.99431	.10716	9.3315	53
8	.10684	.99428	.10746	9.3060	52
9	.10713	.99424	.10775	9.2806	51
10	0.10742	0.99421	0.10805	9.2553	50
11	.10771	.99418	.10834	9.2302	49
12	.10800	.99415	.10863	9.2052	48
13	.10829	.99412	.10893	9.1803	47
14	.10858	.99409	.10922	9.1555	46
15	0.10887	0.99406	0.10952	9.1309	45
16	.10916	.99402	.10981	9.1065	44
17	.10945	.99399	.11011	9.0821	43
18	.10973	.99396	.11040	9.0579	42
19	.11002	.99393	.11070	9.0338	41
20	0.11031	0.99390	0.11099	9.0098	40
21	.11060	.99386	.11128	8.9860	39
22	.11089	.99383	.11158	8.9623	38
23	.11118	.99380	.11187	8.9387	37
24	.11147	.99377	.11217	8.9152	36
25	0.11176	0.99374	0.11246	8.8919	35
26	.11205	.99370	.11276	8.8686	34
27	.11234	.99367	.11305	8.8455	33
28	.11263	.99364	.11335	8.8225	32
29	.11291	.99360	.11364	8.7996	31
30	0.11320	0.99357	0.11394	8.7769	30
31	.11349	.99354	.11423	8.7542	29
32	.11378	.99351	.11452	8.7317	28
33	.11407	.99347	.11482	8.7093	27
34	.11436	.99344	.11511	8.6870	26
35	0.11465	0.99341	0.11541	8.6648	25
36	.11494	.99337	.11570	8.6427	24
37	.11523	.99334	.11600	8.6208	23
38	.11552	.99331	.11629	8.5989	22
39	.11580	.99327	.11659	8.5772	21
40	0.11609	0.99324	0.11688	8.5555	20
41	.11638	.99320	.11718	8.5340	19
42	.11667	.99317	.11747	8.5126	18
43	.11696	.99314	.11777	8.4913	17
44	.11725	.99310	.11806	8.4701	16
45	0.11754	0.99307	0.11836	8.4490	15
46	.11783	.99303	.11865	8.4280	14
47	.11812	.99300	.11895	8.4071	13
48	.11840	.99297	.11924	8.3863	12
49	.11869	.99293	.11954	8.3656	11
50	0.11898	0.99290	9.11983	8.3450	10
51	.11927	.99286	.12013	8.3245	9
52	.11956	.99283	.12042	8.3041	8
53	.11985	.99279	.12072	8.2838	7
54	.12014	.99276	.12101	8.2636	6
55	0.12043	0.99272	0.12131	8.2434	5
56	.12071	.99269	.12160	8.2234	4
57	.12100	.99265	.12190	8.2035	3
58	.12129	.99262	.12219	8.1837	2
59	.12158	.99258	.12249	8.1640	1
60	0.12187	0.99255	0.12278	8.1443	0
M	Cosine	Sine	Cotan.	Tan.	M

83°

Natural Trigonometric Functions

7°

M	Sine	Cosine	Tan.	Cotan.	M
0	0.12187	0.99255	0.12278	8.1443	60
1	.12216	.99251	.12308	8.1248	59
2	.12245	.99248	.12338	8.1054	58
3	.12274	.99244	.12367	8.0860	57
4	.12302	.99240	.12397	8.0667	56
5	0.12331	0.99237	0.12426	8.0476	55
6	.12360	.99233	.12456	8.0285	54
7	.12389	.99230	.12485	8.0095	53
8	.12418	.99226	.12515	7.9906	52
9	.12447	.99222	.12544	7.9718	51
10	0.12476	0.99219	0.12574	7.9530	50
11	.12504	.99215	.12603	7.9344	49
12	.12533	.99211	.12633	7.9158	48
13	.12562	.99208	.12662	7.8973	47
14	.12591	.99204	.12692	7.8789	46
15	0.12620	0.99200	0.12722	7.8606	45
16	.12649	.99197	.12751	7.8424	44
17	.12678	.99193	.12781	7.8243	43
18	.12706	.99189	.12810	7.8062	42
19	.12735	.99186	.12840	7.7882	41
20	0.12764	0.99182	0.12869	7.7704	40
21	.12793	.99178	.12899	7.7525	39
22	.12822	.99175	.12929	7.7348	38
23	.12851	.99171	.12958	7.7171	37
24	.12880	.99167	.12988	7.6996	36
25	0.12908	0.99163	0.13017	7.6821	35
26	.12937	.99160	.13047	7.6647	34
27	.12966	.99156	.13076	7.6473	33
28	.12995	.99152	.13106	7.6301	32
29	.13024	.99148	.13136	7.6129	31
30	0.13053	0.99144	0.13165	7.5958	30
31	.13081	.99141	.13195	7.5787	29
32	.13110	.99137	.13224	7.5618	28
33	.13139	.99133	.13254	7.5449	27
34	.13168	.99129	.13284	7.5281	26
35	0.13197	0.99125	0.13313	7.5113	25
36	.13226	.99122	.13343	7.4947	24
37	.13254	.99118	.13372	7.4781	23
38	.13283	.99114	.13402	7.4615	22
39	.13312	.99110	.13432	7.4451	21
40	0.13341	0.99106	0.13461	7.4287	20
41	.13370	.99102	.13491	7.4124	19
42	.13399	.99098	.13521	7.3962	18
43	.13427	.99094	.13550	7.3800	17
44	.13456	.99091	.13580	7.3639	16
45	0.13485	0.99087	0.13609	7.3479	15
46	.13514	.99083	.13639	7.3319	14
47	.13543	.99079	.13669	7.3160	13
48	.13572	.99075	.13698	7.3002	12
49	.13600	.99071	.13728	7.2844	11
50	0.13629	0.99067	0.13758	7.2687	10
51	.13658	.99063	.13787	7.2531	9
52	.13687	.99059	.13817	7.2375	8
53	.13716	.99055	.13846	7.2220	7
54	.13744	.99051	.13876	7.2066	6
55	0.13773	0.99047	0.13906	7.1912	5
56	.13802	.99043	.13935	7.1759	4
57	.13831	.99039	.13965	7.1607	3
58	.13860	.99035	.13995	7.1455	2
59	.13889	.99031	.14024	7.1304	1
60	0.13917	0.99027	0.14054	7.1154	0
M	Cosine	Sine	Cotan.	Tan.	M

82°

Natural Trigonometric Functions

8°

M	Sine	Cosine	Tan.	Cotan.	M
0	0.13917	0.99027	0.14054	7.1154	60
1	.13946	.99023	.14084	7.1004	59
2	.13975	.99019	.14113	7.0855	58
3	.14004	.99015	.14143	7.0706	57
4	.14033	.99011	.14173	7.0558	56
5	0.14061	0.99006	0.14202	7.0410	55
6	.14090	.99002	.14232	7.0264	54
7	.14119	.98998	.14262	7.0117	53
8	.14148	.98994	.14291	6.9972	52
9	.14177	.98990	.14321	6.9827	51
10	0.14205	0.98986	0.14351	6.9682	50
11	.14234	.98982	.14381	6.9538	49
12	.14263	.98978	.14410	6.9395	48
13	.14292	.98973	.14440	6.9252	47
14	.14320	.98969	.14470	6.9110	46
15	0.14349	0.98965	0.14499	6.8969	45
16	.14378	.98961	.14529	6.8828	44
17	.14407	.98957	.14559	6.8687	43
18	.14436	.98953	.14588	6.8548	42
19	.14464	.98948	.14618	6.8408	41
20	0.14493	0.98944	0.14648	6.8269	40
21	.14522	.98940	.14678	6.8131	39
22	.14551	.98936	.14707	6.7994	38
23	.14580	.98931	.14737	6.7856	37
24	.14608	.98927	.14767	6.7720	36
25	0.14637	0.98923	0.14796	6.7584	35
26	.14666	.98919	.14826	6.7448	34
27	.14695	.98914	.14856	6.7313	33
28	.14723	.98910	.14886	6.7179	32
29	.14752	.98906	.14915	6.7045	31
30	0.14781	0.98902	0.14945	6.6912	30
31	.14810	.98897	.14975	6.6779	29
32	.14838	.98893	.15005	6.6646	28
33	.14867	.98889	.15034	6.6514	27
34	.14896	.98884	.15064	6.6383	26
35	0.14925	0.98880	0.15094	6.6252	25
36	.14954	.98876	.15124	6.6122	24
37	.14982	.98871	.15153	6.5992	23
38	.15011	.98867	.15183	6.5863	22
39	.15040	.98863	.15213	6.5734	21
40	0.15069	0.98858	0.15243	6.5606	20
41	.15097	.98854	.15272	6.5478	19
42	.15126	.98849	.15302	6.5350	18
43	.15155	.98845	.15332	6.5223	17
44	.15184	.98841	.15362	6.5097	16
45	0.15212	0.98836	0.15391	6.4971	15
46	.15241	.98832	.15421	6.4846	14
47	.15270	.98827	.15451	6.4721	13
48	.15290	.98823	.15481	6.4596	12
49	.15327	.98818	.15511	6.4472	11
50	0.15356	0.98814	0.15540	6.4348	10
51	.15385	.98809	.15570	6.4225	9
52	.15414	.98805	.15600	6.4103	8
53	.15442	.98800	.15630	6.3980	7
54	.15471	.09796	.15660	6.3859	6
55	0.15500	0.98791	0.15689	6.3737	5
56	.15529	.98787	.15719	6.3617	4
57	.15557	.98782	.15749	6.3496	3
58	.15586	.98778	.15779	6.3376	2
59	.15615	.98773	.15809	6.3257	1
60	0.15643	0.98769	0.15838	6.3138	0
M	Cosine	Sine	Cotan.	Tan.	M

81°

Natural Trigonometric Functions

9°

M	Sine	Cosine	Tan.	Cotan.	M
0	0.15643	0.98769	0.15838	6.3138	60
1	.15672	.98764	.15868	6.3019	59
2	.15701	.98760	.15898	6.2901	58
3	.15730	.98755	.15928	6.2783	57
4	.15758	.98751	.15958	6.2666	56
5	0.15787	0.98746	0.15988	6.2549	55
6	.15816	.98741	.16017	6.2432	54
7	.15845	.98737	.16047	6.2316	53
8	.15873	.98732	.16077	6.2200	52
9	.15902	.98728	.16107	6,2085	51
10	0.15931	0.98723	0.16137	6.1970	50
11	.15959	.98718	.16167	6.1856	49
12	.15988	.98714	.16196	6.1742	48
13	.16017	.98709	.16226	6.1628	47
14	.16046	.98704	.16256	6.1515	46
15	0.16074	0.98700	0.16286	6.1402	45
16	.16103	.98695	.16316	6.1290	44
17	.16132	.98690	.16346	6.1178	43
18	.16160	.98686	.16376	6.1066	42
19	.16189	.98681	.16405	6.0955	41
20	0.16218	0.98676	0.16435	6.0844	40
21	.16246	.98671	.16465	6.0734	39
22	.16275	.98667	.16495	6.0624	38
23	.16304	.98662	.16525	6.0514	37
24	.16333	.98657	.16555	6.0405	36
25	0.16361	0.98652	0.16585	6.0296	35
26	.16390	.98648	.16615	6.0188	34
27	.16419	.98643	.16645	6.0080	33
28	.16447	.98638	.16674	5.9972	32
29	.16476	.98633	.16704	5.9865	31
30	0.16505	0.98629	0.16734	5.9758	30
31	.16533	.98624	.16764	5.9651	29
32	.16562	.98619	.16794	5.9545	28
33	.16591	.98614	.16824	5.9439	27
34	.16620	.98609	.16854	5.9333	26
35	0.16648	0.98604	0.16884	5.9228	25
36	.16677	.98600	.16914	5.9124	24
37	.16706	.98595	.16944	5.9019	23
38	.16734	.98590	.16974	5.8915	22
39	.16763	.98585	.17004	5.8811	21
40	0.16792	0.98580	0.17033	5.8708	20
41	.16820	.98575	.17063	5.8605	19
42	.16849	.98570	.17093	5.8502	18
43	.16878	.98565	.17123	5.8400	17
44	.16906	.98561	.17153	5.8298	16
45	0.16935	0.98556	0.17183	5.8197	15
46	.16964	.98551	.17213	5.8095	14
47	.16992	.98546	.17243	5.7994	13
48	.17021	.98541	.17273	5.7894	12
49	.17050	.98536	.17303	5.7794	11
50	0.17078	0.98531	0.17333	5.7694	10
51	.17107	.98526	.17363	5.7594	9
52	.17136	.98521	.17393	5.7495	8
53	.17164	.98516	.17423	5.7396	7
54	.17193	.98511	.17453	5.7297	6
55	0.17222	0.98506	0.17483	5.7199	5
56	.17250	.98501	.17513	5.7101	4
57	.17279	.98496	.17543	5.7004	3
58	.17308	.98491	.17573	5.6906	2
59	.17336	.98486	.17603	5.6809	1
60	0.17365	0.98481	0.17633	5.6713	0
M	Cosine	Sine	Cotan.	Tan.	M

80°

Natural Trigonometric Functions

10°

M	Sine	Cosine	Tan.	Cotan.	M
0	0.17365	0.98481	0.17633	5.6713	60
1	.17393	.98476	.17663	5.6617	50
2	.17422	.98471	.17693	5.6521	58
3	.17451	.98466	.17723	5.6425	57
4	.17479	.98461	.17753	5.6329	56
5	0.17508	0.98455	0.17783	5.6234	55
6	.17537	.98450	.17813	5.6140	54
7	.17565	.98445	.17843	5.6045	53
8	.17594	.98440	.17873	5.5951	52
9	.17623	.98435	.17903	5.5857	51
10	0.17651	0.98430	0.17933	5.5764	50
11	.17680	.98425	.17963	5.5671	49
12	.17708	.98420	.17993	5.5578	48
13	.17737	.98414	.18023	5.5485	47
14	.17766	.98409	.18053	5.5393	46
15	0.17794	0.98404	0.18083	5.5301	45
16	.17823	.98399	.18113	5.5209	44
17	.17852	.98394	.18143	5.5118	43
18	.17880	.98389	.18173	5.5026	42
19	.17909	.98383	.18203	5.4936	41
20	0.17937	0.98378	0.18233	5.4845	40
21	.17966	.98373	.18263	5.4755	39
22	.17995	.98368	.18293	5.4665	38
23	.18023	.98362	.18323	5.4575	37
24	.18052	.98357	.18353	5.4486	36
25	0.18081	0.98352	0.18384	5.4397	35
26	.18109	.98347	.18414	5.4308	34
27	.18138	.98341	.18444	5.4219	33
28	.18166	.98336	.18474	5.4131	32
29	.18195	.98331	.18504	5.4043	31
30	0.18224	0.98325	0.18534	5.3955	30
31	.18252	.98320	.18564	5.3868	29
32	.18281	.98315	.18594	5.3781	28
33	.18309	.98310	.18624	5.3694	27
34	.18338	.98304	.18654	5.3607	26
35	0.18367	0.98299	0.18684	5.3521	25
36	.18395	.98294	.18714	5.3435	24
37	.18424	.98288	.18745	5.3349	23
38	.18452	.98283	.18775	5.3263	22
39	.18481	.98277	.18805	5.3178	21
40	0.18509	0.98272	0.18835	5.3093	20
41	.18538	.98267	.18865	5.3008	19
42	.18567	.98261	.18895	5.2924	18
43	.18595	.98256	.18925	5.2839	17
44	.18624	.98250	.18955	5.2755	16
45	0.18652	0.98245	0.18986	5.2672	15
46	.18681	.98240	.19016	5.2588	14
47	.18710	.98234	.19046	5.2505	13
48	.18738	.98229	.19076	5.2422	12
49	.18767	.98223	.19106	5.2339	11
50	0.18795	0.98218	0.19136	5.2257	10
51	.18824	.98212	.19166	5.2174	9
52	.18852	.98207	.19197	5.2092	8
53	.18881	.98201	.19227	5.2011	7
54	.18910	.98196	.19257	5.1929	6
55	0.18938	0.98190	0.19287	5.1848	5
56	.18967	.98185	.19317	5.1767	4
57	.18995	.98179	.19347	5.1686	3
58	.19024	.98174	.19378	5.1606	2
59	.19052	.98168	.19408	5.1526	1
60	0.19081	0.98163	0.19438	5.1446	0
M	Cosine	Sine	Cotan.	Tan.	M

79°

Natural Trigonometric Functions

11°

M	Sine	Cosine	Tan.	Cotan.	M
0	0.19081	0.98163	0.19438	5.1466	60
1	.19109	.98157	.19468	5.1366	59
2	.19138	.98152	.19498	5.1286	58
3	.19167	.98146	.19529	5.1207	57
4	.19195	.98140	.19559	5.1128	56
5	0.19224	0.98135	0.19589	5.1049	55
6	.19252	.98129	.19619	5.0970	54
7	.19281	.98124	.19649	5.0892	53
8	.19309	.98118	.19680	5.0814	52
9	.19338	.98112	.19710	5.0736	51
10	0.19366	0.98107	0.19740	5.0658	50
11	.19395	.98101	.19770	5.0581	49
12	.19423	.98096	.19801	5.0504	48
13	.19452	.98090	.19831	5.0427	47
14	.19481	.98084	.19861	5.0350	46
15	0.19509	0.98079	0.19891	5.0273	45
16	.19538	.98073	.19921	5.0197	44
17	.19566	.98067	.19952	5.0121	43
18	.19595	.98061	.19982	5.0045	42
19	.19623	.98056	.20012	4.9969	41
20	0.19652	0.98050	0.20042	4.9894	40
21	.19680	.98044	.20073	4.9819	39
22	.19709	.98030	.20103	4.9744	38
23	.19737	.98033	.20133	4.9669	37
24	.19766	.98027	.20164	4.9594	36
25	0.19794	0.98021	0.20194	4.9520	35
26	.19823	.98016	.20224	4.9446	34
27	.19851	.98010	.20254	4.9372	33
28	.19880	.98004	.20285	4.9298	32
29	.19908	.97998	.20315	4.9225	31
30	0.19937	0.97992	0.20345	4.9152	30
31	.19965	.97987	.20376	4.9078	29
32	.19994	.97981	.20406	4.9006	28
33	.20022	.97975	.20436	4.8933	27
34	.20051	.97969	.20466	4.8860	26
35	0.20079	0.97962	0.20497	4.8788	25
36	.20108	.97958	.20527	4.8716	24
37	.20136	.97952	.20557	4.8644	23
38	.20165	.97946	.20588	4.8573	22
39	.20193	.97940	.20618	4.8501	21
40	0.20222	0.97934	0.20648	4.8430	20
41	.20250	.97928	.20679	4.8359	19
42	.20279	.97922	.20709	4.8288	18
43	.20307	.97916	.20739	4.8218	17
44	.20336	.97910	.20770	4.8147	16
45	0.20364	0.97905	0.20800	4.8077	15
46	.20393	.97899	.20830	4.8007	14
47	.20421	.97893	.20861	4.7937	13
48	.20450	.97887	.20891	4.7867	12
49	.20478	.97881	.20921	4.7798	11
50	0.20507	0.97875	0.20952	4.7729	10
51	.20535	.97869	.20982	4.7659	9
52	.20563	.97863	.21013	4.7591	8
53	.20592	.97857	.21043	4.7522	7
54	.20620	.97851	.21073	4.7453	6
55	0.20649	0.97845	0.21104	4.7385	5
56	.20677	.97839	.21134	4.7317	4
57	.20706	.97833	.21164	4.7249	3
58	.20734	.97827	.21195	4.7181	2
59	.20763	.97821	.21225	4.7114	1
60	0.20791	0.97815	0.21256	4.7046	0
M	Cosine	Sine	Cotan.	Tan.	M

78°

Natural Trigonometric Functions

12°

M	Sine	Cosine	Tan.	Cotan.	M
0	0.20791	0.97815	0.21256	4.7046	60
1	.20820	.97809	.21286	4.6979	59
2	.20848	.97803	.21316	4.6912	58
3	.20877	.97797	.21347	4.6845	57
4	.20905	.97791	.21377	4.6779	56
5	0.20933	0.97784	0.21408	4.6712	55
6	.20962	.97778	.21438	4.6646	54
7	.20990	.97772	.21469	4.6580	53
8	.21019	.97766	.21499	4.6514	52
9	.21047	.97760	.21529	4.6448	51
10	0.21076	0.97754	0.21560	4.6382	50
11	.21104	.97748	.21590	4.6317	49
12	.21132	.97742	.21621	4.6252	48
13	.21161	.97735	.21651	4.6187	47
14	.21189	.97729	.21682	4.6122	46
15	0.21218	0.97723	0.21712	4.6057	45
16	.21246	.97717	.21743	4.5993	44
17	.21275	.97711	.21773	4.5928	43
18	.21303	.97705	.21804	4.5864	42
19	.21331	.97698	.21834	4.5800	41
20	0.21360	0.97692	0.21864	4.5736	40
21	.21388	.97686	.21895	4.5673	39
22	.21417	.97680	.21925	4.5609	38
23	.21445	.97673	.21956	4.5546	37
24	.21474	.97667	.21986	4.5483	36
25	0.21502	0.97661	0.22017	4.5420	35
26	.21530	.97655	.22047	4.5357	34
27	.21559	.97648	.22078	4.5294	33
28	.21587	.97642	.22108	4.5232	32
29	.21616	.97636	.22139	4.5169	31
30	0.21644	0.97630	0.22169	4.5107	30
31	.21672	.97623	.22200	4.5045	29
32	.21701	.97617	.22231	4.4983	28
33	.21729	.97611	.22261	4.4922	27
34	.21758	.97604	.22292	4.4860	26
35	0.21786	0.97598	0.22322	4.4799	25
36	.21814	.97592	.22353	4.4737	24
37	.21843	.97585	.22383	4.4676	23
38	.21871	.97579	.22414	4.4615	22
39	.21899	.97573	.22444	4.4555	21
40	0.21928	0.97566	0.22475	4.4494	20
41	.21956	.97560	.22505	4.4434	19
42	.21985	.97553	.22536	4.4373	18
43	.22013	.97547	.22567	4.4313	17
44	.22041	.97541	.22597	4.4253	16
45	0.22070	0.97534	0.22628	4.4194	15
46	.22098	.97528	.22658	4.4134	14
47	.22126	.97521	.22689	4.4075	13
48	.22155	.97515	.22719	4.4015	12
49	.22183	.97508	.22750	4.3956	11
50	0.22212	0.97502	0.22781	4.3897	10
51	.22240	.97496	.22811	4.3838	9
52	.22268	.97489	.22842	4.3779	8
53	.22297	.97483	.22872	4.3721	7
54	.22325	.97476	.22903	4.3662	6
55	0.22353	0.97470	0.22934	4.3604	5
56	.22382	.97463	.22964	4.3546	4
57	.22410	.97457	.22995	4.3488	3
58	.22438	.97450	.23026	4.3430	2
59	.22467	.97444	.23056	4.3372	1
60	0.22495	0.97437	0.23087	4.3315	0
M	Cosine	Sine	Cotan.	Tan.	M

77°

Natural Trigonometric Functions

13°

M	Sine	Cosine	Tan.	Cotan.	M
0	0.22495	0.97437	0.23087	4.3315	60
1	.22523	.97430	.23117	4.3257	59
2	.22552	.97424	.23148	4.3200	58
3	.22580	.97417	.23179	4.3143	57
4	.22608	.97411	.23209	4.3086	56
5	0.22637	0.97404	0.23240	4.3029	55
6	.22665	.97398	.23271	4.2972	54
7	.22693	.97391	.23301	4.2916	53
8	.22722	.97384	.23332	4.2859	52
9	.22750	.97378	.23363	4.2803	51
10	0.22778	0.97371	0.23393	4.2747	50
11	.22807	.97365	.23424	4.2691	49
12	.22835	.97358	.23455	4.2635	48
13	.22863	.97351	.23485	4.2580	47
14	.22892	.97345	.23516	4.2524	46
15	0.22920	0.97338	0.23547	4.2468	45
16	.22948	.97331	.23578	4.2413	44
17	.22977	.97325	.23608	4.2358	43
18	.23005	.97318	.23639	4.2303	42
19	.23033	.97311	.23670	4.2248	41
20	0.23062	0.97304	0.23700	4.2193	40
21	.23090	.97298	.23731	4.2139	39
22	.23118	.97291	.23762	4.2084	38
23	.23146	.97284	.23793	4.2030	37
24	.23175	.97278	.23823	4.1976	36
25	0.23203	0.97271	0.23854	4.1922	35
26	.23231	.97264	.23885	4.1868	34
27	.23260	.97257	.23916	4.1814	33
28	.23288	.97251	.23946	4.1760	32
29	.23316	.97244	.23977	4.1706	31
30	0.23345	0.97237	0.24008	4.1653	30
31	.23373	.97230	.24039	4.1600	29
32	.23401	.97223	.24069	4.1547	28
33	.23429	.97217	.24100	4.1493	27
34	.23458	.97210	.24131	4.1441	26
35	0.23486	0.97203	0.24162	4.1388	25
36	.23514	.97196	.24193	4.1335	24
37	.23542	.97189	.24223	4.1282	23
38	.23571	.97182	.24254	4.1230	22
39	.23599	.97176	.24285	4.1178	21
40	0.23627	0.97169	0.24316	4.1126	20
41	.23656	.97162	.24347	4.1074	19
42	.23684	.97155	.24377	4.1022	18
43	.23712	.97148	.24408	4.0970	17
44	.23740	.97141	.24439	4.0918	16
45	0.23769	0.97134	0.24470	4.0867	15
46	.23797	.97127	.24501	4.0815	14
47	.23825	.97120	.24532	4.0764	13
48	.23853	.97113	.24562	4.0713	12
49	.23882	.97106	.24593	4.0662	11
50	0.23910	0.97100	0.24624	4.0611	10
51	.23938	.97093	.24655	4.0560	9
52	.23966	.97086	.24686	4.0509	8
53	.23995	.97079	.24717	4.0459	7
54	.24023	.97072	.24747	4.0408	6
55	0.24051	0.97065	0.24778	4.0358	5
56	.24079	.97058	.24809	4.0308	4
57	.24108	.97051	.24840	4.0257	3
58	.24136	.97044	.24871	4.0207	2
59	.24164	.97037	.24902	4.0158	1
60	0.24192	0.97030	0.24933	4.0108	0
M	Cosine	Sine	Cotan.	Tan.	M

76°

Natural Trigonometric Functions

14°

M	Sine	Cosine	Tan.	Cotan.	M
0	0.24192	0.97030	0.24933	4.0108	60
1	.24220	.97023	.24964	4.0058	59
2	.24249	.97015	.24995	4.0009	58
3	.24277	.97008	.25026	3.9959	57
4	.24305	.97001	.25056	3.9910	56
5	0.24333	0.96994	0.25087	3.9861	55
6	.24362	.96987	.25118	3.9812	54
7	.24390	.96980	.25149	3.9763	53
8	.24418	.96973	.25180	3.9714	52
9	.24446	.96966	.25211	3.9665	51
10	0.24474	0.96959	0.25242	3.9617	50
11	.24503	.96952	.25273	3.9568	49
12	.24531	.96945	.25304	3.9520	48
13	.24559	.96937	.25335	3.9471	47
14	.24587	.96930	.25366	3.9423	46
15	0.24615	0.96923	0.25397	3.9375	45
16	.24644	.96916	.25428	3.9327	44
17	.24672	.96909	.25459	3.9279	43
18	.24700	.96902	.25490	3.9232	42
19	.24728	.96894	.25521	3.9184	41
20	0.24756	0.96887	0.25552	3.9136	40
21	.24784	.96880	.25583	3.9089	39
22	.24813	.96873	.25614	3.9042	38
23	.24841	.96866	.25645	3.8995	37
24	.24869	.96858	.25676	3.8947	36
25	0.24897	0.96851	0.25707	3.8900	35
26	.24925	.96844	.25738	3.8854	34
27	.24954	.96837	.25769	3.8807	33
28	.24982	.96829	.25800	3.8760	32
29	.25010	.96822	.25831	3.8714	31
30	0.25038	0.96815	0.25862	3.8667	30
31	.25066	.96807	.25893	3.8621	29
32	.25094	.96800	.25924	3.8575	28
33	.25122	.96793	.25955	3.8528	27
34	.25151	.96786	.25986	3.8482	26
35	0.25179	0.96778	0.26017	3.8436	25
36	.25207	.96771	.26048	3.8391	24
37	.25235	.96764	.26079	3.8345	23
38	.25263	.96756	.26110	3.8299	22
39	.25291	.96749	.26141	3.8254	21
40	0.25320	0.96742	0.26172	3.8208	20
41	.25348	.96734	.26203	3.8163	19
42	.25376	.96727	.26235	3.8118	18
43	.25404	.96719	.26266	3.8073	17
44	.25432	.96712	.26297	3.8028	16
45	0.25460	0.96705	0.26328	3.7983	15
46	.25488	.96697	.26359	3.7938	14
47	.25516	.96690	.26390	3.7893	13
48	.25545	.96682	.26421	3.7848	12
49	.25573	.96675	.26452	3.7804	11
50	0.25601	0.96667	0.26483	3.7760	10
51	.25629	.96660	.26515	3.7715	9
52	.25657	.96653	.26546	3.7671	8
53	.25685	.96643	.26577	3.7627	7
54	.25713	.96638	.26608	3.7583	6
55	0.25741	0.96630	0.26639	3.7539	5
56	.25769	.96623	.26670	3.7495	4
57	.25798	.96615	.26701	3.7451	3
58	.25826	.96608	.26733	3.7408	2
59	.25854	.96600	.26764	3.7364	1
60	0.25882	0.96593	0.26795	3.7321	0
M	Cosine	Sine	Cotan.	Tan.	*M*

75°

Natural Trigonometric Functions

15°

M	Sine	Cosine	Tan.	Cotan.	M
0	0.25882	0.96593	0.26795	3.7321	60
1	.25910	.96585	.26826	3.7277	59
2	.25938	.96578	.26857	3.7234	58
3	.25966	.96570	.26888	3.7191	57
4	.25994	.96562	.26920	3.7148	56
5	0.26022	0.96555	0.26951	3.7105	55
6	.26050	.96547	.26982	3.7062	54
7	.26079	.96540	.27013	3.7019	53
8	.26107	.96532	.27044	3.6976	52
9	.26135	.96524	.27076	3.6933	51
10	0.26163	0.96517	0.27107	3.6891	50
11	.26191	.96509	.27138	3.6848	49
12	.26219	.96502	.27169	3.6806	48
13	.26247	.96494	.27201	3.6764	47
14	.26275	.96486	.27232	3.6722	46
15	0.26303	0.96479	0.27263	3.6680	45
16	.26331	.96471	.27294	3.6638	44
17	.26359	.96463	.27326	3.6596	43
18	.26387	.96456	.27357	3.6554	42
19	.26415	.96448	.27388	3.6512	41
20	0.26443	0.96440	0.27419	3.6470	40
21	.26471	.96433	.27451	3.6429	39
22	.26500	.96425	.27482	3.6387	38
23	.26528	.96417	.27513	3.6346	37
24	.26556	.96410	.27545	3.6305	36
25	0.26584	0.96402	0.27576	3.6264	35
26	.26612	.96394	.27607	3.6222	34
27	.26640	.96386	.27638	3.6181	33
28	.26668	.96379	.27670	3.6140	32
29	.26696	.96371	.27701	3.6100	31
30	0.26724	0.96363	0.27732	3.6059	30
31	.26752	.96355	.27764	3.6018	29
32	.26780	.96347	.27795	3.5978	28
33	.26808	.96340	.27826	3.5937	27
34	.26836	.96332	.27558	3.5897	26
35	0.26864	0.96324	0.27889	3.5856	25
36	.26892	.96316	.27921	3.5816	24
37	.26920	.96308	.27952	3.5776	23
38	.26948	.96301	.27983	3.5736	22
39	.26976	.96293	.28015	3.5696	21
40	0.27004	0.96285	0.28046	3.5656	20
41	.27032	.96277	.28077	3.5616	19
42	.27060	.96269	.28109	3.5576	18
43	.27088	.96261	.28140	3.5536	17
44	.27116	.96253	.28172	3.5497	16
45	0.27144	0.96246	0.28203	3.5457	15
46	.27172	.96238	.28234	3.5418	14
47	.27200	.96230	.28266	3.5379	13
48	.27228	.96222	.28297	3.5339	12
49	.27256	.96214	.28329	3.5300	11
50	0.27284	0.96206	0.28360	3.5261	10
51	.27312	.96198	.28391	3.5222	9
52	.27340	.96190	.28423	3.5183	8
53	.27368	.96182	.28454	3.5144	7
54	.27396	.96174	.28486	3.5105	6
55	0.27424	0.96166	0.27517	3.5067	5
56	.27452	.96158	.28549	3.5028	4
57	.27480	.96150	.28580	3.4989	3
58	.27508	.96142	.28612	3.4951	2
59	.27536	.96134	.28643	3.4912	1
60	0.27564	0.96126	0.28675	3.4874	0
M	Cosine	Sine	Cotan.	Tan.	*M*

74°

Natural Trigonometric Functions

16°

M	Sine	Cosine	Tan.	Cotan.	M
0	0.27564	0.96126	0.28675	3.4874	60
1	.27592	.96118	.28706	3.4836	59
2	.27620	.96110	.28738	3.4798	58
3	.27648	.96102	.28769	3.4760	57
4	.27676	.96094	.28801	3.4722	56
5	0.27704	0.96086	0.28832	3.4684	55
6	.27731	.96078	.28864	3.4646	54
7	.27759	.96070	.28895	3.4608	53
8	.27787	.96062	.28927	3.4570	52
9	.27815	.96054	.28958	3.4533	51
10	0.27843	0.96046	0.28990	3.4495	50
11	.27871	.96037	.29021	3.4458	49
12	.27899	.96029	.29053	3.4420	48
13	.27927	.96021	.29084	3.4383	47
14	.27955	.96013	.29116	3.4346	46
15	0.27983	0.96005	0.29147	3.4308	45
16	.28011	.95997	.29179	3.4271	44
17	.28039	.95989	.29210	3.4234	43
18	.28067	.95981	.29242	3.4197	42
19	.28095	.95972	.29274	3.4160	41
20	0.28123	0.95964	0.29305	3.4124	40
21	.28150	.95956	.29337	3.4087	39
22	.28178	.95948	.29368	3.4050	38
23	.28206	.95940	.29400	3.4014	37
24	.28234	.95931	.29432	3.3977	36
25	0.28262	0.95923	0.29463	3.3941	35
26	.28290	.95915	.29495	3.3904	34
27	.28318	.95907	.29526	3.3868	33
28	.28346	.95898	.29558	3.3832	32
29	.28374	.95890	.29590	3.3796	31
30	0.28402	0.95882	0.29621	3.3759	30
31	.28429	.95874	.29653	3.3723	29
32	.28457	.95865	.29685	3.3687	28
33	.28485	.95857	.29716	3.3652	27
34	.28513	.95849	.29748	3.3616	26
35	0.28541	0.95841	0.29780	3.3580	25
36	.28569	.95832	.29811	3.3544	24
37	.28597	.95824	.29843	3.3509	23
38	.28625	.95816	.29875	3.3473	22
39	.28652	.95807	.29906	3.3438	21
40	0.28680	0.95799	0.29938	3.3402	20
41	.28708	.95791	.29970	3.3367	19
42	.28736	.95782	.30001	3.3332	18
43	.28764	.95774	.30033	3.3297	17
44	.28792	.95766	.30065	3.3261	16
45	0.28820	0.95757	0.30097	3.3226	15
46	.28847	.95749	.30128	3.3191	14
47	.28875	.95740	.30160	3.3156	13
48	.28903	.95732	.30192	3.3122	12
49	.28931	.95724	.30224	3.3087	11
50	0.28959	0.95715	0.30255	3.3052	10
51	.28987	.95707	.30287	3.3017	9
52	.29015	.95698	.30319	3.2983	8
53	.29042	.95690	.30351	3.2948	7
54	.29070	.95681	.30382	3.2914	6
55	0.29098	0.95673	0.30414	3.2878	5
56	.29126	.95664	.30446	3.2845	4
57	.29154	.95656	.30478	3.2811	3
58	.29182	.95647	.30509	3.2777	2
59	.29209	.95639	.30541	3.2743	1
60	0.29237	0.95630	0.30573	3.2709	0
M	Cosine	Sine	Cotan.	Tan.	M

73°

Natural Trigonometric Functions

17°

M	Sine	Cosine	Tan.	Cotan.	M
0	0.29237	0.95630	0.30573	3.2709	60
1	.29265	.95622	.30605	3.2675	59
2	.29293	.95613	.30637	3.2641	58
3	.29321	.95605	.30669	3.2607	57
4	.29348	.95596	.30700	3.2573	56
5	0.29376	0.95588	0.30732	3.2539	55
6	.29404	.95579	.30764	3.2506	54
7	.29432	.95571	.30796	3.2472	53
8	.29460	.95562	.30828	3.2438	52
9	.29487	.95554	.30860	3.2405	51
10	0.29515	0.95545	0.30891	3.2371	50
11	.29543	.95536	.30923	3.2338	49
12	.29571	.95528	.30955	3.2305	48
13	.29599	.95519	.30987	3.2272	47
14	.29626	.95511	.31019	3.2238	46
15	0.29654	0.95502	0.31051	3.2205	45
16	.29682	.95493	.31083	3.2172	44
17	.29710	.95485	.31115	3.2139	43
18	.29737	.95476	.31147	3.2106	42
19	.29765	.95467	.31178	3.2073	41
20	0.29793	0.95459	0.31210	3.2041	40
21	.29821	.95450	.31242	3.2008	39
22	.29849	.95441	.31274	3.1975	38
23	.29876	.95433	.31306	3.1943	37
24	.29904	.95424	.31338	3.1910	36
25	0.29932	0.05415	0.31370	3.1878	35
26	.29960	.95407	.31402	3.1845	34
27	.29987	.95398	.31434	3.1813	33
28	.30015	.95389	.31466	3.1780	32
29	.30043	.95380	.31498	3.1748	31
30	0.30071	0.95372	0.31530	3.1716	30
31	.30098	.95363	.31562	3.1684	29
32	.30126	.95354	.31594	3.1652	28
33	.30154	.95345	.31626	3.1620	27
34	.30182	.95337	.31658	3.1588	26
35	0.30209	0.95328	0.31690	3.1556	25
36	.30237	.95319	.31722	3.1524	24
37	.30265	.95310	.31754	3.1492	23
38	.30292	.95301	.31786	3.1460	22
39	.30320	.95293	.31818	3.1429	21
40	0.30348	0.95284	0.31850	3.1397	20
41	.30376	.95275	.31882	3.1366	19
42	.30403	.95266	.31914	3.1334	18
43	.30431	.95257	.31946	3.1303	17
44	.30459	.95248	.31978	3.1271	16
45	0.30486	0.95240	0.32010	3.1240	15
46	.30514	.95231	.32042	3.1209	14
47	.30542	.95222	.32074	3.1178	13
48	.30570	.95213	.32106	3.1146	12
49	.30597	.95204	.32139	3.1115	11
50	0.30625	0.95195	0.32171	3.1084	10
51	.30653	.95186	.32203	3.1053	9
52	.30680	.95177	.32235	3.1022	8
53	.30708	.95168	.32267	3.0991	7
54	.30736	.95159	.32299	3.0961	6
55	0.30763	0.95150	0.32331	3.0930	5
56	.30791	.95142	.32363	3.0899	4
57	.30819	.95133	.32396	3.0868	3
58	.30846	.95124	.32428	3.0838	2
59	.30874	.95115	.32460	3.0807	1
60	0.30902	0.95106	0.32492	3.0777	0
M	Cosine	Sine	Cotan.	Tan.	M

72°

Natural Trigonometric Functions

18°

M	Sine	Cosine	Tan.	Cotan.	M
0	0.30902	0.95106	0.32492	3.0777	60
1	.30929	.95097	.32524	3.0746	59
2	.30957	.95088	.32556	3.0716	58
3	.30985	.95079	.32588	3.0686	57
4	.31012	.95070	.32621	3.0655	56
5	0.31040	0.95061	0.32653	3.0625	55
6	.31068	.95052	.32685	3.0595	54
7	.31095	.95043	.32717	3.0565	53
8	.31123	.95033	.32749	3.0535	52
9	.31151	.95024	.32782	3.0505	51
10	0.31178	0.95015	0.32814	3.0475	50
11	.31206	.95006	.32846	3.0445	49
12	.31233	.94997	.32878	3.0415	48
13	.31261	.94988	.32911	3.0385	47
14	.31289	.94979	.32943	3.0356	46
15	0.31316	0.94970	0.32975	3.0326	45
16	.31344	.94961	.33007	3.0296	44
17	.31372	.94952	.33040	3.0267	43
18	.31399	.94943	.33072	3.0237	42
19	.31427	.94933	.33104	3.0208	41
20	0.31454	0.94924	0.33136	3.0178	40
21	.31482	.94915	.33169	3.0149	39
22	.31510	.94906	.33201	3.0120	38
23	.31537	.94897	.33233	3.0090	37
24	.31565	.94888	.33266	3.0061	36
25	0.31593	0.94878	0.33298	3.0032	35
26	.31620	.94869	.33330	3.0003	34
27	.31648	.94860	.33363	2.9974	33
28	.31675	.94851	.33395	2.9945	32
29	.31703	.94842	.33427	2.9916	31
30	0.31730	0.94832	0.33460	2.9887	30
31	.31758	.94823	.33492	2.9858	29
32	.31786	.94814	.33524	2.9829	28
33	.31813	.94805	.33557	2.9800	27
34	.31841	.94795	.33589	2.9772	26
35	0.31868	0.94786	0.33621	2.9743	25
36	.31896	.94777	.33654	2.9714	24
37	.31923	.94768	.33686	2.9686	23
38	.31951	.94758	.33718	2.9657	22
39	.31979	.94749	.33751	2.9629	21
40	0.32006	0.94740	0.33783	2.9600	20
41	.32034	.94730	.33816	2.9572	19
42	.32061	.94721	.33848	2.9544	18
43	.32089	.94712	.33881	2.9515	17
44	.32116	.94702	.33913	2.9487	16
45	0.32144	0.94693	0.33945	2.9459	15
46	.32171	.94684	.33978	2.9431	14
47	.32199	.94674	.34010	2.9403	13
48	.32227	.94665	.34043	2.9375	12
49	.32254	.94656	.34075	2.9347	11
50	0.32282	0.94646	0.34108	2.9319	10
51	.32309	.94637	.34140	2.9291	9
52	.32337	.94627	.34173	2.9263	8
53	.32364	.94618	.34205	2.9235	7
54	.32392	.94609	.34238	2.9208	6
55	0.32419	0.94599	0.34270	2.9180	5
56	.32447	.94590	.34303	2.9152	4
57	.32474	.94580	.34335	2.9125	3
58	.32502	.94571	.34368	2.9097	2
59	.32529	.94561	.34400	2.9070	1
60	0.32557	0.94552	0.34433	2.9042	0
M	Cosine	Sine	Cotan.	Tan.	M

71°

Natural Trigonometric Functions

19°

M	Sine	Cosine	Tan.	Cotan.	M
0	0.32557	0.94552	0.34433	2.9042	60
1	.32584	.94542	.34465	2.9015	59
2	.32612	.94533	.34498	2.8987	58
3	.32639	.94523	.34530	2.8960	57
4	.32667	.94514	.34563	2.8933	56
5	0.32694	0.94504	0.34596	2.8905	55
6	.32722	.94495	.34628	2.8878	54
7	.32749	.94485	.34661	2.8851	53
8	.32777	.94476	.34693	2.8824	52
9	.32804	.94466	.34726	2.8797	51
10	0.32832	0.94457	0.34758	2.8770	50
11	.32859	.94447	.34791	2.8743	49
12	.32887	.94438	.34824	2.8716	48
13	.32914	.94428	.34856	2.8689	47
14	.32942	.94418	.34889	2.8662	46
15	0.32969	0.94409	0.34922	2.8636	45
16	.32997	.94399	.34954	2.8609	44
17	.33024	.94390	.34987	2.8582	43
18	.33051	.94380	.35020	2.8556	42
19	.33079	.94370	.35052	2.8529	41
20	0.33106	0.94361	0.35085	2.8502	40
21	.33134	.94351	.35118	2.8476	39
22	.33161	.94342	.35150	2.8449	38
23	.33189	.94332	.35183	2.8423	37
24	.33216	.94322	.35216	2.8397	36
25	0.33244	0.94313	0.35248	2.8370	35
26	.33271	.94303	.35281	2.8344	34
27	.33298	.94293	.35314	2.8318	33
28	.33326	.94284	.35346	2.8291	32
29	.33353	.94274	.35379	2.8265	31
30	0.33381	0.94264	0.35412	2.8239	30
31	.33408	.94254	.35445	2.8213	29
32	.33436	.94245	.35477	2.8187	28
33	.33463	.94235	.35510	2.8161	27
34	.33490	.94225	.35543	2.8135	26
35	0.33518	0.94215	0.35576	2.8109	25
36	.33545	.94206	.35608	2.8083	24
37	.33573	.94196	.35641	2.8057	23
38	.33600	.94186	.35674	2.8032	22
39	.33627	.94176	.35707	2.8006	21
40	0.33655	0.94167	0.35740	2.7980	20
41	.33682	.94157	.35772	2.7955	19
42	.33710	.94147	.35805	2.7929	18
43	.33737	.94137	.35838	2.7903	17
44	.33764	.94127	.35871	2.7878	16
45	0.33792	0.94118	0.35904	2.7852	15
46	.33819	.94108	.35937	2.7827	14
47	.33846	.94098	.35969	2.7801	13
48	.33874	.94088	.36002	2.7776	12
49	.33901	.94078	.36035	2.7751	11
50	0.33929	0.94068	0.36068	2.7725	10
51	.33956	.94058	.36101	2.7700	9
52	.33983	.94049	.36134	2.7675	8
53	.34011	.94039	.36167	2.7650	7
54	.34038	.94029	.36199	2.7625	6
55	0.34065	0.94019	0.36232	2.7600	5
56	.34093	.94009	.36265	2.7575	4
57	.34120	.93999	.36298	2.7550	3
58	.34147	.93989	.36331	2.7525	2
59	.34175	.93979	.36364	2.7500	1
60	0.34202	0.93969	0.36397	2.7475	0
M	Cosine	Sine	Cotan.	Tan.	M

70°

Natural Trigonometric Functions

20°

M	Sine	Cosine	Tan.	Cotan.	M
0	0.34202	0.93969	0.36397	2.7478	60
1	.34229	.93959	.36430	2.7450	59
2	.34257	.93949	.36463	2.7425	58
3	.34284	.93939	.36496	2.7400	57
4	.34311	.93929	.36529	2.7376	56
5	0.34339	0.93919	0.36562	2.7351	55
6	.34366	.93909	.36595	2.7326	54
7	.34393	.93899	.36628	2.7302	53
8	.34421	.93889	.36661	2.7277	52
9	.34448	.93879	.36694	2.7253	51
10	0.34475	0.93869	0.36727	2.7228	50
11	.34503	.93859	.36760	2.7204	49
12	.34530	.93849	.36793	2.7179	48
13	.34557	.93839	.36826	2.7155	47
14	.34584	.93829	.36859	2.7130	46
15	0.34612	0.93819	0.36892	2.7106	45
16	.34639	.93809	.36925	2.7082	44
17	.34666	.93799	.36958	2.7058	43
18	.34694	.93789	.36991	2.7034	42
19	.34721	.93779	.37024	2.7009	41
20	0.34748	0.93769	0.37057	2.6985	40
21	.34775	.93759	.37090	2.6961	39
22	.34803	.93748	.37123	2.6937	38
23	.34830	.93738	.37157	2.6913	37
24	.34857	.93728	.37190	2.6889	36
25	0.34884	0.93718	0.37223	2.6865	35
26	.34912	.93708	.37256	2.6841	34
27	.34939	.93698	.37289	2.6818	33
28	.34966	.93688	.37322	2.6794	32
29	.34993	.93677	.37355	2.6770	31
30	0.35021	0.93667	0.37388	2.6746	30
31	.35048	.93657	.37422	2.6723	29
32	.35075	.93647	.37455	2.6699	28
33	.35102	.93637	.37488	2.6675	27
34	.35130	.93626	.37521	2.6652	26
35	0.35157	0.93616	0.37554	2.6628	25
36	.35184	.93606	.37588	2.6605	24
37	.35211	.93596	.37621	2.6581	23
38	.35239	.93585	.37654	2.6558	22
39	.35266	.93575	.37687	2.6534	21
40	0.35293	0.93565	0.37720	2.6511	20
41	.35320	.93555	.37754	2.6488	19
42	.35347	.93544	.37787	2.6464	18
43	.35375	.93534	.37820	2.6441	17
44	.35402	.93524	.37853	2.6418	16
45	0.35429	0.93514	0.37887	2.6395	15
46	.35456	.93503	.37920	2.6371	14
47	.35484	.93493	.37953	2.6348	13
48	.35511	.93483	.37986	2.6328	12
49	.35538	.93472	.38020	2.6302	11
50	0.35565	0.93462	0.38053	2.6279	10
51	.35592	.93452	.38086	2.6256	9
52	.35619	.93441	.38120	2.6233	8
53	.35647	.93431	.38153	2.6210	7
54	.35674	.93420	.38186	2.6187	6
55	0.35701	0.93410	0.38220	2.6165	5
56	.35728	.93400	.38253	2.6142	4
57	.35755	.93389	.38286	2.6119	3
58	.35782	.93379	.38320	2.6096	2
59	.35810	.93368	.38353	2.6074	1
60	0.35837	0.93358	0.38386	2.6051	0
M	Cosine	Sine	Cotan.	Tan.	M

69°

Natural Trigonometric Functions

21°

M	Sine	Cosine	Tan.	Cotan.	M
0	0.35837	0.93358	0.38386	2.6051	60
1	.35864	.93348	.38420	2.6028	59
2	.35891	.93337	.38453	2.6006	58
3	.35918	.93327	.38487	2.5983	57
4	.35945	.93316	.38520	2.5961	56
5	0.35973	0.93306	0.38553	2.5938	55
6	.36000	.93295	.38587	2.5916	54
7	.36027	.03285	.38620	2.5893	53
8	.36054	.93274	.38654	2.5871	52
9	.36081	.93264	.38687	2.5848	51
10	0.36108	0.93253	0.38721	2.5826	50
11	.36135	.93243	.38754	2.5804	49
12	.36162	.93232	.38787	2.5782	48
13	.36190	.93222	.38821	2.5759	47
14	.36217	.93211	.38854	2.5737	46
15	0.36244	0.93201	0.38888	2.5715	45
16	.36271	.93190	.38921	2.5693	44
17	.36298	.93180	.38955	2.5671	43
18	.36325	.93169	.38988	2.5649	42
19	.36352	.93159	.39022	2.5627	41
20	0.36379	0.93148	0.39055	2.5605	40
21	.36406	.93137	.39089	2.5583	39
22	.36434	.93127	.39122	2.5561	38
23	.36461	.93116	.39156	2.5539	37
24	.36488	.93106	.39190	2.5517	36
25	0.36515	0.93095	0.39223	2.5495	35
26	.36542	.93084	.39257	2.5473	34
27	.36569	.93074	.39290	2.5452	33
28	.36596	.93063	.39324	2.5430	32
29	.36623	.93052	.39357	2.5408	31
30	0.36650	0.93042	0.39391	2.5386	30
31	.36677	.93031	.39425	2.5365	29
32	.36704	.93020	.39458	2.5343	28
33	.36731	.93010	.39492	2.5322	27
34	.36758	.92999	.39526	2.5300	26
35	0.36785	0.92988	0.39559	2.5279	25
36	.36812	.92978	.39593	2.5257	24
37	.36839	.92967	.39626	2.5236	23
38	.36867	.92956	.39660	2.5214	22
39	.36894	.92945	.39694	2.5193	21
40	0.36921	0.92935	0.39727	2.5172	20
41	.36948	.92924	.39761	2.5150	19
42	.36975	.92913	.39795	2.5129	18
43	.37002	.92902	.39829	2.5108	17
44	.37029	.92892	.39862	2.5086	16
45	0.37056	0.92881	0.39896	2.5065	15
46	.37083	.92870	.39930	2.5044	14
47	.37110	.92859	.39963	2.5023	13
48	.37137	.92849	.39997	2.5002	12
49	.37164	.92838	.40031	2.4981	11
50	0.37191	0.92827	0.40065	2.4960	10
51	.37218	.92816	.40098	2.4939	9
52	.37245	.92805	.40132	2.4918	8
53	.37272	.92794	.40166	2.4897	7
54	.37299	.92784	.40200	2.4876	6
55	0.37326	0.92773	0.40234	2.4855	5
56	.37353	.92762	.40267	2.4834	4
57	.37380	.92751	.40301	2.4813	3
58	.37407	.92740	.40335	2.4792	2
59	.37434	.92729	.40369	2.4772	1
60	0.37461	0.92718	0.40403	2.4751	0
M	Cosine	Sine	Cotan.	Tan.	M

68°

Natural Trigonometric Functions

22°

M	Sine	Cosine	Tan.	Cotan.	M
0	0.37461	0.92718	0.40403	2.4751	60
1	.37488	.92707	.40436	2.4730	59
2	.37515	.92697	.40470	2.4709	58
3	.37542	.92686	.40504	2.4689	57
4	.37569	.92675	.40538	2.4668	56
5	0.37595	0.92664	0.40572	2.4648	55
6	.37622	.92653	.40606	2.4627	54
7	.37649	.92642	.40640	2.4606	53
8	.37676	.92631	.40674	2.4586	52
9	.37703	.92620	.40707	2.4566	51
10	0.37730	0.92609	0.40741	2.4545	50
11	.37757	.92598	.40775	2.4525	49
12	.37784	.92587	.40809	2.4504	48
13	.37811	.92576	.40843	2.4484	47
14	.37838	.92565	.40877	2.4464	46
15	0.37865	0.92554	0.40911	2.4443	45
16	.37892	.92543	.40945	2.4423	44
17	.37919	.92532	.40979	2.4403	43
18	.37946	.92521	.41013	2.4383	42
19	.37973	.92510	.41047	2.4362	41
20	0.37999	0.92499	0.41081	2.4342	40
21	.38026	.92488	.41115	2.4322	39
22	.38053	.92477	.41149	2.4302	38
23	.38080	.92466	.41183	2.4282	37
24	.38107	.92455	.41217	2.4262	36
25	0.38134	0.92444	0.41251	2.4242	35
26	.38161	.92432	.41285	2.4222	34
27	.38188	.92421	.41319	2.4202	33
28	.38215	.92410	.41353	2.4181	32
29	.38241	.92399	.41387	2.4162	31
30	0.38268	0.92388	0.41421	2.4142	30
31	.38295	.92377	.41455	2.4122	29
32	.38322	.92366	.41490	2.4102	28
33	.38349	.92355	.41524	2.4083	27
34	.38376	.92343	.41558	2.4063	26
35	0.38403	0.92332	0.41592	2.4043	25
36	.38430	.92321	.41626	2.4023	24
37	.38456	.92310	.41660	2.4004	23
38	.38483	.92299	.41694	2.3984	22
39	.38510	.92287	.41728	2.3964	21
40	0.38537	0.92276	0.41763	2.3945	20
41	.38564	.92265	.41797	2.3925	19
42	.38591	.92254	.41831	2.3906	18
43	.38617	.92243	.41865	2.3886	17
44	.38644	.92231	.41899	2.3867	16
45	0.38671	0.92220	0.41933	2.3847	15
46	.38698	.92209	.41968	2.3828	14
47	.38725	.92198	.42002	2.3808	13
48	.38752	.92186	.42036	2.3789	12
49	.38778	.92175	.42070	2.3770	11
50	0.38805	0.92164	0.42105	2.3750	10
51	.38872	.92152	.42139	2.3731	9
52	.38859	.92141	.42173	2.3712	8
53	.38886	.92130	.42207	2.3693	7
54	.38912	.92119	.42242	2.3673	6
55	0.38939	0.92107	0.42276	2.3654	5
56	.38966	.92096	.42310	2.3635	4
57	.38993	.92085	.42345	2.3616	3
58	.39020	.92073	.42379	2.3597	2
59	.39046	.92062	.42413	2.3578	1
60	0.39073	0.92050	0.42447	2.3559	0
M	Cosine	Sine	Cotan.	Tan.	M

67°

Natural Trigonometric Functions

23°

M	Sine	Cosine	Tan.	Cotan.	M
0	0.39073	0.92050	0.42447	2.3559	60
1	.39100	.92039	.42482	2.3539	59
2	.39127	.92028	.42516	2.3520	58
3	.39153	.92016	.42551	2.3501	57
4	.39180	.92005	.42583	2.3483	56
5	0.39207	0.91994	0.42619	2.3464	55
6	.39234	.91982	.42654	2.3445	54
7	.39260	.91971	.42688	2.3426	53
8	.39287	.91959	.42772	2.3407	52
9	.39314	.91948	.42757	2.3388	51
10	0.39341	0.91956	0.42791	2.3369	50
11	.39367	.91925	.42826	2.3351	49
12	.39394	.91914	.42860	2.3332	48
13	.39421	.91902	.42894	2.3313	47
14	.39448	.91891	.42929	2.3204	46
15	0.39474	0.91879	0.42963	2.3276	45
16	.39501	.91868	.42998	2.3257	44
17	.39528	.91856	.43032	2.3238	43
18	.39555	.91845	.43067	2.3220	42
19	.39581	.91833	.43101	2.3201	41
20	0.39608	0.91822	0.43136	2.3183	40
21	.39635	.91810	.43170	2.3164	39
22	.39661	.91799	.43205	2.3146	38
23	.39688	.91787	.43239	2.3127	37
24	.39715	.91775	.43274	2.3109	36
25	0.39741	0.91764	0.43308	2.3090	35
26	.39768	.91752	.43343	2.3072	34
27	.39795	.91741	.43378	2.3053	33
28	.39822	.91729	.43412	2.3035	32
29	.39848	.91718	.43447	2.3017	31
30	0.39875	0.91706	0.43481	2.2998	30
31	.39902	.91694	.43516	2.2980	29
32	.39928	.91683	.43550	2.2962	28
33	.39955	.91671	.43585	2.2944	27
34	.39982	.91660	.43620	2.2925	26
35	0.40008	0.91648	0.43654	2.2907	25
36	.40035	.91636	.43689	2.2889	24
37	.40062	.91625	.43724	2.2871	23
38	.40088	.91613	.43758	2.2853	22
39	.40115	.91601	.43793	2.2835	21
40	0.40141	0.91590	0.43828	2.2817	20
41	.40168	.91578	.43862	2.2709	19
42	.40195	.91566	.43897	2.2781	18
43	.40221	.91555	.43932	2.2763	17
44	.40248	.91543	.43966	2.2745	16
45	0.40275	0.91531	0.44001	2.2727	15
46	.40301	.91519	.44036	2.2709	14
47	.40328	.91508	.44071	2.2691	13
48	.40355	.91496	.44105	2.2673	12
49	.40381	.91484	.44140	2.2655	11
50	0.40408	0.91472	0.44175	2.2637	10
51	.40434	.91461	.44210	2.2620	9
52	.40461	.91449	.44244	2.2602	8
53	.40488	.91437	.44279	2.2584	7
54	.40514	.91425	.44314	2.2566	6
55	0.40541	0.91414	0.44349	2.2549	5
56	.40567	.91402	.44384	2.2531	4
57	.40594	.91390	.44418	2.2513	3
58	.40621	.91378	.44453	2.2496	2
59	.40641	.91366	.44488	2.2478	1
60	0.40674	0.91355	0.44523	1.2460	0
M	Cosine	Sine	Cotan.	Tan.	M

66°

Natural Trigonometric Functions

24°

M	Sine	Cosine	Tan.	Cotan.	*M*
0	0.40674	0.91355	0.44523	2.2460	60
1	.40700	.91343	.44558	2.2443	59
2	.40727	.91331	.44593	2.2425	58
3	.40753	.91319	.44627	2.2408	57
4	.40780	.91307	.44662	2.2390	56
5	0.40806	0.91295	0.44697	2.2373	55
6	.40833	.91283	.44732	2.2355	54
7	.40860	.91272	.44767	2.2338	53
8	.40886	.91260	.44802	2.2320	52
9	.40913	.91248	.44837	2.2303	51
10	0.40939	0.91236	0.44872	2.2286	50
11	.40966	.91224	.44907	2.2268	49
12	.40992	.91212	.44942	2.2251	48
13	.41019	.91200	.44977	2.2234	47
14	.41045	.91188	.45012	2.2216	46
15	0.41072	0.91176	0.45047	2.2199	45
16	.41098	.91164	.45082	2.2182	44
17	.41125	.91152	.45117	2.2165	43
18	.41151	.91140	.45152	2.2148	42
19	.41178	.91128	.45187	2.2130	41
20	0.41204	0.91116	0.45222	2.2113	40
21	.41231	.91104	.45257	2.2096	39
22	.41257	.91092	.45292	2.2079	38
23	.41284	.91080	.45327	2.2062	37
24	.41310	.91068	.45362	2.2045	36
25	0.41337	0.91056	0.45397	2.2028	35
26	.41363	.91044	.45432	2.2011	34
27	.41390	.91032	.45467	2.1994	33
28	.41416	.91020	.45502	2.1977	32
29	.41443	.91008	.45537	2.1960	31
30	0.41469	0.90996	0.45573	2.1943	30
31	.41496	.90984	.45608	2.1926	29
32	.41522	.90972	.45643	2.1909	28
33	.41549	.90960	.45678	2.1892	27
34	.41575	.90948	.45713	2.1876	26
35	0.41602	0.90936	0.45748	2.1859	25
36	.41628	.90924	.45784	2.1842	24
37	.41655	.90911	.45819	2.1825	23
38	.41681	.90899	.45854	2.1808	22
39	.41707	.90887	.45889	2.1792	21
40	0.41734	0.90875	0.45924	2.1775	20
41	.41760	.90863	.45960	2.1758	19
42	.41787	.90851	.45995	2.1742	18
43	.41813	.90839	.46030	2.1725	17
44	.41840	.90826	.46065	2.1708	16
45	0.41866	0.90814	0.46101	2.1692	15
46	.41892	.90802	.46136	2.1675	14
47	.41919	.90790	.46171	2.1659	13
48	.41945	.90778	.46206	2.1642	12
49	.41972	.90766	.46242	2.1625	11
50	0.41998	0.90753	0.46277	2.1609	10
51	.42024	.90741	.46312	2.1592	9
52	.42051	.90729	.46348	2.1576	8
53	.42077	.90717	.46383	2.1560	7
54	.42104	.90704	.46418	2.1543	6
55	0.42130	0.90692	0.46454	2.1527	5
56	.42156	.90680	.46489	2.1510	4
57	.42183	.90668	.46525	2.1494	3
58	.42209	.90655	.46560	2.1478	2
59	.42235	.90643	.46595	2.1461	1
60	0.42262	0.90631	0.46631	2.1445	0
M	Cosine	Sine	Cotan.	Tan.	*M*

65°

Natural Trigonometric Functions

25°

M	Sine	Cosine	Tan.	Cotan.	*M*
0	0.42262	0.90631	0.46631	2.1445	60
1	.42288	.90618	.46666	2.1429	59
2	.42315	.90606	.46702	2.1413	58
3	.42341	.90594	.46737	2.1396	57
4	.42367	.90582	.46772	2.1380	56
5	0.42394	0.90569	0.46808	2.1364	55
6	.42420	.90557	.46843	2.1348	54
7	.42446	.90545	.46879	2.1332	53
8	.42473	.90532	.46914	2.1315	52
9	.42499	.90520	.46950	2.1299	51
10	0.42525	0.90507	0.46985	2.1283	50
11	.42552	.90495	.47021	2.1267	49
12	.42578	.90483	.47056	2.1251	48
13	.42604	.90470	.47092	2.1235	47
14	.42631	.90458	.47128	2.1219	46
15	0.42657	0.90446	0.47163	2.1203	45
16	.42683	.90433	.47199	2.1187	44
17	.42709	.90421	.47234	2.1171	43
18	.42736	.90408	.47270	2.1155	42
19	.42762	.90396	.47305	2.1139	41
20	0.42788	0.90383	0.47341	2.1123	40
21	.42815	.90371	.47377	2.1107	39
22	.42841	.90358	.47412	2.1092	38
23	.42867	.90346	.47448	2.1076	37
24	.42894	.90334	.47483	2.1060	36
25	0.42920	0.90321	0.47519	2.1044	35
26	.42946	.90309	.47555	2.1028	34
27	.42972	.90296	.47590	2.1013	33
28	.42999	.90284	.47626	2.0997	32
29	.43025	.90271	.47662	2.0981	31
30	0.43051	0.90259	0.47698	2.0965	30
31	.43077	.90246	.47733	2.0950	29
32	.43104	.90233	.47769	2.0934	28
33	.43130	.90221	.47805	2.0918	27
34	.43156	.90208	.47840	2.0903	26
35	0.43182	0.90196	0.47876	2.0887	25
36	.43209	.90183	.47912	2.0872	24
37	.43235	.90171	.47948	2.0856	23
38	.43261	.90158	.47984	2.0840	22
39	.43287	.90146	.48019	2.0825	21
40	0.43313	0.90133	0.48055	2.0809	20
41	.43340	.90120	.48091	2.0794	19
42	.43366	.90108	.48127	2.0778	18
43	.43392	.90095	.48163	2.0763	17
44	.43418	.90082	.48198	2.0748	16
45	0.43445	0.90070	0.48234	2.0732	15
46	.43471	.90067	.48270	2.0717	14
47	.43497	.90045	.48306	2.0701	13
48	.43523	.90032	.48342	2.0686	12
49	.43549	.90019	.48378	2.0671	11
50	0.43575	0.90007	0.48414	2.0655	10
51	.43602	.89994	.48450	2.0640	9
52	.43628	.89981	.48486	2.0625	8
53	.43654	.89968	.48521	2.0609	7
54	.43680	.89956	.48557	2.0594	6
55	0.43706	0.89943	0.48593	2.0579	5
56	.43733	.89930	.48629	2.0564	4
57	.43759	.89918	.48665	2.0549	3
58	.43785	.89905	.48701	2.0533	2
59	.43811	.89892	.48737	2.0518	1
60	0.43837	0.89879	0.48773	2.0503	0
M	Cosine	Sine	Cotan.	Tan.	*M*

64°

Natural Trigonometric Functions

26°

M	Sine	Cosine	Tan.	Cotan.	M
0	0.43837	0.89879	0.48773	2.0508	60
1	.43863	.89867	.48809	2.0488	59
2	.43889	.89854	.48845	2.0473	58
3	.43916	.89841	.48881	2.0458	57
4	.43942	.89828	.48917	2.0443	56
5	0.43968	0.89816	0.48953	2.0428	55
6	.43994	.89803	.48989	2.0413	54
7	.44020	.89790	.49026	2.0398	53
8	.44046	.89777	.49062	2.0383	52
9	.44072	.89764	.49098	2.0368	51
10	0.44098	0.89752	0.49134	2.0353	50
11	.44124	.89739	.49170	2.0338	49
12	.44151	.89726	.49206	2.0323	48
13	.44177	.89713	.49242	2.0308	47
14	.44203	.89700	.49278	2.0293	46
15	0.44229	0.89687	0.49315	2.0278	45
16	.44255	.89674	.49351	2.0263	44
17	.44281	.89662	.49387	2.0248	43
18	.44307	.89649	.49423	2.0233	42
19	.44333	.89636	.49459	2.0219	41
20	0.44359	0.89623	0.49495	2.0204	40
21	.44385	.89610	.49532	2.0189	39
22	.44411	.89597	.49568	2.0174	38
23	.44437	.89584	.49604	2.0160	37
24	.44464	.89571	.49640	2.0145	36
25	0.44490	0.89558	0.49677	2.0130	35
26	.44516	.89545	.49713	2.0115	34
27	.44542	.89532	.49749	2.0101	33
28	.44568	.89519	.49786	2.0086	32
29	.44594	.89506	.49822	2.0072	31
30	0.44620	0.89493	0.49858	2.0057	30
31	.44646	.89480	.49894	2.0042	29
32	.44672	.89467	.49931	2.0028	28
33	.44698	.89454	.49967	2.0013	27
34	.44724	.89441	.50004	1.9999	26
35	0.44750	0.89428	0.50040	1.9984	25
36	.44776	.89415	.50076	1.9970	24
37	.44802	.89402	.50113	1.9955	23
38	.44828	.89389	.50149	1.9941	22
39	.44854	.89376	.50185	1.9926	21
40	0.44880	0.89363	0.50222	1.9912	20
41	.44906	.89350	.50258	1.9897	19
42	.44932	.89337	.50295	1.9883	18
43	.44958	.89324	.50331	1.9868	17
44	.44984	.89311	.50368	1.9854	16
45	0.45010	0.89298	0.50404	1.9840	15
46	.45036	.89285	.50441	1.9825	14
47	.45062	.89272	.50477	1.9811	13
48	.45088	.89259	.50514	1.9797	12
49	.45114	.89245	.50550	1.9782	11
50	0.45140	0.89232	0.50587	1.9768	10
51	.45166	.89219	.50623	1.9754	9
52	.45192	.89206	.50660	1.9740	8
53	.45218	.89193	.50696	1.9725	7
54	.45243	.89180	.50733	1.9711	6
55	0.45269	0.89167	0.50769	1.9697	5
56	.45295	.89153	.50806	1.9683	4
57	.45321	.89140	.50843	1.9669	3
58	.45347	.89127	.50879	1.9654	2
59	.45373	.89114	.50916	1.9640	1
60	0.45399	0.89101	0.50953	1.9626	0
M	Cosine	Sine	Cotan.	Tan.	M

63°

Natural Trigonometric Functions

27°

M	Sine	Cosine	Tan.	Cotan.	M
0	0.45399	0.89101	0.50953	1.9626	60
1	.45425	.89087	.50989	1.9612	59
2	.45451	.89074	.51026	1.9598	58
3	.45477	.89061	.51063	1.9584	57
4	.45503	.89048	.51099	1.9570	56
5	0.45529	0.89035	0.51136	1.9556	55
6	.45554	.89021	.51173	1.9542	54
7	.45580	.89008	.51209	1.9528	53
8	.45606	.88995	.51246	1.9514	52
9	.45632	.88981	.51283	1.9500	51
10	0.45658	0.88968	0.51319	1.9486	50
11	.45684	.88955	.51356	1.9472	49
12	.45710	.88942	.51393	1.9458	48
13	.45736	.88928	.51430	1.9444	47
14	.45762	.88915	.51467	1.9430	46
15	0.45787	0.88902	0.51503	1.9416	45
16	.45813	.88888	.51540	1.9402	44
17	.45839	.88875	.51577	1.9388	43
18	.45865	.88862	.51614	1.9375	42
19	.45891	.88848	.51651	1.9361	41
20	0.45917	0.88835	0.51688	1.9347	40
21	.45942	.88822	.51724	1.9333	39
22	.45968	.88808	.51761	1.9319	38
23	.45934	.88795	.51798	1.9306	37
24	.46029	.88782	.51835	1.9292	36
25	0.46046	0.88768	0.51872	1.9278	35
26	.46072	.88755	.51909	1.9265	34
27	.46097	.88741	.51946	1.9251	33
28	.46123	.88728	.51983	1.9237	32
29	.46149	.88715	.52020	1.9223	31
30	0.46175	0.88701	0.52057	1.9210	30
31	.46201	.88688	.52094	1.9196	29
32	.46226	.88674	.52131	1.9183	28
33	.46252	.88661	.52168	1.9169	27
34	.46278	.88647	.52205	1.9155	26
35	0.46304	0.88634	0.52242	1.9142	25
36	.46330	.88620	.52279	1.9128	24
37	.46355	.88607	.52316	1.9115	23
38	.46381	.88593	.52353	1.9101	22
39	.46407	.88580	.52390	1.9088	21
40	0.46433	0.88566	0.52437	1.9074	20
41	.46458	.88553	.52464	1.9061	19
42	.46484	.88539	.52501	1.9047	18
43	.46510	.88526	.52538	1.9034	17
44	.46536	.88512	.52575	1.9020	16
45	0.46561	0.88499	0.52613	1.9007	15
46	.46587	.88485	.52650	1.8993	14
47	.46613	.88472	.52687	1.8980	13
48	.46639	.88458	.52724	1.8967	12
49	.46664	.88445	.52761	1.8953	11
50	0.46690	0.88431	0.52798	1.8940	10
51	.46716	.88417	.52836	1.8927	9
52	.46742	.88404	.52873	1.8913	8
53	.46767	.88390	.52910	1.8900	7
54	.46793	.88377	.52947	1.8887	6
55	0.46819	0.88363	0.52985	1.8873	5
56	.46844	.88349	.53022	1.8860	4
57	.46870	.88336	.53059	1.8847	3
58	.46896	.88322	.53096	1.8834	2
59	.46921	.88308	.53134	1.8820	1
60	0.46947	0.88295	0.53171	1.8807	0
M	Cosine	Sine	Cotan.	Tan.	M

62°

Natural Trigonometric Functions

28°

M	Sine	Cosine	Tan.	Cotan.	M
0	0.46947	0.88295	0.53171	1.8807	60
1	.46973	.88281	.53208	1.8794	59
2	.46999	.88267	.53246	1.8781	58
3	.47024	.88254	.53283	1.8768	57
4	.47050	.88240	.53320	1.8755	56
5	0.47076	0.88226	0.53358	1.8741	55
6	.47101	.88213	.53395	1.8728	54
7	.47127	.88199	.53432	1.8715	53
8	.47153	.88185	.53470	1.8702	52
9	.47178	.88172	.53507	1.8689	51
10	0.47204	0.88158	0.53545	1.8676	50
11	.47229	.88144	.53582	1.8663	49
12	.47255	.88130	.53620	1.8650	48
13	.47281	.88117	.53657	1.8637	47
14	.47306	.88103	.53694	1.8624	46
15	0.47332	0.88089	0.53732	1.8611	45
16	.47358	.88075	.53769	1.8598	44
17	.47383	.88062	.53807	1.8585	43
18	.47409	.88048	.53844	1.8572	42
19	.47434	.88034	.53882	1.8559	41
20	0.47460	0.88020	0.53920	1.8546	40
21	.47486	.88006	.53957	1.8533	39
22	.47511	.87993	.53995	1.8520	38
23	.47537	.87979	.54032	1.8507	37
24	.47562	.87965	.54070	1.8495	36
25	0.47588	0.87951	0.54107	1.8482	35
26	.47614	.87937	.54145	1.8469	34
27	.47639	.87923	.54183	1.8456	33
28	.47665	.87909	.54220	1.8443	32
29	.47690	.87896	.54258	1.8430	31
30	0.47716	0.87882	0.54296	1.8418	30
31	.47741	.87882	.54333	1.8405	29
32	.47767	.87854	.54371	1.8392	28
33	.47793	.87840	.54409	1.8379	27
34	.47818	.87826	.54446	1.8367	26
35	0.47844	0.87812	0.54484	1.8354	25
36	.47869	.87798	.54522	1.8341	24
37	.47895	.87784	.54560	1.8329	23
38	.47920	.87770	.54597	1.8316	22
39	.47946	.87756	.54635	1.8303	21
40	0.47971	0.87743	0.54673	1.8291	20
41	.47997	.87729	.54711	1.8278	19
42	.48022	.87715	.54748	1.8265	18
43	.48048	.87701	.54786	1.8253	17
44	.48073	.87687	.54824	1.8240	16
45	0.48099	0.87673	0.54862	1.8228	15
46	.48124	.87659	.54900	1.8215	14
47	.48150	.87645	.54938	1.8202	13
48	.48175	.87631	.54975	1.8190	12
49	.48201	.87617	.55013	1.8177	11
50	0.48226	0.87603	0.55051	1.8165	10
51	.48252	.87589	.55089	1.8152	9
52	.48277	.87575	.55127	1.8140	8
53	.48303	.87561	.55165	1.8127	7
54	.48328	.87546	.55203	1.8115	6
55	0.48354	0.87532	0.55241	1.8103	5
56	.48379	.87518	.55270	1.8090	4
57	.48405	.87504	.55317	1.8078	3
58	.48430	.87490	.55355	1.8065	2
59	.48456	.87476	.55393	1.8053	1
60	0.48481	0.87462	0.55431	1.8040	0
M	Cosine	Sine	Cotan.	Tan.	M

61°

Natural Trigonometric Functions

29°

M	Sine	Cosine	Tan.	Cotan.	M
0	0.48481	0.87462	0.55431	1.8040	60
1	.48506	.87448	.55469	1.8028	59
2	.48532	.87434	.55507	1.8016	58
3	.48557	.87420	.55545	1.8003	57
4	.48583	.87406	.55583	1.7991	56
5	0.48608	0.87391	0.55621	1.7979	55
6	.48634	.87377	.55659	1.7966	54
7	.48659	.87363	.55697	1.7954	53
8	.48684	.87349	.55736	1.7942	52
9	.48710	.87335	.55774	1.7930	51
10	0.48735	0.87321	0.55812	1.7917	50
11	.48761	.87306	.55850	1.7905	49
12	.48786	.87292	.55888	1.7893	48
13	.48811	.87278	.55926	1.7881	47
14	.48837	.87264	.55964	1.7868	46
15	0.48862	0.87250	0.56003	1.7856	45
16	.48888	.87235	.56041	1.7844	44
17	.48913	.87221	.56079	1.7832	43
18	.48938	.87207	.56117	1.7820	42
19	.48964	.87193	.56156	1.7808	41
20	0.48989	0.87178	0.56194	1.7796	40
21	.49014	.87164	.56232	1.7783	39
22	.49040	.87150	.56270	1.7771	38
23	.49065	.87136	.56309	1.7759	37
24	.49090	.87121	.56347	1.7747	36
25	0.49116	0.87107	0.56385	1.7735	35
26	.49141	.87093	.56424	1.7723	34
27	.49166	.87079	.56462	1.7711	33
28	.49192	.87064	.56501	1.7699	32
29	.49217	.87050	.56539	1.7687	31
30	0.49242	0.87036	0.56577	1.7675	30
31	.49268	.87021	.56616	1.7663	29
32	.49293	.87007	.56654	1.7651	28
33	.49318	.86993	.56693	1.7639	27
34	.49344	.86978	.56731	1.7627	26
35	0.49369	0.86964	0.56769	1.7615	25
36	.49394	.86949	.56808	1.7603	24
37	.49419	.86935	.56846	1.7591	23
38	.49445	.86921	.56885	1.7579	22
39	.49470	.86906	.56923	1.7567	21
40	0.49459	0.86892	0.56962	1.7556	20
41	.49521	.86878	.57000	1.7544	19
42	.49546	.86863	.57039	1.1532	18
43	.49571	.86849	.57078	1.7520	17
44	.49596	.86834	.57116	1.7508	16
45	0.49622	0.86820	0.57155	1.7496	15
46	.49647	.86805	.57193	1.7485	14
47	.49672	.86791	.57232	1.7473	13
48	.49697	.86777	.57271	1.7461	12
49	.49723	.86762	.57309	1.7449	11
50	0.49748	0.86748	0.57348	1.7437	10
51	.49773	.86733	.57386	1.7426	9
52	.49798	.86719	.57425	1.7414	8
53	.49824	.86704	.57464	1.7402	7
54	.49849	.86690	.57503	1.7391	6
55	0.49874	0.86675	0.57541	1.7379	5
56	.49899	.86661	.57580	1.7367	4
57	.49924	.86646	.57619	1.7355	3
58	.49950	.86632	.57657	1.7344	2
59	.49975	.86617	.57696	1.7332	1
60	0.50000	0.86603	0.57735	1.7321	0
M	Cosine	Sine	Cotan.	Tan.	M

60°

Natural Trigonometric Functions

30°

M	Sine	Cosine	Tan.	Cotan.	M
0	0.50000	0.86603	0.57735	1.7321	60
1	.50025	.86588	.57774	1.7309	59
2	.50050	.86573	.57813	1.7297	58
3	.50076	.86559	.57851	1.7286	57
4	.50101	.86544	.57890	1.7274	56
5	0.50126	0.86530	0.57929	1.7262	55
6	.50151	.86515	.57968	1.7251	54
7	.50176	.86501	.58007	1.7239	53
8	.50201	.86486	.58046	1.7228	52
9	.50227	.86471	.58085	1.7216	51
10	0.50252	.086457	0.58124	1.7205	50
11	.50277	.86442	.58162	1.7193	49
12	.50302	.86427	.58201	1.7182	48
13	.50327	.86413	.58240	1.7170	47
14	.50352	.86398	.58279	1.7159	46
15	0.50377	0.86384	0.58318	1.7147	45
16	.50403	.86369	.58357	1.7136	44
17	.50428	.86354	.58396	1.1724	43
18	.50453	.86340	.58435	1.7113	42
19	.50478	.86325	.58474	1.7102	41
20	0.50503	0.86310	0.58513	1.7090	40
21	.50528	.86295	.58552	1.7079	39
22	.50553	.86281	.58591	1.7067	38
23	.50578	.86266	.58631	1.7056	37
24	.50603	.86251	.58670	1.7045	36
25	0.50628	0.86237	0.58709	1.7033	35
26	.50654	.86222	.58748	1.7022	34
27	.50679	.86207	.58787	1.7011	33
28	.50704	.86192	.58826	1.6999	32
29	.50729	.86178	.58865	1.6988	31
30	0.50754	0.86163	0.58905	1.6977	30
31	.50779	.86148	.58944	1.6965	29
32	.50804	.86133	.58983	1.6954	28
33	.50829	.86119	.59022	1.6943	27
34	.50854	.86104	.59061	1.6932	26
35	0.50879	0.86089	0.59101	1.6920	25
36	.50904	.86074	.59140	1.6909	24
37	.50929	.86059	.59179	1.6898	23
38	.50954	.86045	.59218	1.6887	22
39	.50979	.86030	.59258	1.6875	21
40	0.51004	0.86015	0.59297	1.6864	20
41	.51029	.86000.	.59336	1.6853	19
42	.51054	.85985	.59376	1.6842	18
43	.51079	.85970	.59415	1.6831	17
44	.51104	.85956	.59454	1.6820	16
45	0.51129	0.85941	0.59494	1.6808	15
46	.51154	.85926	.59533	1.6797	14
47	.51179	.85911	.59573	1.6786	13
48	.51204	.85896	.59612	1.6775	12
49	.51229	.85881	.59651	1.6764	11
50	0.51254	0.85866	0.59691	1.6753	10
51	.51279	.85851	.59730	1.6742	9
52	.51304	.85836	.59770	1.6731	8
53	.51329	.85821	.59809	1.6720	7
54	.51354	.85806	.59849	1.6709	6
55	0.51379	0.85792	0.59888	1.6698	5
56	.51404	.85777	.59928	1.6687	4
57	.51429	.85762	.59967	1.6676	3
58	.51454	.85747	.60007	1.6665	2
59	.51479	.85732	.60046	1.6654	1
60	0.51504	0.85717	0.60086	1.6643	0
M	Cosine	Sine	Cotan.	Tan.	M

59°

Natural Trigonometric Functions

31°

M	Sine	Cosine	Tan.	Cotan.	M
0	0.51504	0.85717	0.60086	1.6643	60
1	.51529	.85702	.60126	1.6632	59
2	.51554	.85687	.60165	1.6621	58
3	.51579	.85672	.60205	1.6610	57
4	.51604	.85657	.60245	1.6599	56
5	0.51628	0.85642	0.60284	1.6588	55
6	.51653	.85627	.60324	1.6577	54
7	.51678	.85612	.60364	1.6566	53
8	.51703	.85597	.60403	1.6555	52
9	.51728	.85582	.60443	1.6545	51
10	0.51753	0.85567	0.60483	1.6534	50
11	.51778	.85551	.60522	1.6523	49
12	.51803	.85536	.60562	1.6512	48
13	.51828	.85521	.60602	1.6501	47
14	.51852	.85506	.60642	1.6490	46
15	0.51877	0.85491	0.60681	1.6479	45
16	.51902	.85476	.60721	1.6469	44
17	.51927	.85461	.60761	1.6458	43
18	.51952	.85446	.60801	1.6447	42
19	.51977	.85431	.60841	1.6436	41
20	0.52002	0.85416	0.60881	1.6426	40
21	.52026	.85401	.60921	1.6415	39
22	.52051	.85385	.60960	1.6404	38
23	.52076	.85370	.61000	1.6393	37
24	.52101	.85355	.61040	1.6383	36
25	0.52126	0.85340	0.61080	1.6372	35
26	.52151	.85325	.61120	1.6361	34
27	.52175	.85310	.61160	1.6351	33
28	.52200	.85294	.61200	1.6340	32
29	.52225	.85279	.61240	1.6329	31
30	0.52250	0.85264	0.61280	1.6319	30
31	.52275	.85249	.61320	1.6308	29
32	.52299	.85234	.61360	1.6297	28
33	.52324	.85218	.61400	1.6287	27
34	.52349	.85203	.61440	1.6276	26
35	0.52374	0.85188	0.61480	1.6265	25
36	.52399	.85173	.61520	1.6255	24
37	.52423	.85157	.61561	1.6244	23
38	.52448	.85142	.61601	1.6234	22
39	.52473	.85127	.61641	1.6223	21
40	0.52498	0.85112	0.61681	1.6212	20
41	.52522	.85096	.61721	1.6202	19
42	.52547	.85081	.61761	1.6191	18
43	.52572	.85066	.61801	1.6181	17
44	.52597	.85051	.61842	1.6170	16
45	0.52621	0.85035	0.61882	1.6160	15
46	.52646	.85020	.61922	1.6149	14
47	.52671	.85005	.61962	1.6139	13
48	.52696	.84989	.62003	1.6128	12
49	.52720	.84974	.62043	1.6118	11
50	0.52745	0.84959	0.62083	1.6107	10
51	.52770	.84943	.62124	1.6097	9
52	.52794	.84928	.62164	1.6087	8
53	.52819	.84913	.62204	1.6076	7
54	.52844	.84897	.62245	1.6066	6
55	0.52869	0.84882	0.62285	1.6055	5
56	.52893	.84866	.62325	1.6045	4
57	.52918	.84851	.62366	1.6034	3
58	.52943	.84836	.62406	1.6024	2
59	.52967	.84820	.62446	1.6014	1
60	0.52992	0.84805	0.62487	1.6003	0
M	Cosine	Sine	Cotan.	Tan.	M

58°

Natural Trigonometric Functions

32°

M	Sine	Cosine	Tan.	Cotan.	M
0	0.52992	0.84805	0.62487	1.6003	60
1	.53017	.84789	.62527	1.5993	59
2	.53041	.84774	.62568	1.5983	58
3	.53066	.84759	.62608	1.5972	57
4	.53091	.84743	.62649	1.5962	56
5	0.53115	0.84728	0.62689	1.5952	55
6	.53140	.84712	.62730	1.5941	54
7	.53164	.84697	.62770	1.5931	53
8	.53189	.84681	.62811	1.5921	52
9	.53214	.84666	.62852	1.5911	51
10	0.53238	0.84650	0.62892	1.5900	50
11	.53263	.84635	.62933	1.5890	49
12	.53288	.84619	.62973	1.5880	48
13	.53312	.84604	.63014	1.5869	47
14	.53337	.84588	.63055	1.5859	46
15	0.53361	0.84573	0.63095	1.5849	45
16	.53386	.84557	.63136	1.5839	44
17	.53411	.84542	.63177	1.5829	43
18	.53435	.84526	.63217	1.5818	42
19	.53460	.84511	.63258	1.5808	41
20	0.53484	0.84495	0.63299	1.5798	40
21	.53509	.84480	.63340	1.5788	39
22	.53534	.84464	.63380	1.5778	38
23	.53558	.84448	.63421	1.5768	37
24	.53583	.84433	.63462	1.5757	36
25	0.53607	0.84417	0.63503	1.5747	35
26	.53632	.84402	.63544	1.5737	34
27	.53656	.84386	.63584	1.5727	33
28	.53681	.84370	.63625	1.5717	32
29	.53705	.84355	.63666	1.5707	31
30	0.53730	0.84339	0.63707	1.5697	30
31	.53754	.84324	.63748	1.5687	29
32	.53779	.84308	.63789	1.5677	28
33	.53804	.84292	.63830	1.5667	27
34	.53828	.84277	.63871	1.5657	26
35	0.53853	0.84261	0.63912	1.5647	25
36	.53877	.84245	.63953	1.5637	24
37	.53902	.84230	.63994	1.5627	23
38	.53926	.84214	.64035	1.5617	22
39	.53951	.84198	.64076	1.5607	21
40	0.53975	0.84182	0.64117	1.5597	20
41	.54000	.84167	.64158	1.5587	19
42	.54024	.84151	.64199	1.5577	18
43	.54049	.84135	.64240	1.5567	17
44	.54073	.84120	.64281	1.5557	16
45	0.54097	0.84104	0.64322	1.5547	15
46	.54122	.84088	.64363	1.5537	14
47	.54146	.84072	.64404	1.5527	13
48	.54171	.84057	.64446	1.5517	12
49	.54195	.84041	.64487	1.5507	11
50	0.54220	0.84025	0.64528	1.5497	10
51	.54244	.84009	.64569	1.5487	9
52	.54269	.83994	.64610	1.5477	8
53	.54293	.83978	.64652	1.5468	7
54	.54317	.83962	.64693	1.5458	6
55	0.54342	0.83946	0.64734	1.5448	5
56	.54366	.83930	.64775	1.5438	4
57	.54391	.83915	.64817	1.5428	3
58	.54415	.83899	.64858	1.5418	2
59	.54440	.83883	.64899	1.5408	1
60	0.54464	0.83867	0.64941	1.5399	0
M	Cosine	Sine	Cotan.	Tan.	M

57°

Natural Trigonometric Functions

33°

M	Sine	Cosine	Tan.	Cotan.	M
0	0.54464	0.83867	0.64941	1.5399	60
1	.54488	.83851	.64982	1.5389	59
2	.54513	.83835	.65024	1.5379	58
3	.54537	.83819	.65065	1.5369	57
4	.54561	.83804	.65106	1.5359	56
5	0.54586	0.83788	0.65148	1.5350	55
6	.54610	.83772	.65189	1.5340	54
7	.54135	.83756	.65231	1.5330	53
8	.54659	.83740	.65272	1.5320	52
9	.54683	.83724	.65314	1.5311	51
10	0.54708	0.83708	0.65355	1.5301	50
11	.54732	.83692	.65397	1.5291	49
12	.54756	.83676	.65438	1.5282	48
13	.54781	.83660	.65480	1.5272	47
14	.54805	.83645	.65521	1.5262	46
15	0.54829	0.83629	0.65563	1.5253	45
16	.54854	.83613	.65604	1.5243	44
17	.54878	.83597	.65646	1.5233	43
18	.54902	.83581	.65688	1.5224	42
19	.54927	.83565	.65729	1.5214	41
20	0.54951	0.83549	0.65771	1.5204	40
21	.54975	.83533	.65813	1.5195	39
22	.54999	.83517	.65854	1.5185	38
23	.55024	.83501	.65896	1.5175	37
24	.55048	.83485	.65938	1.5166	36
25	0.55072	0.83469	0.65980	1.5156	35
26	.55097	.83453	.66021	1.5147	34
27	.55121	.83437	.66063	1.5137	33
28	.55145	.83421	.66105	1.5127	32
29	.55169	.83405	.66147	1.5118	31
30	0.55194	0.83389	0.66189	1.5108	30
31	.55218	.83373	.66230	1.5099	29
32	.55242	.83356	.66272	1.5089	28
33	.55266	.83340	.66314	1.5080	27
34	.55291	.83324	.66356	1.5070	26
35	0.55315	0.83308	0.66398	1.5061	25
36	.55339	.83292	.66440	1.5051	24
37	.55363	.83276	.66482	1.5042	23
38	.55388	.83260	.66524	1.5032	22
39	.55412	.83244	.66566	1.5023	21
40	0.55436	0.83228	0.66608	1.5013	20
41	.55460	.83212	.66650	1.5004	19
42	.55484	.83195	.66692	1.4994	18
43	.55509	.83179	.66734	1.4985	17
44	.55533	.83163	.66776	1.4975	16
45	0.55557	0.83147	0.66818	1.4966	15
46	.55581	.83131	.66860	1.4957	14
47	.55605	.83115	.66902	1.4947	13
48	.55630	.83098	.66944	1.4938	12
49	.55654	.83082	.66986	1.4928	11
50	0.55678	0.83066	0.67028	1.4919	10
51	.55702	.83050	.67071	1.4910	9
52	.55726	.83034	.67113	1.4900	8
53	.55750	.83017	.67155	1.4891	7
54	.55775	.83001	.67197	1.4882	6
55	0.55799	0.82985	0.67239	1.4872	5
56	.55823	.82969	.67282	1.4863	4
57	.55847	.82953	.67324	1.4854	3
58	.55871	.82936	.67366	1.4844	2
59	.55895	.82920	.67409	1.4835	1
60	0.55919	0.82904	0.67451	1.4826	0
M	Cosine	Sine	Cotan.	Tan.	M

56°

Natural Trigonometric Functions

34°

M	Sine	Cosine	Tan.	Cotan.	M
0	0.55919	0.82904	0.67451	1.4826	60
1	.55943	.82887	.67493	1.4816	59
2	.55968	.82871	.67536	1.4807	58
3	.55992	.82855	.67578	1.4798	57
4	.56016	.82839	.67620	1.4788	56
5	0.56040	0.82822	0.67663	1.4779	55
6	.56064	.82806	.67705	1.4770	54
7	.56088	.82790	.67748	1.4761	53
8	.56112	.82773	.67790	1.4751	52
9	.56136	.82757	.67832	1.4742	51
10	0.56160	0.82741	0.67875	1.4733	50
11	.56184	.82724	.67917	1.4724	49
12	.56208	.82708	.67960	1.4715	48
13	.56232	.82692	.68002	1.4705	47
14	.56256	.82675	.68045	1.4696	46
15	0.56280	0.82659	0.68088	1.4687	45
16	.56305	.82643	.68130	1.4678	44
17	.56329	.82626	.68173	1.4669	43
18	.56353	.82610	.68215	1.4659	42
19	.56377	.82593	.68258	1.4650	41
20	0.56401	0.82577	0.68301	1.4641	40
21	.56425	.82561	.68343	1.4632	39
22	.56449	.82544	.68386	1.4623	38
23	.56473	.82528	.68429	1.4614	37
24	.56497	.82511	.68471	1.4605	36
25	0.56521	0.82495	0.68514	1.4596	35
26	.56545	.82478	.68557	1.4586	34
27	.56569	.82462	.68600	1.4577	33
28	.56593	.82446	.68642	1.4568	32
29	.56617	.82429	.68685	1.4559	31
30	0.56641	0.82413	0.68728	1.4550	30
31	.56665	.82396	.68771	1.4541	29
32	.56689	.82380	.68814	1.4532	28
33	.56713	.82363	.68857	1.4523	27
34	.56736	.82347	.68900	1.4514	26
35	0.56760	0.82330	0.68942	1.4505	25
36	.56784	.82314	.68985	1.4496	24
37	.56808	.82297	.69028	1.4487	23
38	.56832	.82281	.69071	1.4478	22
39	.56856	.82264	.69114	1.4469	21
40	0.56880	0.82248	0.69157	1.4460	20
41	.56904	.82231	.69200	1.4451	19
42	.56928	.82214	.69243	1.4442	18
43	.56952	.82198	.69286	1.4433	17
44	.56976	.82181	.69329	1.4424	16
45	0.57000	0.82165	0.69372	1.4415	15
46	.57024	.82148	.69416	1.4406	14
47	.57047	.82132	.69459	1.4397	13
48	.57071	.82115	.69502	1.4388	12
49	.57095	.82098	.69545	1.4379	11
50	0.57119	0.82082	0.69588	1.4370	10
51	.57143	.82065	.69631	1.4361	9
52	.57167	.82048	.69675	1.4352	8
53	.57191	.82032	.69718	1.4344	7
54	.57215	.82015	.69761	1.4335	6
55	0.57238	0.81999	0.69804	1.4326	5
56	.57262	.81982	.69847	1.4317	4
57	.57286	.81965	.69891	1.4308	3
58	.57310	.81949	.69934	1.4299	2
59	.57334	.81932	.69977	1.4290	1
60	0.57358	0.81915	0.70021	1.4281	0
M	Cosine	Sine	Cotan.	Tan.	M

55°

Natural Trigonometric Functions

35°

M	Sine	Cosine	Tan.	Cotan.	M
0	0.57358	0.81915	0.70021	1.4281	60
1	.57381	.81899	.70064	1.4273	59
2	.57405	.81882	.70107	1.4264	58
3	.57429	.81865	.70151	1.4255	57
4	.57453	.81848	.70194	1.4246	56
5	0.57477	0.81832	.70238	1.4237	55
6	.57501	.81815	.70281	1.4229	54
7	.57524	.81798	.70325	1.4220	53
8	.57548	.81782	.70368	1.4211	52
9	.57572	.81765	.70412	1.4202	51
10	0.57596	0.81748	0.70455	1.4193	50
11	.57619	.81731	.70499	1.4185	49
12	.57643	.81714	.70542	1.4176	48
13	.57667	.81698	.70586	1.4167	47
14	.57691	.81681	.70629	1.4158	46
15	0.57715	0.81664	0.70673	1.4150	45
16	.57738	.81647	.70717	1.4141	44
17	.57762	.81631	.70760	1.4132	43
18	.57786	.81614	.70804	1.4124	42
19	.57810	.81597	.70848	1.4115	41
20	0.57833	0.81580	0.70891	1.4106	40
21	.57857	.81563	.70935	1.4097	39
22	.57881	.81546	.70979	1.4089	38
23	.57904	.81530	.71023	1.4080	37
24	.57928	.81513	.71066	1.4071	36
25	0.57952	0.81496	0.71110	1.4063	35
26	.57976	.81479	.71154	1.4054	34
27	.57999	.81462	.71198	1.4045	33
28	.58023	.81445	.71242	1.4037	32
29	.58047	.81428	.71285	1.4028	31
30	0.58070	0.81412	0.71329	1.4019	30
31	.58094	.81395	.71373	1.4011	29
32	.58118	.81378	.71417	1.4002	28
33	.58141	.81361	.71461	1.3994	27
34	.58165	.81344	.71505	1.3985	26
35	0.58189	0.81327	0.71549	1.3976	25
36	.58212	.81310	.71593	1.3968	24
37	.58236	.81293	.71637	1.3959	23
38	.58260	.81276	.71681	1.3951	22
39	.58283	.81259	.71725	1.3942	21
40	0.58307	0.81242	0.71769	1.3934	20
41	.58330	.81225	.71813	1.3925	19
42	.58354	.81208	.71857	1.3916	18
43	.58378	.81191	.71901	1.3908	17
44	.58401	.81174	.71946	1.3899	16
45	0.58425	0.81157	0.71990	1.3891	15
46	.58449	.81140	.72034	1.3882	14
47	.58472	.81123	.72078	1.3874	13
48	.58496	.81106	.72122	1.3865	12
49	.58519	.81089	.72167	1.3857	11
50	0.58543	0.81072	0.72211	1.3848	10
51	.58567	.81055	.72255	1.3840	9
52	.58590	.81038	.72299	1.3831	8
53	.58614	.81021	.72344	1.3823	7
54	.58637	.81004	.72388	1.3814	6
55	0.58661	0.80987	0.72432	1.3806	5
56	.58684	.80970	.72477	1.3798	4
57	.58708	.80953	.72521	1.3789	3
58	.58731	.80936	.72565	1.3781	2
59	.58755	.80919	.72610	1.3772	1
60	0.58779	0.80902	0.72654	1.3764	0
M	Cosine	Sine	Cotan.	Tan.	M

54°

Natural Trigonometric Functions

36°

M	Sine	Cosine	Tan.	Cotan.	M
0	0.58779	0.80902	0.72654	1.3764	60
1	.58802	.80885	.72699	1.3755	59
2	.58826	.80867	.72743	1.3747	58
3	.58849	.80850	.72788	1.3739	57
4	.58873	.80833	.72832	1.3730	56
5	0.58896	0.80816	0.72877	1.3722	55
6	.58920	.80799	.72921	1.3713	54
7	.58943	.80782	.72966	1.3705	53
8	.58967	.80765	.73010	1.3697	52
9	.58990	.80748	.73055	1.3688	51
10	0.59014	0.80730	0.73100	1.3680	50
11	.59037	.80713	.73144	1.3672	49
12	.59061	.80696	.73189	1.3663	48
13	.59084	.80679	.73234	1.3655	47
14	.59108	.80662	.73278	1.3647	46
15	0.59131	0.80644	0.73323	1.3638	45
16	.59154	.80627	.73368	1.3630	44
17	.59178	.80610	.73413	1.3622	43
18	.59201	.80593	.73457	1.3613	42
19	.59225	.80576	.73502	1.3605	41
20	0.59248	0.80558	0.73547	1.3597	40
21	.59272	.80541	.73592	1.3588	39
22	.59295	.80524	.73637	1.3580	38
23	.59318	.80507	.73681	1.3572	37
24	.59342	.80489	.73726	1.3564	36
25	0.59365	0.80472	0.73771	1.3555	35
26	.59389	.80455	.73816	1.3547	34
27	.59412	.80438	.73861	1.3539	33
28	.59436	.80420	.73906	1.3531	32
29	.59459	.80403	.73951	1.3522	31
30	0.59482	0.80386	0.73996	1.3514	30
31	.59506	.80368	.74041	1.3506	29
32	.59529	.80351	.74086	1.3498	28
33	.59552	.80334	.74131	1.3490	27
34	.59576	.80316	.74176	1.3481	26
35	0.59599	0.80299	0.74221	1.3473	25
36	.59622	.80282	.74267	1.3465	24
37	.59646	.80264	.74312	1.3457	23
38	.59669	.80247	.74357	1.3449	22
39	.59693	.80230	.74402	1.3440	21
40	0.59716	0.80212	0.74447	1.3432	20
41	.59739	.80195	.74492	1.3424	19
42	.59763	.80178	.74538	1.3416	18
43	.59786	.80160	.74583	1.3408	17
44	.59809	.80143	.74628	1.3400	16
45	0.59832	0.80125	0.74674	1.3392	15
46	.59856	.80108	.74719	1.3384	14
47	.59879	.80091	.74764	1.3375	13
48	.59902	.80073	.74810	1.3367	12
49	.59926	.80056	.74855	1.3359	11
50	0.59949	0.80038	0.74900	1.3351	10
51	.59972	.80021	.74946	1.3343	9
52	.59995	.80003	.74991	1.3335	8
53	.60019	.79986	.75037	1.3327	7
54	.60042	.79968	.75082	1.3319	6
55	0.60065	0.79951	0.75128	1.3311	5
56	.60089	.79934	.75173	1.3303	4
57	.60112	.79916	.75219	1.3295	3
58	.60135	.79899	.75264	1.3287	2
59	.60158	.79881	.75310	1.3278	1
60	0.60182	0.79864	0.75355	1.3270	0
M	Cosine	Sine	Cotan.	Tan.	M

53°

Natural Trigonometric Functions

37°

M	Sine	Cosine	Tan.	Cotan.	M
0	0.60182	0.79864	0.75355	1.3270	60
1	.60205	.79846	.75401	1.3262	59
2	.60228	.79829	.75447	1.3254	58
3	.60251	.79811	.75492	1.3246	57
4	.60274	.79793	.75538	1.3238	56
5	0.60298	0.79776	0.75584	1.3230	55
6	.60321	.79758	.75629	1.3222	54
7	.60344	.79741	.75675	1.3214	53
8	.60367	.79723	.75721	1.3206	52
9	.60390	.79706	.75767	1.3198	51
10	0.60414	0.79688	0.75812	1.3190	50
11	.60437	.79671	.75858	1.3182	49
12	.60460	.79653	.75904	1.3175	48
13	.60483	.79635	.75950	1.3167	47
14	.60506	.79618	.75996	1.3159	46
15	0.60529	0.79600	0.76042	1.3151	45
16	.60553	.79583	.76088	1.3143	44
17	.60576	.79565	.76134	1.3135	43
18	.60599	.79547	.76180	1.3127	42
19	.60622	.79530	.76226	1.3119	41
20	0.60645	0.79512	0.76272	1.3111	40
21	.60668	.79494	.76318	1.3103	39
22	.60691	.79477	.76364	1.3095	38
23	.60714	.79459	.76410	1.3087	37
24	.60738	.79441	.76456	1.3079	36
25	0.60761	0.79424	0.76502	1.3072	35
26	.60784	.79406	.76548	1.3064	34
27	.60807	.79388	.76594	1.3056	33
28	.60830	.79371	.76640	1.3048	32
29	.60853	.79353	.76686	1.3040	31
30	0.60876	0.79335	0.76733	1.3032	30
31	.60899	.79318	.76779	1.3024	29
32	.60922	.79300	.76825	1.3017	28
33	.60945	.79282	.76871	1.3009	27
34	.60968	.79264	.76819	1.3001	26
35	0.60991	0.79247	0.76964	1.2993	25
36	.61015	.79229	.77010	1.2985	24
37	.61038	.79211	.77057	1.2977	23
38	.61061	.79193	.77103	1.2970	22
39	.61084	.79176	.77149	1.2962	21
40	0.61107	0.79158	0.77196	1.2954	20
41	.61130	.79140	.77242	1.2946	19
42	.61153	.79122	.77289	1.2938	18
43	.61176	.79105	.77335	1.2931	17
44	.61199	.79087	.77382	1.2923	16
45	0.61222	0.79067	0.77428	1.2915	15
46	.61245	.79051	.77475	1.2907	14
47	.61268	.79033	.77521	1.2900	13
48	.61291	.79016	.77568	1.2892	12
49	.61314	.78998	.77615	1.2884	11
50	0.61337	0.78980	0.77661	1.2876	10
51	.61360	.78962	.77708	1.2869	9
52	.61383	.78944	.77754	1.2861	8
53	.61406	.78926	.77801	1.2853	7
54	.61429	.78908	.77848	1.2846	6
55	0.61451	0.78891	0.77895	1.2838	5
56	.61474	.78873	.77941	1.2830	4
57	.61497	.78855	.77988	1.2822	3
58	.61520	.78837	.78035	1.2815	2
59	.61543	.78819	.78082	1.2807	1
60	0.61566	0.78801	0.78129	1.2799	0
M	Cosine	Sine	Cotan.	Tan.	M

52°

Natural Trigonometric Functions

38°

M	Sine	Cosine	Tan.	Cotan.	M
0	0.61566	0.78801	0.78129	1.2799	60
1	.61589	.78783	.78175	1.2792	59
2	.61612	.78765	.78222	1.2784	58
3	.61635	.78747	.78269	1.2776	57
4	.61658	.78729	.78316	1.2769	56
5	0.61681	0.78711	0.78363	1.2761	55
6	.61704	.78694	.78410	1.2753	54
7	.61726	.78676	.78457	1.2746	53
8	.61749	.78658	.78504	1.2738	52
9	.61772	.78640	.78551	1.2731	51
10	0.61795	0.78622	0.78598	1.2723	50
11	.61818	.78604	.78645	1.2715	49
12	.61841	.78586	.78692	1.2708	48
13	.61864	.78568	.78739	1.2700	47
14	.61887	.78550	.78786	1.2693	46
15	0.61909	0.78532	0.78834	1.2685	45
16	.61932	.78514	.78881	1.2677	44
17	.61955	.78496	.78928	1.2670	43
18	.61978	.78478	.78975	1.2662	42
19	.62001	.78460	.79022	1.2655	41
20	0.62024	0.78442	0.79070	1.2647	40
21	.62046	.78424	.79117	1.2640	39
22	.62069	.78405	.79164	1.2632	38
23	.62092	.78387	.79212	1.2624	37
24	.62115	.78369	.79259	1.2617	36
25	0.62138	0.78351	0.79306	1.2609	35
26	.62160	.78333	.79354	1.2602	34
27	.62183	.78315	.79401	1.2594	33
28	.62206	.78297	.79449	1.2587	32
29	.62229	.78279	.79496	1.2579	31
30	0.62251	0.78261	0.79544	1.2572	30
31	.62274	.78243	.79591	1.2564	29
32	.62297	.78225	.79639	1.2557	28
33	.62320	.78206	.79686	1.2549	27
34	.62342	.78188	.79734	1.2542	26
35	0.62365	0.78170	0.79781	1.2534	25
36	.62388	.78152	.79829	1.2527	24
37	.62411	.78134	.79877	1.2519	23
38	.62433	.78116	.79924	1.2512	22
39	.62456	.78098	.79972	1.2504	21
40	0.62479	0.78079	0.80020	1.2497	20
41	.62502	.78061	.80067	1.2489	19
42	.62524	.78043	.80115	1.2482	18
43	.62547	.78025	.80163	1.2475	17
44	.62570	.78007	.80211	1.2467	16
45	0.62592	0.77988	0.80258	1.2460	15
46	.62615	.77970	.80306	1.2452	14
47	.62638	.77952	.80354	1.2445	13
48	.62660	.77934	.80402	1.2437	12
49	.62683	.77916	.80450	1.2430	11
50	0.62706	0.77897	0.80498	1.2423	10
51	.62728	.77879	.80546	1.2415	9
52	.62751	.77861	.80594	1.2408	8
53	.62774	.77843	.80642	1.2401	7
54	.62796	.77824	.80690	1.2393	6
55	0.62819	0.77806	0.80738	1.2386	5
56	.62842	.77788	.80786	1.2378	4
57	.62864	.77769	.80834	1.2371	3
58	.62887	.77751	.80882	1.2364	2
59	.62909	.77733	.80930	1.2356	1
60	0.62932	0.77715	0.80978	1.2349	0
M	Cosine	Sine	Cotan.	Tan.	M

51°

Natural Trigonometric Functions

39°

M	Sine	Cosine	Tan.	Cotan.	M
0	0.62932	0.77715	0.80978	1.2349	60
1	.62955	.77696	.81027	1.2342	59
2	.62977	.77678	.81075	1.2334	58
3	.63000	.77660	.81123	1.2327	57
4	.63022	.77641	.81171	1.2320	56
5	0.63045	0.77623	0.81220	1.2312	55
6	.63068	.77605	.81268	1.2305	54
7	.63090	.77586	.81316	1.2298	53
8	.63113	.77568	.81364	1.2290	52
9	.63135	.77550	.81413	1.2283	51
10	0.63158	0.77531	0.81461	1.2276	50
11	.63180	.77513	.81510	1.2268	49
12	.63203	.77494	.81558	1.2261	48
13	.63225	.77476	.81606	1.2254	47
14	.63248	.77458	.81655	1.2247	46
15	0.63271	0.77439	0.81703	1.2239	45
16	.63293	.77421	.81752	1.2232	44
17	.63316	.77402	.81800	1.2225	43
18	.63338	.77384	.81849	1.2218	42
19	.63361	.77366	.81898	1.2210	41
20	0.63383	0.77347	0.81946	1.2203	40
21	.63406	.77329	.81995	1.2196	39
22	.63428	.77310	.82044	1.2189	38
23	.63451	.77292	.82092	1.2181	37
24	.63473	.77273	.82141	1.2174	36
25	0.63496	0.77255	0.82190	1.2167	35
26	.63518	.77236	.82238	1.2160	34
27	.63540	.77218	.82287	1.2153	33
28	.63563	.77199	.82336	1.2145	32
29	.63585	.77181	.82385	1.2138	31
30	0.63608	0.77162	0.82434	1.2131	30
31	.63630	.77144	.82483	1.2124	29
32	.63653	.77125	.82531	1.2117	28
33	.63675	.77107	.82580	1.2109	27
34	.63698	.77088	.82629	1.2102	26
35	0.63720	0.77070	0.82678	1.2095	25
36	.63742	.77051	.82727	1.2088	24
37	.63765	.77033	.82776	1.2081	23
38	.63787	.77014	.82825	1.2074	22
39	.63810	.76996	.82874	1.2066	21
40	0.63832	0.76977	0.82923	1.2059	20
41	.63854	.76959	.82972	1.2052	19
42	.63877	.76940	.83022	1.2045	18
43	.63899	.76921	.83071	1.2038	17
44	.63922	.76903	.83120	1.2031	16
45	0.63944	0.76884	0.83169	1.2024	15
46	.63966	.76866	.83218	1.2017	14
47	.63989	.76847	.82368	1.2009	13
48	.64011	.76828	.83317	1.2002	12
49	.64033	.76810	.83366	1.1995	11
50	0.64056	0.76791	0.83415	1.1988	10
51	.64078	.76772	.83465	1.1981	9
52	.64100	.76754	.83514	1.1974	8
53	.64123	.76735	.83564	1.1967	7
54	.64145	.76717	.83613	1.1960	6
55	0.64167	0.76698	0.83662	1.1953	5
56	.64190	.76679	.83712	1.1946	4
57	.64212	.76661	.83761	1.1939	3
58	.64234	.76642	.83811	1.1932	2
59	.64256	.76623	.83860	1.1925	1
60	0.64279	0.76604	0.83910	1.1918	0
M	Cosine	Sine	Cotan.	Tan.	M

50°

Natural Trigonometric Functions

40°

M	Sine	Cosine	Tan.	Cotan.	M
0	0.64279	0.76604	0.83910	1.1918	60
1	.64301	.76586	.83960	1.1910	59
2	.64323	.76567	.84009	1.1903	58
3	.64346	.76548	.84059	1.1896	57
4	.64368	.76530	.84108	1.1889	56
5	0.64390	0.76511	0.84158	1.1882	55
6	.64412	.76492	.84208	1.1875	54
7	.64435	.76473	.84258	1.1868	53
8	.64457	.76455	.84307	1.1861	52
9	.64479	.76436	.84357	1.1854	51
10	0.64501	0.76417	0.84407	1.1847	50
11	.64524	.76398	.84457	1.1840	49
12	.64546	.76380	.84507	1.1833	48
13	.64568	.76361	.84556	1.1826	47
14	.64590	.76342	.84606	1.1819	46
15	0.64612	0.76323	0.84656	1.1812	45
16	.64635	.76304	.84706	1.1806	44
17	.64657	.76286	.84756	1.1799	43
18	.64679	.76267	.84806	1.1792	42
19	.64701	.76248	.84856	1.1785	41
20	0.64723	0.76229	0.84906	1.1778	40
21	.64746	.76210	.84956	1.1771	39
22	.64768	.76192	.85006	1.1764	38
23	.64790	.76173	.85057	1.1757	37
24	.64812	.76154	.85107	1.1750	36
25	0.64834	0.76135	0.85157	1.1743	35
26	.64856	.76116	.85207	1.1736	34
27	.64878	.76097	.85257	1.1729	33
28	.64901	.76078	.85308	1.1722	32
29	.64923	.76059	.85358	1.1715	31
30	0.64945	0.76041	0.85408	1.1708	30
31	.64967	.76022	.85458	1.1702	29
32	.64989	.76003	.85509	1.1695	28
33	.65011	.75984	.85559	1.1688	27
34	.65033	.75965	.85609	1.1681	26
35	0.65055	0.75946	0.85660	1.1674	25
36	.65077	.75927	.85710	1.1667	24
37	.65100	.75908	.85761	1.1660	23
38	.65122	.75889	.85811	1.1653	22
39	.65144	.75870	.85862	1.1647	21
40	0.65166	0.75851	0.85912	1.1640	20
41	.65188	.75832	.85963	1.1633	19
42	.65210	.75813	.86014	1.1626	18
43	.65232	.75794	.86064	1.1619	17
44	.65254	.75775	.86115	1.1612	16
45	0.65276	0.75756	0.86166	1.1606	15
46	.65298	.75738	.86216	1.1599	14
47	.65320	.75719	.86267	1.1592	13
48	.65342	.75700	.86318	1.1585	12
59	.65364	.75680	.86368	1.1578	11
50	0.65386	0.75661	0.86419	1.1571	10
51	.65408	.75642	.86470	1.1565	9
52	.65430	.75623	.86521	1.1558	8
53	.65452	.75604	.86572	1.1551	7
54	.65474	.75585	.86623	1.1544	6
55	0.65496	0.75566	0.86674	1.1538	5
56	.65518	.75547	.86725	1.1531	4
57	.65540	.75528	.86776	1.1524	3
58	.65562	.75509	.86827	1.1517	2
59	.65584	.75490	.86878	1.1510	1
60	0.65606	0.75471	0.86929	1.1504	0
M	Cosine	Sine	Cotan.	Tan.	M

49°

Natural Trigonometric Functions

41°

M	Sine	Cosine	Tan.	Cotan.	M
0	0.65606	0.75471	0.86929	1.1504	60
1	.65628	.75452	.86980	1.1497	59
2	.65650	.75433	.87031	1.1490	58
3	.65672	.75414	.87082	1.1483	57
4	.65694	.75395	.87133	1.1477	56
5	0.65716	0.75375	0.87184	1.1470	55
6	.65738	.75356	.87236	1.1463	54
7	.65759	.75337	.87287	1.1456	53
8	.65781	.75318	.87338	1.1450	52
9	.65803	.75299	.87389	1.1443	51
10	0.65825	0.75280	0.87441	1.1436	50
11	.65847	.75261	.87492	1.1430	49
12	.65869	.75241	.87543	1.1423	48
13	.65891	.75222	.87595	1.1416	47
14	.65913	.75203	.87646	1.1410	46
15	0.65935	0.75184	0.87698	1.1403	45
16	.65956	.75165	.87749	1.1396	44
17	.65978	.75146	.87801	1.1389	43
18	.66000	.75126	.87852	1.1383	42
19	.66022	.75107	.87904	1.1376	41
20	0.66044	0.75088	0.87955	1.1369	40
21	.66066	.75069	.88007	1.1363	39
22	.66088	.75050	.88059	1.1356	38
23	.66109	.75030	.88110	1.1349	37
24	.66131	.75011	.88162	1.1343	36
25	0.66153	0.74992	0.88214	1.1336	35
26	.66175	.74973	.88265	1.1329	34
27	.66197	.74953	.88317	1.1323	33
28	.66218	.74934	.88369	1.1316	32
29	.66240	.74915	.88421	1.1310	31
30	0.66262	0.74896	0.88473	1.1303	30
31	.66284	.74876	.88524	1.1296	29
32	.66306	.74857	.88576	1.1290	28
33	.66327	.74838	.88628	1.1283	27
34	.66349	.74818	.88680	1.1276	26
35	0.66371	0.74799	0.88732	1.1270	25
36	.66393	.74780	.88784	1.1263	24
37	.66414	.74760	.88836	1.1257	23
38	.66436	.74741	.88888	1.1250	22
39	.66458	.74722	.88940	1.1243	21
40	0.66480	0.74703	0.88992	1.1237	20
41	.66501	.74683	.89045	1.1230	19
42	.66523	.74664	.89097	1.1224	18
43	.66545	.74644	.89149	1.1217	17
44	.66566	.74625	.89201	1.1211	16
45	0.66588	0.74606	0.89253	1.1204	15
46	.66610	.74586	.89306	1.1197	14
47	.66632	.74567	.89358	1.1191	13
48	.66653	.74548	.89410	1.1184	12
49	.66675	.74528	.89463	1.1178	11
50	0.66697	0.74509	0.89515	1.1171	10
51	.66718	.74489	.89567	1.1165	9
52	.66740	.74470	.89620	1.1158	8
53	.66762	.74451	.89672	1.1152	7
54	.66783	.74431	.89725	1.1145	6
55	0.66805	0.74412	0.89777	1.1139	5
56	.66827	.74392	.89830	1.1132	4
57	.66848	.74373	.89883	1.1126	3
58	.66870	.74353	.89935	1.1119	2
59	.66891	.74334	.89988	1.1113	1
60	0.66913	0.74314	0.90040	1.1106	0
M	Cosine	Sine	Cotan.	Tan.	M

48°

Natural Trigonometric Functions

42°

M	Sine	Cosine	Tan.	Cotan.	M
0	0.66913	0.74314	0.90040	1.1106	60
1	.66935	.74295	.90093	1.1100	59
2	.66956	.74276	.90146	1.1093	58
3	.66978	.74256	.90199	1.1087	57
4	.66999	.74237	.90251	1.1080	56
5	0.67021	0.74217	0.90304	1.1074	55
6	.67043	.74198	.90357	1.1067	54
7	.67064	.74178	.90410	1.1061	53
8	.67086	.74159	.90463	1.1054	52
9	.67107	.74139	.90516	1.1048	51
10	0.67129	0.74120	0.90569	1.1041	50
11	.67151	.74100	.90621	1.1035	49
12	.67172	.74080	.90674	1.1028	48
13	.67194	.74061	.90727	1.1022	47
14	.67215	.74041	.90781	1.1016	46
15	0.67237	0.74022	0.90834	1.1009	45
16	.67258	.74002	.90887	1.1003	44
17	.67280	.73983	.90940	1.0996	43
18	.67301	.73963	.90993	1.0990	42
19	.67323	.73944	.91046	1.0983	41
20	0.67344	0.73924	0.91099	1.0977	40
21	.67366	.73904	.91153	1.0971	39
22	.67387	.73885	.91206	1.0964	38
23	.67409	.73865	.91259	1.0958	37
24	.67430	.73846	.91313	1.0951	36
25	0.67452	0.73826	0.91366	1.0945	35
26	.67473	.73806	.91419	1.0939	34
27	.67495	.73787	.91473	1.0932	33
28	.67516	.73767	.91526	1.0926	32
29	.67538	.73747	.91580	1.0919	31
30	0.67559	0.73728	0.91633	1.0913	30
31	.67580	.73708	.91687	1.0907	29
32	.67602	.73688	.91740	1.0900	28
33	.67623	.73669	.91794	1.0894	27
34	.67645	.73649	.91847	1.0888	26
35	0.67666	0.73629	0.91901	1.0881	25
36	.67688	.73610	.91955	1.0875	24
37	.67709	.73590	.92008	1.0869	23
38	.67730	.73570	.92062	1.0862	22
39	.67752	.73551	.92116	1.0856	21
40	0.67773	0.73531	0.92170	1.0850	20
41	.67795	.73511	.92224	1.0843	19
42	.67816	.73491	.92277	1.0837	18
43	.67837	.73472	.92331	1.0831	17
44	.67859	.73542	.92385	1.0824	16
45	0.67880	0.73432	0.92439	1.0818	15
46	.67901	.73413	.92493	1.0812	14
47	.67923	.73393	.92547	1.0805	13
48	.67944	.73373	.92601	1.0799	12
49	.67965	.73353	.92655	1.0793	11
50	0.67987	0.73333	0.92709	1.0786	10
51	.68008	.73314	.92763	1.0780	9
52	.68029	.73294	.92817	1.0774	8
53	.68051	.73274	.92872	1.0768	7
54	.68072	.73254	.92926	1.0761	6
55	0.68093	0.73234	0.92980	1.0755	5
56	.68115	.73215	.93034	1.0749	4
57	.68136	.73195	.93088	1.0742	3
58	.68157	.73175	.93143	1.0736	2
59	.68179	.73155	.93197	1.0730	1
60	0.68200	0.73135	0.93252	1.0724	0
M	Cosine	Sine	Cotan.	Tan.	M

47°

Natural Trigonometric Functions

43°

M	Sine	Cosine	Tan.	Cotan.	M
0	0.68200	0.73135	0.93252	1.0724	60
1	.68221	.73116	.93306	1.0717	59
2	.68242	.73096	.93360	1.0711	58
3	.68264	.73076	.93415	1.0705	57
4	.68285	.73056	.93469	1.0699	56
5	0.68306	0.73036	0.93524	1.0692	55
6	.68327	.73016	.93578	1.0686	54
7	.68349	.72996	.93633	1.0680	53
8	.68370	.72976	.93688	1.0674	52
9	.68391	.72957	.93742	1.0668	51
10	0.68412	0.72937	0.93797	1.0661	50
11	.68434	.72917	.93852	1.0655	49
12	.68455	.72897	.93906	1.0649	48
13	.68476	.72877	.93961	1.0643	47
14	.68497	.72857	.94016	1.0637	46
15	0.68518	0.72837	0.94071	1.0630	45
16	.68539	.72817	.94125	1.0624	44
17	.68561	.72797	.94180	1.0618	43
18	.68582	.72777	.94235	1.0612	42
19	.68603	.72757	.94290	1.0606	41
20	0.68624	0.72737	0.94345	1.0599	40
21	.68645	.72717	.94400	1.0593	39
22	.68666	.72697	.94455	1.0587	38
23	.68688	.72677	.94510	1.0581	37
24	.68709	.72657	.94565	1.0575	36
25	0.68730	0.72637	0.94620	1.0569	35
26	.68751	.72617	.94676	1.0562	34
27	.68772	.72597	.94731	1.0556	33
28	.68793	.72577	.94786	1.0550	32
29	.68814	.72557	.94841	1.0544	31
30	0.68835	0.72537	0.94896	1.0538	30
31	.68857	.72517	.94952	1.0532	29
32	.68878	.72497	.95007	1.0526	28
33	.68899	.72477	.95062	1.0519	27
34	.68920	.72457	.95118	1.0513	26
35	0.68941	0.72437	0.95173	1.0507	25
36	.68962	.72417	.95229	1.0501	24
37	.68983	.72397	.95284	1.0495	23
38	.69004	.72377	.95340	1.0489	22
39	.69025	.72357	.95395	1.0483	21
40	0.69046	0.72337	0.95451	1.0477	20
41	.69067	.72317	.95506	1.0470	19
42	.69088	.72297	.95562	1.0464	18
43	.69109	.72277	.95618	1.0458	17
44	.69130	.72257	.95673	1.0452	16
45	0.69151	0.72236	0.95729	1.0446	15
46	.69172	.72216	.95785	1.0440	14
47	.69193	.72196	.95841	1.0434	13
48	.69214	.72176	.95897	1.0428	12
49	.69235	.72156	.95952	1.0422	11
50	0.69256	0.72136	0.96008	1.0416	10
51	.69277	.72116	.96064	1.0410	9
52	.69298	.72095	.96120	1.0404	8
53	.69319	.72075	.96176	1.0398	7
54	.69340	.72055	.96232	1.0392	6
55	0.69361	0.72035	0.96288	1.0385	5
56	.69382	.72015	.96344	1.0379	4
57	.69403	.71995	.96400	1.0373	3
58	.69224	.71974	.96457	1.0367	2
59	.69445	.71954	.96513	1.0361	1
60	0.69466	0.71934	0.96569	1.0355	0
M	Cosine	Sine	Cotan.	Tan.	M

46°

M	Sine	Cosine	Tan.	Cotan.	M
0	0.69466	0.71934	0.96569	1.0355	60
1	.69487	.71914	.96625	1.0349	59
2	.69508	.71894	.96681	1.0343	58
3	.69529	.71873	.96738	1.0337	57
4	.69549	.71853	.96794	1.0331	56
5	0.69570	0.71833	0.96850	1.0325	55
6	.69591	.71813	.96907	1.0319	54
7	.69612	.71792	.96963	1.0313	53
8	.69633	.71772	.97020	1.0307	52
9	.69654	.71752	.97076	1.0301	51
10	0.69675	0.71732	0.97133	1.0295	50
11	.69696	.71711	.97189	1.0289	49
12	.69717	.71691	.97246	1.0283	48
13	.69737	.71671	.97302	1.0277	47
14	.69758	.71650	.97359	1.0271	46
15	0.69779	0.71630	0.97416	1.0265	45
16	.69800	.71610	.97472	1.0259	44
17	.69821	.71590	.97529	1.0253	43
18	.69842	.71569	.97586	1.0247	42
19	.69862	.71549	.97643	1.0241	41
20	0.69883	0.71529	0.97700	1.0235	40
21	.69904	.71508	.97756	1.0230	39
22	.69925	.71488	.97813	1.0224	38
23	.69946	.71468	.97870	1.0218	37
24	.69966	.71447	.97927	1.0212	36
25	0.69987	0.71427	0.97984	1.0206	35
26	.70008	.71407	.98041	1.0200	34
27	.70029	.71386	.98098	1.0194	33
28	.70049	.71366	.98155	1.0188	32
29	.70070	.71345	.98213	1.0182	31
30	0.70091	0.71325	0.98270	1.0176	30
31	.70112	.71305	.98327	1.0170	29
32	.70132	.71284	.98384	1.0164	28
33	.70153	.71264	.98441	1.0158	27
34	.70174	.71243	.98499	1.0152	26
35	0.70195	0.71223	0.98556	1.0147	25
36	.70215	.71203	.98613	1.0141	24
37	.70236	.71182	.98671	1.0135	23
38	.70257	.71162	.98728	1.0129	22
39	.70277	.71141	.98786	1.0123	21
40	0.70298	0.71121	0.98843	1.0117	20
41	.70319	.71100	.98901	1.0111	19
42	.70339	.71080	.98958	1.0105	18
43	.70360	.71059	.99016	1.0099	17
44	.70381	.71039	.99073	1.0094	16
45	0.70401	0.71019	0.99131	1.0088	15
46	.70422	.70998	.99189	1.0082	14
47	.70443	.70978	.99247	1.0076	13
48	.70463	.70957	.99304	1.0070	12
49	.70484	.70937	.99362	1.0064	11
50	0.70505	0.70916	0.99420	1.0058	10
51	.70525	.70896	.99478	1.0052	9
52	.70546	.70875	.99536	1.0047	8
53	.70567	.70855	.99594	1.0041	7
54	.70587	.70834	.99652	1.0035	6
55	0.70608	0.70813	0.99710	1.0029	5
56	.70628	.70793	.99768	1.0023	4
57	.70649	.70772	.99826	1.0017	3
58	.70670	.70752	.99884	1.0012	2
59	.70690	.70731	.99942	1.0006	1
60	0.70711	0.70711	1.00000	1.0000	0
M	Cosine	Sine	Cotan.	Tan.	M

STADIA REDUCTIONS B

Horizontal Distances and Elevations from Stadia Readings

	0°		1°		2°		3°	
Minutes	Hor. Dist.	Diff. Elev.	Hor. Dist.	Diff. Elev.	Hor. Dist.	Diff. Elev.	Hor. Dist.	Diff. Elev.
0	100.00	0.00	99.97	1.74	99.88	3.49	99.73	5.23
2	100.00	0.06	99.97	1.80	99.87	3.55	99.72	5.28
4	100.00	0.12	99.97	1.86	99.87	3.60	99.71	5.34
6	100.00	0.17	99.96	1.92	99.87	3.66	99.71	5.40
8	100.00	0.23	99.96	1.98	99.86	3.72	99.70	5.46
10	100.00	0.29	99.96	2.04	99.86	3.78	99.69	5.52
12	100.00	0.35	99.96	2.09	99.85	3.84	99.69	5.57
14	100.00	0.41	99.95	2.15	99.85	3.89	99.68	5.63
16	100.00	0.47	99.95	2.21	99.84	3.95	99.68	5.69
18	100.00	0.52	99.95	2.27	99.84	4.01	99.67	5.75
20	100.00	0.58	99.95	2.33	99.83	4.07	99.66	5.80
22	100.00	0.64	99.94	2.38	99.83	4.13	99.66	5.86
24	100.00	0.70	99.94	2.44	99.82	4.18	99.65	5.92
26	99.99	0.76	99.94	2.50	99.82	4.24	99.64	5.98
28	99.99	0.81	99.93	2.56	99.81	4.30	99.63	6.04
30	99.99	0.87	99.93	2.62	99.81	4.36	99.63	6.09
32	99.99	0.93	99.93	2.67	99.80	4.42	99.62	6.15
34	99.99	0.99	99.93	2.73	99.80	4.47	99.61	6.21
36	99.99	1.05	99.92	2.79	99.79	4.53	99.61	6.27
38	99.99	1.11	99.92	2.85	99.79	4.59	99.60	6.32
40	99.99	1.16	99.92	2.91	99.78	4.65	99.59	6.38
42	99.99	1.22	99.91	2.97	99.78	4.71	99.58	6.44
44	99.98	1.28	99.91	3.02	99.77	4.76	99.58	6.50
46	99.98	1.34	99.90	3.08	99.77	4.82	99.57	6.56
48	99.98	1.40	99.90	3.14	99.76	4.88	99.56	6.61
50	99.98	1.45	99.90	3.20	99.76	4.94	99.55	6.67
52	99.98	1.51	99.89	3.26	99.75	4.99	99.55	6.73
54	99.98	1.57	98.89	3.31	99.74	5.05	99.54	6.79
56	99.97	1.63	99.89	3.37	99.74	5.11	99.53	6.84
58	99.97	1.69	99.88	3.43	99.73	5.17	99.52	6.90
60	99.97	1.74	99.88	3.49	99.73	5.23	99.51	6.96
$C = 0.75$	0.75	0.01	0.75	0.02	0.75	0.03	0.75	0.05
$C = 1.00$	1.00	0.01	1.00	0.03	1.00	0.04	1.00	0.06
$C = 1.25$	1.25	0.02	1.25	0.03	1.25	0.05	1.25	0.08

Horizontal Distances and Elevations from Stadia Readings—*Continued*

	4°		5°		6°		7°	
Minutes	Hor. Dist.	Diff. Elev.	Hor. Dist.	Diff. Elev.	Hor. Dist.	Diff. Elev.	Hor. Dist.	Diff. Elev.
0	99.51	6.96	99.24	8.68	98.91	10.40	98.51	12.10
2	99.51	7.02	99.23	8.74	98.90	10.45	98.50	12.15
4	99.50	7.07	99.22	8.80	98.88	10.51	98.49	12.21
6	99.49	7.13	99.21	8.85	98.87	10.57	98.47	12.27
8	99.48	7.19	99.20	8.91	98.86	10.62	98.46	12.32
10	99.47	7.25	99.19	8.97	98.85	10.68	98.44	12.38
12	99.46	7.30	99.18	9.03	98.83	10.74	98.43	12.43
14	99.46	7.36	99.17	9.08	98.82	10.79	98.41	12.49
16	99.45	7.42	99.16	9.14	98.81	10.85	98.40	12.55
18	99.44	7.48	99.15	9.20	98.80	10.91	98.39	12.60
20	99.43	7.53	99.14	9.25	98.78	10.96	98.37	12.66
22	99.42	7.59	99.13	9.31	98.77	11.02	98.36	12.72
24	99.41	7.65	99.11	9.37	98.76	11.08	98.34	12.77
26	99.40	7.71	99.10	9.43	98.74	11.13	98.33	12.83
28	99.39	7.76	99.09	9.48	98.73	11.19	98.31	12.88
30	99.38	7.82	99.08	9.54	98.72	11.25	98.30	12.94
32	99.38	7.88	99.07	9.60	98.71	11.30	98.28	13.00
34	99.37	7.94	99.06	9.65	98.69	11.36	98.27	13.05
36	99.36	7.99	99.05	9.71	98.68	11.42	98.25	13.11
38	99.35	8.05	99.04	9.77	98.67	11.47	98.24	13.17
40	99.34	8.11	99.03	9.83	98.65	11.53	98.22	13.22
42	99.33	8.17	99.01	9.88	98.64	11.59	98.20	13.28
44	99.32	8.22	99.00	9.94	98.63	11.64	98.19	13.33
46	99.31	8.28	98.99	10.00	98.61	11.70	98.17	13.39
48	99.30	8.34	98.98	10.05	98.60	11.76	98.16	13.45
50	99.29	8.40	98.97	10.11	98.58	11.81	98.14	13.50
52	99.28	8.45	98.96	10.17	98.57	11.87	98.13	13.56
54	99.27	8.51	98.94	10.22	98.56	11.93	98.11	13.61
56	99.26	8.57	98.93	10.28	98.54	11.98	98.10	13.67
58	99.25	8.63	98.92	10.34	98.53	12.04	98.08	13.73
60	99.24	8.68	98.91	10.40	98.51	12.10	98.06	13.78
$C = 0.75$	0.75	0.06	0.75	0.07	0.75	0.08	0.74	0.10
$C = 1.00$	1.00	0.08	1.00	0.10	0.99	0.11	0.99	0.13
$C = 1.25$	1.25	0.10	1.24	0.12	1.24	0.14	1.24	0.16

Horizontal Distances and Elevations from Stadia Readings—*Continued*

	8°		9°		10°		11°	
Minutes	Hor. Dist.	Diff. Elev.	Hor. Dist.	Diff. Elev.	Hor. Dist.	Diff. Elev.	Hor. Dist.	Diff. Elev.
0	98.06	13.78	97.55	15.45	96.98	17.10	96.36	18.73
2	98.05	13.84	97.53	15.51	96.96	17.16	96.34	18.78
4	98.03	13.89	97.52	15.56	96.94	17.21	96.32	18.84
6	98.01	13.95	97.50	15.62	96.92	17.26	96.29	18.89
8	98.00	14.01	97.48	15.67	96.90	17.32	96.27	18.95
10	97.98	14.06	97.46	15.73	96.88	17.37	96.25	19.00
12	97.97	14.12	97.44	15.78	96.86	17.43	96.23	19.05
14	97.95	14.17	97.43	15.84	96.84	17.48	96.21	19.11
16	97.93	14.23	97.41	15.89	96.82	17.54	96.18	19.16
18	97.92	14.28	97.39	15.95	96.80	17.59	96.16	19.21
20	97.90	14.34	97.37	16.00	96.78	17.65	96.14	19.27
22	97.88	14.40	97.35	16.06	96.76	17.70	96.12	19.32
24	97.87	14.45	97.33	16.11	96.74	17.76	96.09	19.38
26	97.85	14.51	97.31	16.17	96.72	17.81	96.07	19.43
28	97.83	14.56	97.29	16.22	96.70	17.86	96.05	19.48
30	97.82	14.62	97.28	16.28	96.68	17.92	96.03	19.54
32	97.80	14.67	97.26	16.33	96.66	17.97	96.00	19.59
34	97.78	14.73	97.24	16.39	96.64	18.03	95.98	19.64
36	97.76	14.79	97.22	16.44	96.62	18.08	95.96	19.70
38	97.75	14.84	97.20	16.50	96.60	18.14	95.93	19.75
40	97.73	14.90	97.18	16.55	96.57	18.19	95.91	19.80
42	97.71	14.95	97.16	16.61	96.55	18.24	95.89	19.86
44	97.69	15.01	97.14	16.66	96.53	18.30	95.86	19.91
46	97.68	15.06	97.12	16.72	96.51	18.35	95.84	19.96
48	97.66	15.12	97.10	16.77	96.49	18.41	95.82	20.02
50	97.64	15.17	97.08	16.83	96.47	18.46	95.79	20.07
52	97.62	15.23	97.06	16.88	96.45	18.51	95.77	20.12
54	97.61	15.28	97.04	16.94	96.42	18.57	95.75	20.18
56	97.59	15.34	97.02	16.99	96.40	18.62	95.72	20.23
58	97.57	15.40	97.00	17.05	96.38	18.68	95.70	20.28
60	97.55	15.45	96.98	17.10	96.36	18.73	95.68	20.34
$C = 0.75$	0.74	0.11	0.74	0.12	0.74	0.14	0.73	0.15
$C = 1.00$	0.99	0.15	0.99	0.17	0.98	0.18	0.98	0.20
$C = 1.25$	1.24	0.18	1.23	0.21	1.23	0.23	1.22	0.25

Horizontal Distances and Elevations from Stadia Readings—*Continued*

	12°		13°		14°		15°	
Minutes	Hor. Dist.	Diff. Elev.	Hor. Dist.	Diff. Elev.	Hor. Dist.	Diff. Elev.	Hor. Dist.	Diff. Elev.
0	95.68	20.34	94.94	21.92	94.15	23.47	93.30	25.00
2	95.65	20.39	94.91	21.97	94.12	23.52	93.27	25.05
4	95.63	20.44	94.89	22.02	94.09	23.58	93.24	25.10
6	95.61	20.50	94.86	22.08	94.07	23.63	93.21	25.15
8	95.58	20.55	94.84	22.13	94.04	23.68	93.18	25.20
10	95.56	20.60	94.81	22.18	94.01	23.73	93.16	25.25
12	95.53	20.66	94.79	22.23	93.98	23.78	93.13	25.30
14	95.51	20.71	94.76	22.28	93.95	23.83	93.10	25.35
16	95.49	20.76	94.73	22.34	93.93	23.88	93.07	25.40
18	95.46	20.81	94.71	22.39	93.90	23.93	93.04	25.45
20	95.44	20.87	94.68	22.44	93.87	23.99	93.01	25.50
22	95.41	20.92	94.66	22.49	93.84	24.04	92.98	25.55
24	95.39	20.97	94.63	22.54	93.82	24.09	92.95	25.60
26	95.36	21.03	94.60	22.60	93.79	24.14	92.92	25.65
28	95.34	21.08	94.58	22.65	93.76	24.19	92.89	25.70
30	95.32	21.13	94.55	22.70	93.73	24.24	92.86	25.75
32	95.29	21.18	94.52	22.75	93.70	24.29	92.83	25.80
34	95.27	21.24	94.50	22.80	93.67	24.34	92.80	25.85
36	95.24	21.29	94.47	22.85	93.65	24.39	92.77	25.90
38	95.22	21.34	94.44	22.91	93.62	24.44	92.74	25.95
40	95.19	21.39	94.42	22.96	93.59	24.49	92.71	26.00
42	95.17	21.45	94.39	23.01	93.56	24.55	92.68	26.05
44	95.14	21.50	94.36	23.06	93.53	24.60	92.65	26.10
46	95.12	21.55	94.34	23.11	93.50	24.65	92.62	26.15
48	95.09	21.60	94.31	23.16	93.47	24.70	92.59	26.20
50	95.07	21.66	94.28	23.22	93.45	24.75	92.56	26.25
52	95.04	21.71	94.26	23.27	93.42	24.80	92.53	26.30
54	95.02	21.76	94.23	23.32	93.39	24.85	92.49	26.35
56	94.99	21.81	94.20	23.37	93.36	24.90	92.46	26.40
58	94.97	21.87	94.17	23.42	93.33	24.95	92.43	26.45
60	94.94	21.92	94.15	23.47	93.30	25.00	92.40	26.50
$C = 0.75$	0.73	0.16	0.73	0.18	0.73	0.19	0.72	0.20
$C = 1.00$	0.98	0.22	0.97	0.23	0.97	0.25	0.96	0.27
$C = 1.25$	1.22	0.27	1.21	0.29	1.21	0.31	1.20	0.33

Horizontal Distances and Elevations from Stadia Readings—*Continued*

	16°		17°		18°		19°	
Minutes	Hor. Dist.	Diff. Elev.	Hor. Dist.	Diff. Elev.	Hor. Dist.	Diff. Elev.	Hor. Dist.	Diff. Elev.
0	92.40	26.50	91.45	27.96	90.45	29.39	89.40	30.78
2	92.37	26.55	91.42	28.01	90.42	29.44	89.36	30.83
4	92.34	26.59	91.39	28.06	90.38	29.48	89.33	30.87
6	92.31	26.64	91.35	28.10	90.35	29.53	89.29	30.92
8	92.28	26.69	91.32	28.15	90.31	29.58	89.26	30.97
10	92.25	26.74	91.29	28.20	90.28	29.62	89.22	31.01
12	92.22	26.79	91.26	28.25	90.24	29.67	89.18	31.06
14	92.19	26.84	91.22	28.30	90.21	29.72	89.15	31.10
16	92.15	26.89	91.19	28.34	90.18	29.76	89.11	31.15
18	92.12	26.94	91.16	28.39	90.14	29.81	89.08	31.19
20	92.09	26.99	91.12	28.44	90.11	29.86	89.04	31.24
22	92.06	27.04	91.09	28.49	90.07	29.90	89.00	31.28
24	92.03	27.09	91.06	28.54	90.04	29.95	88.97	31.33
26	92.00	27.13	91.02	28.58	90.00	30.00	88.93	31.38
28	91.97	27.18	90.99	28.63	89.97	30.04	88.89	31.42
30	91.93	27.23	90.96	28.68	89.93	30.09	88.86	31.47
32	91.90	27.28	90.92	28.73	89.90	30.14	88.82	31.51
34	91.87	27.33	90.89	28.77	89.86	30.18	88.78	31.56
36	91.84	27.38	90.86	28.82	89.83	30.23	88.75	31.60
38	91.81	27.43	90.82	28.87	89.79	30.28	88.71	31.65
40	91.77	27.48	90.79	28.92	89.76	30.32	88.67	31.69
42	91.74	27.52	90.76	28.96	89.72	30.37	88.64	31.74
44	91.71	27.57	90.72	29.01	89.69	30.41	88.60	31.78
46	91.68	27.62	90.69	29.06	89.65	30.46	88.56	31.83
48	91.65	27.67	90.66	29.11	89.61	30.51	88.53	31.87
50	91.61	27.72	90.62	29.15	89.58	30.55	88.49	31.92
52	91.58	27.77	90.59	29.20	89.54	30.60	88.45	31.96
54	91.55	27.81	90.55	29.25	89.51	30.65	88.41	32.01
56	91.52	27.86	90.52	29.30	89.47	30.69	88.38	32.05
58	91.48	27.91	90.49	29.34	89.44	30.74	88.34	32.09
60	91.45	27.96	90.45	29.39	89.40	30.78	88.30	32.14
$C = 0.75$	0.72	0.21	0.72	0.23	0.71	0.24	0.71	0.25
$C = 1.00$	0.96	0.28	0.95	0.30	0.95	0.32	0.94	0.33
$C = 1.25$	1.20	0.36	1.19	0.38	1.19	0.40	1.18	0.42

Horizontal Distances and Elevations from Stadia Readings—*Continued*

	20°		21°		22°		23°	
Minutes	Hor. Dist.	Diff. Elev.	Hor. Dist.	Diff. Elev.	Hor. Dist.	Diff. Elev.	Hor. Dist.	Diff. Elev.
0	88.30	32.14	87.16	33.46	85.97	34.73	84.73	35.97
2	88.26	32.18	87.12	33.50	85.93	34.77	84.69	36.01
4	88.23	32.23	87.08	33.54	85.89	34.82	84.65	36.05
6	88.19	32.27	87.04	33.59	85.85	34.86	84.61	36.09
8	88.15	32.32	87.00	33.63	85.80	34.90	84.57	36.13
10	88.11	32.36	86.96	33.67	85.76	34.94	84.52	36.17
12	88.08	32.41	86.92	33.72	85.72	34.98	84.48	36.21
14	88.04	32.45	86.88	33.76	85.68	35.02	84.44	36.25
16	88.00	32.49	86.84	33.80	85.64	35.07	84.40	36.29
18	87.96	32.54	86.80	33.84	85.60	35.11	84.35	36.33
20	87.93	32.58	86.77	33.89	85.56	35.15	84.31	36.37
22	87.89	32.63	86.73	33.93	85.52	35.19	84.27	36.41
24	87.85	32.67	86.69	33.97	85.48	35.23	84.23	36.45
26	87.81	32.72	86.65	34.01	85.44	35.27	84.18	36.49
28	87.77	32.76	86.61	34.06	85.40	35.31	84.14	36.53
30	87.74	32.80	86.57	34.10	85.36	35.36	84.10	36.57
32	87.70	32.85	86.53	34.14	85.31	35.40	84.06	36.61
34	87.66	32.89	86.49	34.18	85.27	35.44	84.01	36.65
36	87.62	32.93	86.45	34.23	85.23	35.48	83.97	36.69
38	87.58	32.98	86.41	34.27	85.19	35.52	83.93	36.73
40	87.54	33.02	86.37	34.31	85.15	35.56	83.89	36.77
42	87.51	33.07	86.33	34.35	85.11	35.60	83.84	36.80
44	87.47	33.11	86.29	34.40	85.07	35.64	83.80	36.84
46	87.43	33.15	86.25	34.44	85.02	35.68	83.76	36.88
48	87.39	33.20	86.21	34.48	84.98	35.72	83.72	36.92
50	87.35	33.24	86.17	34.52	84.94	35.76	83.67	36.96
52	87.31	33.28	86.13	34.57	84.90	35.80	83.63	37.00
54	87.27	33.33	86.09	34.61	84.86	35.85	83.59	37.04
56	87.24	33.37	86.05	34.65	84.82	35.89	83.54	37.08
58	87.20	33.41	86.01	34.69	84.77	35.93	83.50	37.12
60	87.16	33.46	85.97	34.73	84.73	35.97	83.46	37.16
$C = 0.75$	0.70	0.26	0.70	0.27	0.69	0.29	0.69	0.30
$C = 1.00$	0.94	0.35	0.93	0.37	0.92	0.38	0.92	0.40
$C = 1.25$	1.17	0.44	1.16	0.46	1.15	0.48	1.15	0.50

Horizontal Distances and Elevations from Stadia Readings—*Continued*

	24°		25°		26°		27°	
Minutes	Hor. Dist.	Diff. Elev.	Hor. Dist.	Diff. Elev.	Hor. Dist.	Diff. Elev.	Hor. Dist.	Diff. Elev.
0	83.46	37.16	82.14	38.30	80.78	39.40	79.39	40.45
2	83.41	37.20	82.09	38.34	80.74	39.44	79.34	40.49
4	83.37	37.23	82.05	38.38	80.69	39.47	79.30	40.52
6	83.33	37.27	82.01	38.41	80.65	39.51	79.25	40.55
8	83.28	37.31	81.96	38.45	80.60	39.54	79.20	40.59
10	83.24	37.35	81.92	38.49	80.55	39.58	79.15	40.62
12	83.20	37.39	81.87	38.53	80.51	39.61	79.11	40.66
14	83.15	37.43	81.83	38.56	80.46	39.65	79.06	40.69
16	83.11	37.47	81.78	38.60	80.41	39.69	79.01	40.72
18	83.07	37.51	81.74	38.64	80.37	39.72	78.96	40.76
20	83.02	37.54	81.69	38.67	80.32	39.76	78.92	40.79
22	82.98	37.58	81.65	38.71	80.28	39.79	78.87	40.82
24	82.93	37.62	81.60	38.75	80.23	39.83	78.82	40.86
26	82.89	37.66	81.56	38.78	80.18	39.86	78.77	40.89
28	82.85	37.70	81.51	38.82	80.14	39.90	78.73	40.92
30	82.80	37.74	81.47	38.86	80.09	39.93	78.68	40.96
32	82.76	37.77	81.42	38.89	80.04	39.97	78.63	40.99
34	82.72	37.81	81.38	38.93	80.00	40.00	78.58	41.02
36	82.67	37.85	81.33	38.97	79.95	40.04	78.54	41.06
38	82.63	37.89	81.28	39.00	79.90	40.07	78.49	41.09
40	82.58	37.93	81.24	39.04	79.86	40.11	78.44	41.12
42	82.54	37.96	81.19	39.08	79.81	40.14	78.39	41.16
44	82.49	38.00	81.15	39.11	79.76	40.18	78.34	41.19
46	82.45	38.04	81.10	39.15	79.72	40.21	78.30	41.22
48	82.41	38.08	81.06	39.18	79.67	40.24	78.25	41.26
50	82.36	38.11	81.01	39.22	79.62	40.28	78.20	41.29
52	82.32	38.15	80.97	39.26	79.58	40.31	78.15	41.32
54	82.27	38.19	80.92	39.29	79.53	40.35	78.10	41.35
56	82.23	38.23	80.87	39.33	79.48	40.38	78.06	41.39
58	82.18	38.26	80.83	39.36	79.44	40.42	78.01	41.42
60	82.14	38.30	80.78	39.40	79.39	40.45	77.96	41.45
$C = 0.75$	0.68	0.31	0.68	0.32	0.67	0.33	0.67	0.35
$C = 1.00$	0.91	0.41	0.90	0.43	0.89	0.45	0.89	0.46
$C = 1.25$	1.14	0.52	1.13	0.54	1.12	0.56	1.11	0.58

Horizontal Distances and Elevations from Stadia Readings—*Continued*

	28°		29°		30°	
Minutes	Hor. Dist.	Diff. Elev.	Hor. Dist.	Diff. Elev.	Hor. Dist.	Diff. Elev.
0	77.96	41.45	76.50	42.40	75.00	43.30
2	77.91	41.48	76.45	42.43	74.95	43.33
4	77.86	41.52	76.40	42.46	74.90	43.36
6	77.81	41.55	76.35	42.49	74.85	43.39
8	77.77	41.58	76.30	42.53	74.80	43.42
10	77.72	41.61	76.25	42.56	74.75	43.45
12	77.67	41.65	76.20	42.59	74.70	43.47
14	77.62	41.68	76.15	42.62	74.65	43.50
16	77.57	41.71	76.10	42.65	74.60	43.53
18	77.52	41.74	76.05	42.68	74.55	43.56
20	77.48	41.77	76.00	42.71	74.49	43.59
22	77.42	41.81	75.95	42.74	74.44	43.62
24	77.38	41.84	75.90	42.77	74.39	43.65
26	77.33	41.87	75.85	42.80	74.34	43.67
28	77.28	41.90	75.80	42.83	74.29	43.70
30	77.23	41.93	75.75	42.86	74.24	43.73
32	77.18	41.97	75.70	42.89	74.19	43.76
34	77.13	42.00	75.65	42.92	74.14	43.79
36	77.09	42.03	75.60	42.95	74.09	43.82
38	77.04	42.06	75.55	42.98	74.04	43.84
40	76.99	42.09	75.50	43.01	73.99	43.87
42	76.94	42.12	75.45	43.04	73.93	43.90
44	76.89	42.15	75.40	43.07	73.88	43.93
46	76.84	42.19	75.35	43.10	73.83	43.95
48	76.79	42.22	75.30	43.13	73.78	43.98
50	76.74	42.25	75.25	43.16	73.73	44.01
52	76.69	42.28	75.20	43.18	73.68	44.04
54	76.64	42.31	75.15	43.21	73.63	44.07
56	76.59	42.34	75.10	43.24	73.58	44.09
58	76.55	42.37	75.05	43.27	73.52	44.12
60	76.50	42.40	75.00	43.30	73.47	44.15
$C = 0.75$	0.66	0.36	0.65	0.37	0.65	0.38
$C = 1.00$	0.88	0.48	0.87	0.49	0.86	0.51
$C = 1.25$	1.10	0.60	1.09	0.62	1.08	0.63

INDEX

INDEX